ELECTRONIC TEST EQUIPMENT

Theory and Applications

Electronic Test Equipment
Theory and Applications

TJ Byers

Intertext Publications, Inc.
McGraw-Hill Book Company
New York, N.Y.

Library of Congress Catalog Number 87-81303

10 9 8 7 6 5 4 3 2 1

ISBN 0-07-009522-1

Intertext Publications, Inc.
McGraw-Hill Book Company
11 West 19th Street
New York, NY 10011

DEDICATION

This book is dedicated to my parents for their love and support over the years.

ACKNOWLEDGEMENTS

The following companies and persons contributed toward the writing of this book: Audio Precision, Inc.; Brian Fenton, Editor, *Radio-Electronics*; Hewlett-Packard Company; Maron Production Inc.; RC Electronics Inc.; Tektronix, Inc.; Unisys.

Artwork by Ted Garcia; Ana Maria Calderon assisted in its production.

This book was composed using Xerox Ventura Publisher by Context, Inc., San Diego, California.

Contents

Preface

It can be argued that electronics is the science of measurement. Whether the end product is an automatic coffee maker, a VCR, or a mainframe computer, it took countless numbers of precise electronic measurements and accurate analysis of the results to produce that item. And, it will continue to take more measurements and decisions based on those measurements to keep that component maintained over its lifetime. Everywhere, electronics are metering quantities, monitoring materials, and controlling processes — all under precise measurement control.

That is why, regardless of your discipline, an understanding of electronic test equipment is essential. This couldn't be more true than of today's professional, whose very existence depends on the precise handling of measurements made using electronic test equipment.

The purpose of this book is to assist the professional in understanding the underlying theory and subsequent application of test instruments.

The book is a unique balance of theory and application. It is important to understand how something operates, but it's often more important to know how to put those features to work.

The book begins with a through analysis of the venerable analog meter and its applications. Covered are voltmeters of every type — from basic analog movements to electronic voltmeters — along with complete analysis of ammeters, resistance meters, and simple-to-complex volt-ohm meters (VOM). Particular emphasis has been placed on instrument-induced measurement errors; their cause and correction.

Chapter 2 is devoted to digital test instruments. From clocks to voltmeters, digital electronics have invaded our daily lives. This chapter addresses some misconceptions about digital instruments — particularly those related to measurement precision — and emphasizes the theory behind digital instrument design.

Signal sources are the subject of Chapter 3. Without calibrated signal sources accurate measurements are impossible. This chapter explains the complex relationship between the signal generator, the device under test, and the measuring instrument, emphasizing the changes in signal characteristics as it passes through each stage.

The oscilloscope is undoubtedly the most versatile of all laboratory test instruments, and is the topic of the next three chapters. These three chapters provide a comprehensive discussion of all aspects of oscilloscope design and application.

In the final two chapters we examine the latest area of test technology: computer-based testing. Although still in its infancy, computer-based testing is appearing in applications that were once the exclusive domain of the laboratory technician, especially in the fields of automotive and video diagnosis and repair. The in-depth analysis of this wide and varied subject runs the spectrum from computer-assisted instruments to dedicated PC-based systems to GPIB networks.

1. ANALOG INSTRUMENTS

Basic to all scientific investigation, electronic circuit design, circuit evaluation, and electronic servicing are accurate measurements, selective analysis, and mathematical formulation. First and foremost is measurement, defined as the process by which physical, analog quantities are converted to numbers that are used to analyze and formulate conclusions about the world around us.

Converting real-life quantities into numbers that we can deal with is done using test instruments. Test instruments come in a wide range of types and descriptions, each designed with a specific measurement to perform.

The most basic, and most venerable, of all test instruments is the *deflection pointer*, or *analog meter*. There was a time when the analog meter was the only source of reliable electric measurements. Recent advances in digital technology, however, have permitted the design of instruments that eliminate the measuring error introduced by the mechanical components of an analog meter. While it can be argued that deflection meters are not as accurate as modern digital laboratory test instruments, it is doubtful that the overall results obtained by a more precise test instrument would necessarily be more meaningful than those attainable from an inexpensive meter. Consequently, analog meters have secured their place in the laboratory and field as a reliable source of measurements.

Meter Basics

All deflection instruments require three forces for proper operation: a deflecting force, a controlling force, and a damping force. The deflecting force is commonly generated by a magnetic field produced by a current flowing through a coil. This field interacts with a controlling magnetic field to produce a repelling force. When a mechanical pointer is associated with one of the magnetic forces, movement is noted.

The earliest deflection device was Oersted's experimental measuring apparatus which he called a *galvanometer*. Essentially, it was a compass needle placed below a wire in which current was to be measured. With no current flowing, compass needle and wire were aligned in the north-south direction. When current passed through the wire, the compass needle was deflected toward the east-west direction by an amount proportional to the current flowing in the wire. In this device, the magnetic field of the earth is the controlling force, and the magnetic field generated by current flowing through the wire is the deflecting force. Damping, in the form of friction within the needle pivot, prevents the pointer from oscillating (or, as it is sometimes called, seeking).

In later years, Lord Kelvin improved the sensitivity of the device by wrapping turns of the wire around the body of the compass, thus intensifying the strength of the deflecting force at lower current levels.

Most meters today are of the D'Arsonval design. The D'Arsonval movement consists of a moving coil or pivoted coil suspended in the force field of a permanent magnet. The magnet is much larger than the coil and shaped so

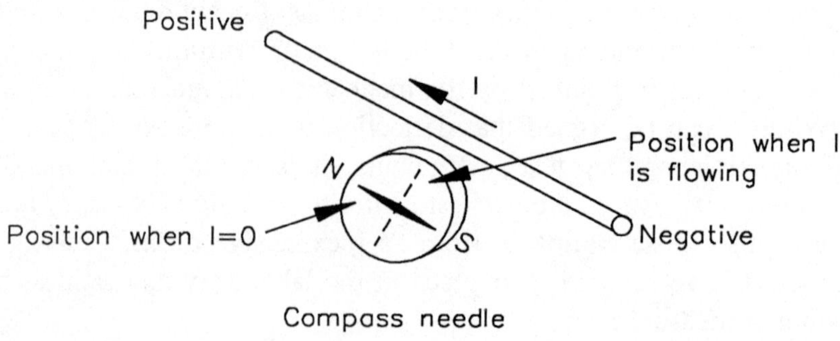

Figure 1-1 Oersted's galvanometer

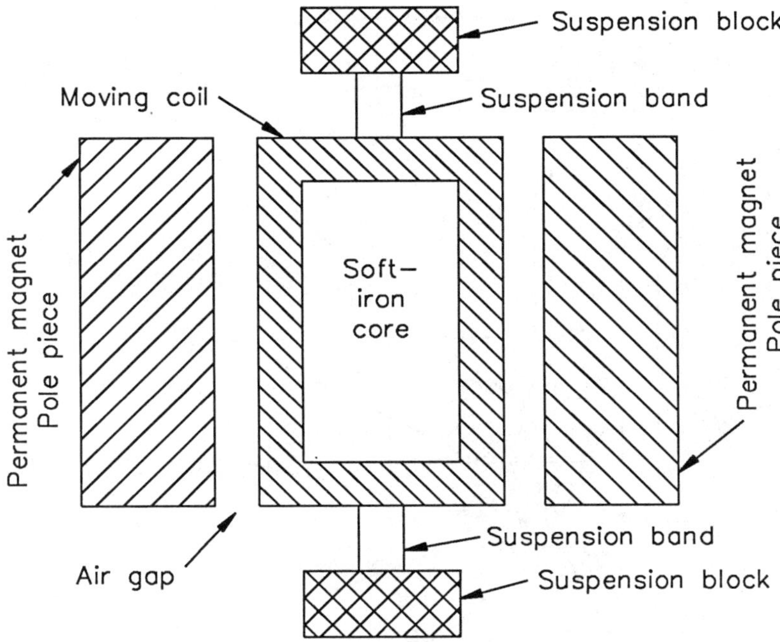

Figure 1-2 Taut-band suspension meter movement

that the coil can be suspended between its poles. Two types of suspension systems are used with D'Arsonval designs, taut-band and jeweled pivots.

The taut-band suspension type suspends the coil between two strips of flat metal stretched between a mounting block, as shown in Figure 1-2. This method prevents any movement of the coil except for rotation around the suspension axis. The inherent elastic properties of the metal provide the restoring torque for the controlling force.

In the pivoted suspension design, the coil is mounted on a shaft aligned perpendicular to its axis, and the shaft in turn is suspended top and bottom on a jeweled bearing so that it is rigidly held with a minimum of friction. In this configuration, the coil is free to spin around the suspension axis in much the same way that a motor armature rotates within its housing. Orientation and placement of the coil is done using spiral springs attached to the suspended coil. The springs retain the coil in a zero position when no current is flowing through the coil, and provide restoring torque to the movement in the way of a controlling force.

Taut-band meters are generally more accurate than pivoted meters because the suspension has no friction that the coil must overcome, but the lack

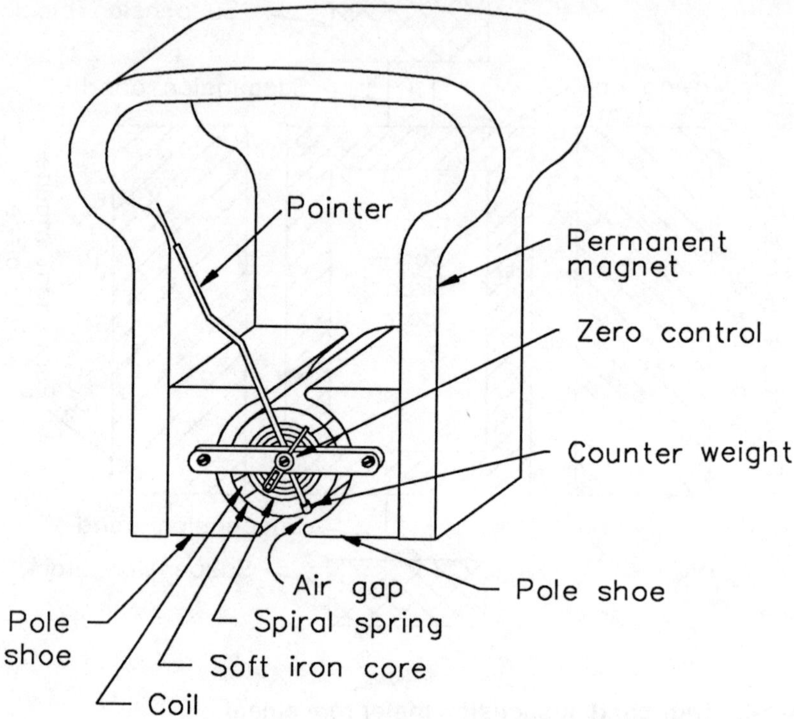

Figure 1-3 Jeweled-bearing D'Arsonal meter movement

of friction comes at the expense of some sensitivity. Jewel-pivoted meters are more sensitive to low currents but are also more sensitive to shock than are taut-band meter movements, and they can easily be damaged in the field.

In both designs, the deflecting force is provided by the interaction of the magnetic field generated by the coil when current flows through it and the opposing magnetic field of the permanent magnet. The coil's magnetic field is concentrated radially by a soft iron cylinder. When current is passed through the coil, horizontal forces called *side thrusts* are exerted on the vertical sides of the coil because of the interaction of the current and the established magnetic field. The strength of the side thrust (F) is proportional to the number of turns in the moving coil (N), the dimensions of the coil in length (L) and width (W), the current flowing through the coil (I), and the strength of the magnetic field, and is equal to:

$$F = NBIL$$

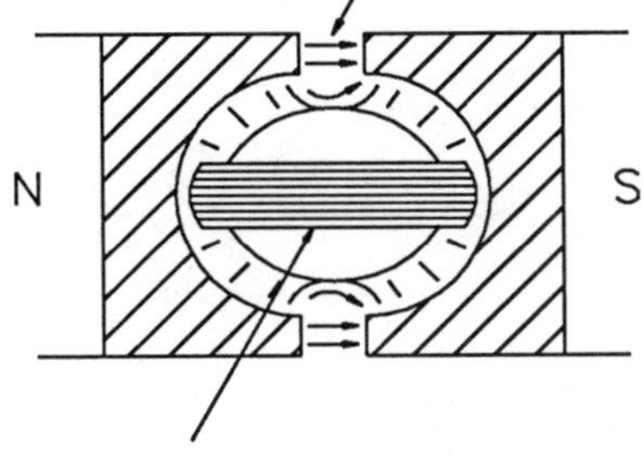

Figure 1-4 Torquing lines of force. Note that the magnetic flux lines of B are radially directed by the soft iron cylinder.

where

N = number of turns

B = magnetic field strength

I = current

L = vertical length of coil

The torque is calculated by multiplying the force times the distance to the point of suspension, and is equal to:

$$T = NBILW$$

where W/2 is the distance to the point of suspension.

This torque will cause the coil to rotate until an equilibrium position is reached at an angle ϕ with its original position. In this position, the torque exerted by the suspension is exactly equal to that due to the current and the field. The angle of rotation is equal to:

$$\phi = \frac{NBLW}{k} I$$

where k is a constant determined by the thickness, the width, and the spring quality of the suspending strip or restoring spring.

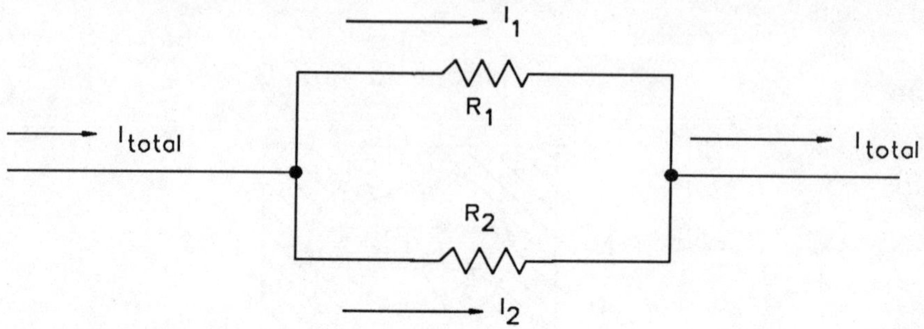

Figure 1-5 Current divider rule

Ammeters

The suspension D'Arsonval-type meter can detect and measure currents as small as 1 microampere, or 1 x 10^{-12} amp. To measure larger currents, the meter must be modified to increase its range. This is easily done using an external *current shunt* across the meter movement. A current shunt is nothing more than a resistor placed in parallel with the meter, thus forming a current divider.

The current divider rule states: The ratio of the currents in any two parallel branches is inversely proportional to the ratio of the resistances of the branches, or

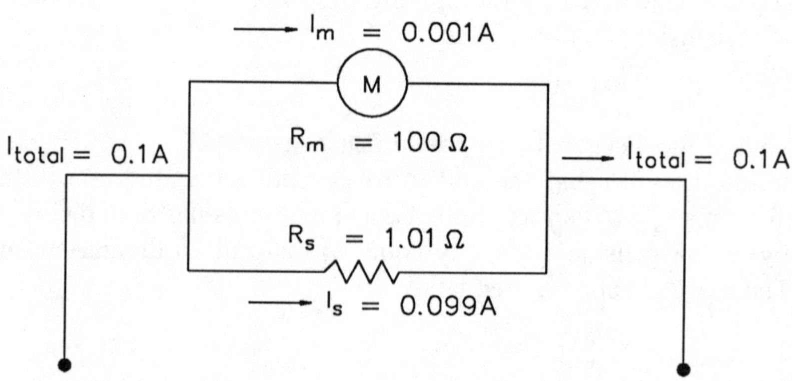

Figure 1-6 Current values in meter shunt

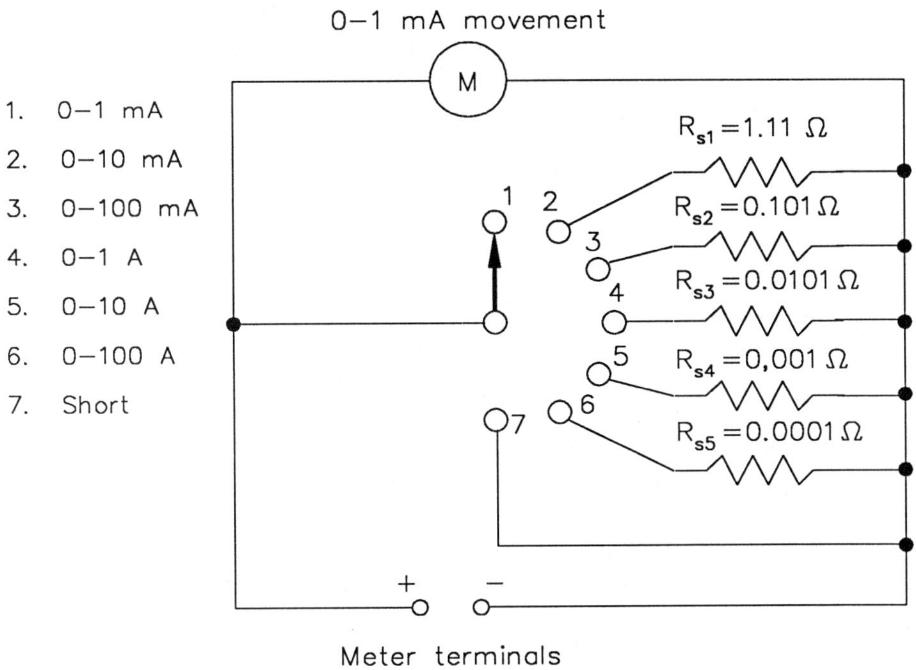

Figure 1-7 Simpson 373 ammeter

$$\frac{I_1}{I_2} = \frac{R_2}{R_1}$$

If the meter movement, with its internal resistance of R_m, is substituted for R_1, and the shunt resistance R_s is substituted for R_2, the equation reads:

$$\frac{I_m}{I_s} = \frac{R_s}{R_m}$$

If $I_s = I_{total} - I_m$, then

$$\frac{I_m}{I_{total} - I_m} = \frac{R_s}{R_m}$$

and

$$R_s = \frac{I_m R_m}{I_{total} - I_m}$$

This last equation can be used to calculate the value of a shunt resistor to increase the current range of the meter to display I_{total}. To allow a 1-mA meter with an internal resistance of 100 ohms to measure 100 mA of current, for example, the shunt resistance is:

$$R_S = \frac{(0.001)(100)}{(0.1) - (0.001)}$$

$$= \frac{0.1}{0.099}$$

$$= 1.01\ \Omega$$

Using this principle, any microammeter, milliammeter, or ammeter may measure a larger current of any magnitude by using suitable shunt resistors to change the range. In fact, this is how commercial instruments measure a wide range of currents using a single meter movement. A schematic diagram of a seven-range commercial ammeter is shown in Figure 1-7. Notice that a rotary switch is used to insert the various shunt resistors across the meter for the range selected.

CURRENT MEASURING ERRORS. An ammeter is always connected in series with the circuit in which current is to be measured. To avoid affecting the current level in the circuit during measurement, the ammeter must have a resistance much lower than the circuit resistance. Ideally, the ammeter would

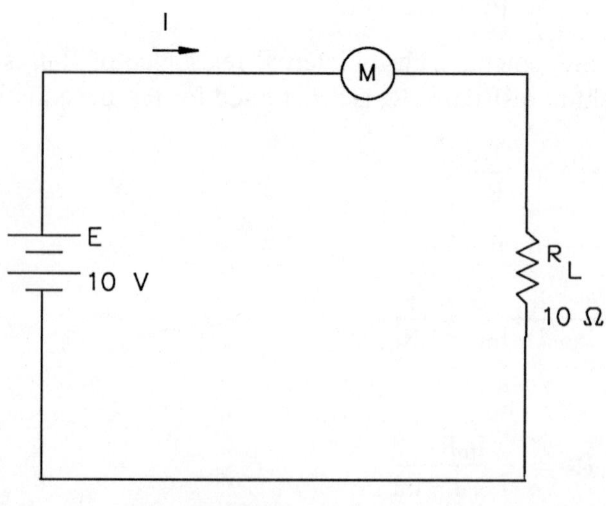

Figure 1-8 Current measuring circuit

have zero resistance, but such a meter is not possible. The effect of meter loading is demonstrated in the following example.

Consider the circuit below, in which an ammeter is connected to measure the current in a 10-ohm load supplied from a 10-volt source. Ohm's law says that under ideal conditions the current flow through the circuit is:

$$I = \frac{E}{R} = \frac{10}{10} = 1A$$

If the resistance of the ammeter is 1 ohm, however, the meter will not register 1 amp, but will indicate a current value of 0.91 amp. The discrepancy between the calculated and measured values is made apparent when the circuit is redrawn to include the resistance of the ammeter, as indicated in Figure 1-9.

Using Ohm's law, we can now conclude that the ammeter is telling the truth. With I = 10 V/11 Ω, the actual current flow is only 0.909 amp. The amount of error introduced by the insertion of the measuring device is equal to:

$$\%\ \text{error} = \frac{\text{ideal value} - \text{measured value}}{\text{ideal value}}\ (100\%)$$

$$\text{error} = \frac{1\ \text{A} - 0.909\ \text{A}}{1\ \text{A}}\ (100\%)$$

$$\text{error} = 9.1\%$$

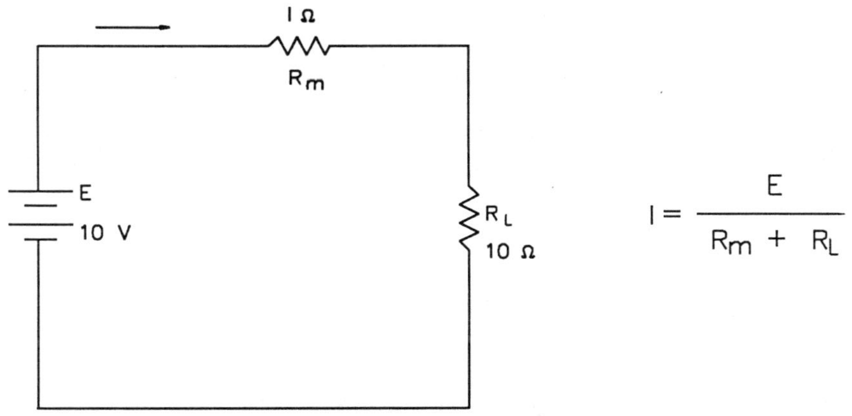

Figure 1-9 Equivalent current measuring circuit

If the resistance of the meter were reduced to 0.1 ohm, then the measured value would be I = 10 V/ 10.1 Ω = 0.99 A, and the amount of error created by the meter would be approximately 1%.

DC Voltmeters

The basic microammeter movement may also be adapted to measure voltage. Because the deflection of the instrument is proportional to the current flowing through the moving coil, Ohm's law states that the deflection of the pointer is also proportional to the voltage across the coil. Therefore, the scale of the meter can be calibrated to indicate voltage rather than current.

The coil resistance of most microammeters is normally quite small, on the order of 1000 ohms or less; thus the coil voltage is also usually very small. Full-scale voltage deflection of a 50-μA meter with a 1000-Ω coil, for example, is:

$$E = IR$$

$$= 5 \times 10^{-5} \times 2 \times 10^{3}$$

$$= 1 \times 10^{-1} \text{ V}$$

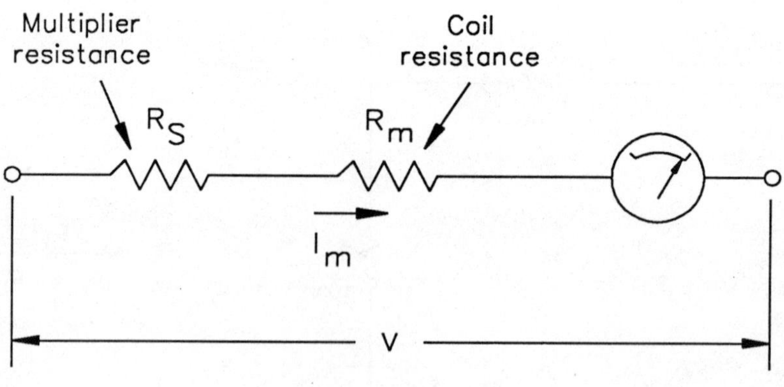

Figure 1-10 Basic voltmeter circuit

or 0.1 volt. Unfortunately, it is not very often that one needs to measure voltage values of 0.1 volt or less.

The range of a voltmeter can easily be increased by connecting a resistance in series with the instrument, as illustrated in Figure 1-10. The resistor limits the current through the meter so that the instrument can measure higher voltages than it can when using its own internal resistance. These resistors, whether mounted in or out of the meter case, are called *multipliers.* The value of the multiplier resistor can be determined by the current for full-scale deflection and by the range of the voltage to be measured. Since the current through the meter is directly proportional to the applied voltage, Ohm's law can be used to calculate its value accordingly:

$$R_S = \frac{E}{I_m} - R_m$$

where

R_m = resistance of the meter coil

R_S = multiplier resistor

I_m = full-scale deflection current of the meter

Consider a microammeter with an internal resistance R_m of 100 Ω and a full-scale current value of 100 A. For this meter to indicate the range of voltages from zero to 1 volt, the value of the multiplier resistor is:

$$R_S = \frac{1}{1 \times 10^{-4}} - 100$$

$$= 10{,}000 - 100$$

$$= 9.9\ \text{k}\Omega$$

When a 9.9-kΩ resistor is placed in series with this meter movement and the scale is calibrated in volts, the result is a 0- to 1-V voltmeter.

As with the ammeter, multirange voltmeters can be constructed using one meter movement and a number of multiplier resistors and a rotary switch. Two circuit configurations are possible. In Figure 1-11(a), individual multiplier resistors are selected by the rotary switch for insertion in series with the meter movement. In Figure 1-11(b), the multiplier resistors are wired in series, and the junction between each resistor is connected to one of the switch terminals.

Of the two circuits, configuration 1-11(b) is the least expensive to construct because only one resistor (R1) is a nonstandard value. The nonstandard

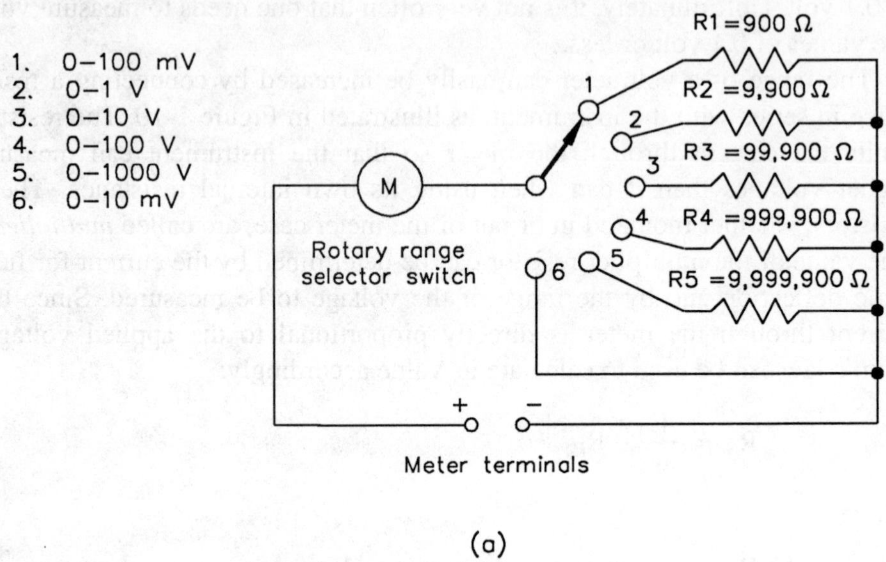

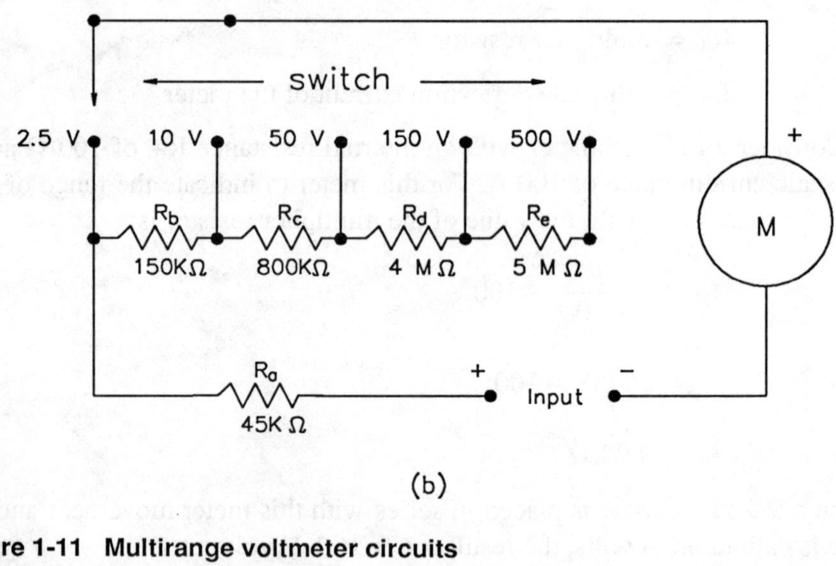

Figure 1-11 Multirange voltmeter circuits

value is required to compensate for the internal resistance of the meter. Once the meter's internal resistance has been brought up to standard value, all other resistor values will fall within the scope of standard (precision) resistor devices, leading to lower construction costs. Configuration 1-11(a), on the other hand, requires each resistor to be a nonstandard value because each is individually affected by the meter's internal resistance.

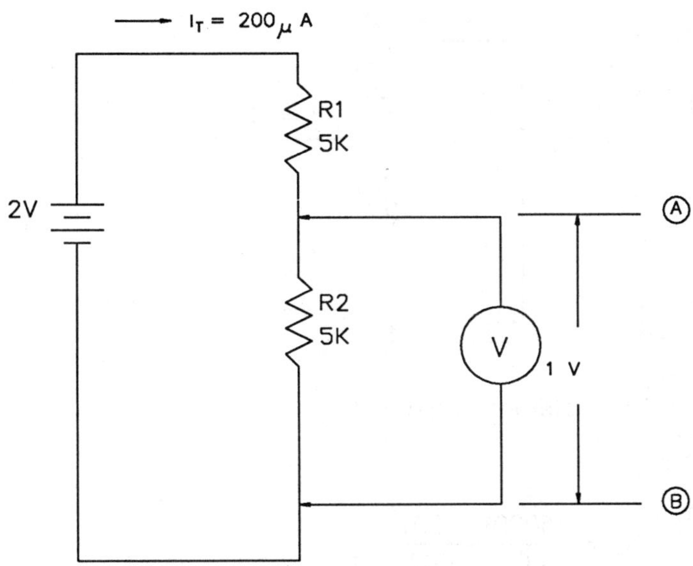

Figure 1-12 Series-resistance voltage divider

VOLTAGE MEASURING ERRORS. In contrast to the ammeter, voltmeters are always connected in parallel with the circuit under test. Because the voltmeter requires a finite amount of current (however small) to operate, the circuit under test is affected. This is known as *voltmeter loading effect.* Such a loading effect tends to lower the voltage across the source, which in turn affects the measurement.

The effect can be described using the example in Figure 1-12. In this example, a voltmeter is used to monitor the voltage across one resistor of a series pair. In circuit 1-12, the potential across points A and B is 1 volt without the voltmeter connected. This is obvious because resistance values are equal, thus dividing the source voltage exactly in half.

When the voltmeter is applied, however, the equivalent configuration of the circuit changes. The voltmeter can be represented by a defined resistance in parallel with R2, as illustrated in Figure 1-13.

Assuming the voltmeter has an effective resistance of 5000 ohms, that being the sum of R_s and R_m, the equivalent resistive load R_L between points A and B can be calculated as being:

$$R_L = \frac{R2(R_s + R_m)}{R2 + (R_s + R_m)}$$

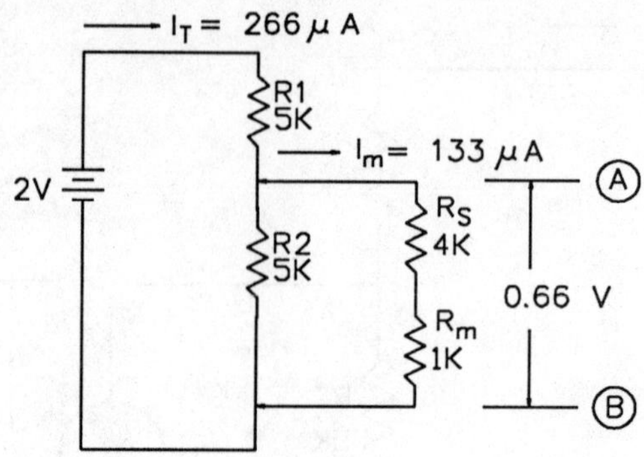

Figure 1-13 Voltmeter loading effect

$$R_L = \frac{(5000)(5000)}{5000 + 5000}$$

$$= 2500\ \Omega$$

Application of Ohm's law, using the new circuit values, yields:

$$I = \frac{2\ \text{V}}{7500\ \Omega}$$

$$= 2.66 \times 10^{-4}\ \text{A}$$

Calculation for the voltage across A and B yields:

$$V_{A\text{-}B} = (2.66 \times 10^{-4})(2500)$$

$$= 0.665\ \text{V}$$

As we can see, the voltmeter will not indicate the correct voltage of the circuit under test, but instead modifies the values to give a false measurement. Instead of indicating the true value of 1 volt, the voltmeter reads 0.66 volt.

The voltmeter loading factor is called the *sensitivity*, and is expressed in *ohms per volt*, where:

$$\text{sensitivity} = \frac{R_S + R_m}{\text{full-scale voltage value}}$$

The higher the sensitivity of the voltmeter, the less current will be drawn and the less circuit loading. In the example above, the sensitivity of the voltmeter involved is:

$$\text{sensitivity} = \frac{5000}{1\ \text{V}}$$

$$= 5000 \text{ ohms per volt}$$

or,

$$= 5000\ \Omega/\text{V}$$

The ideal voltmeter has infinite ohm-per-volt sensitivity in that the circuit parameters are not affected by its presence. Laboratory and service quality analog instruments commonly have a sensitivity of 20,000 Ω/V.

One way to reduce the effects of voltmeter loading is to switch the voltage range to a higher scale. If the voltmeter in Figure 1-13 had a full-scale deflection of 10 volts instead of 1 volt, its internal resistance would be:

$$R_{in} = (\text{full-scale voltage value})(\text{sensitivity})$$

$$= (10)(5000)$$

$$= 50\ \text{k}\Omega$$

This is ten times more resistance than the previous meter setting, and represents a lighter load that has less effect on the circuit under test. When this value is substituted for the values in Figure 1-13, the resultant meter reading is 0.95 volt, an error of only 5 percent.

Unfortunately, analog meters are notoriously inaccurate at the lower end of their scale. Because of frictional forces within the bearings of the coil mounting, a certain amount of torque must be applied before pointer deflection is achieved. As the current through the coil increases, the repelling forces are greater, and the effects of bearing friction are less noticeable. Consequently, readings that fall on the lower one-third of a mechanical meter's scale are less accurate than those taken at half scale or full deflection.

Ohmmeter and Resistance Measurements

Analog meters can also be used to measure resistance. Resistance, as defined by Ohm's law, is the constant of proportionality between the current and the voltage of the current. Consequently, if we know the amount of current flowing through a device and the voltage drop across it, both quantities of which can be measured by a meter, we can calculate its resistance using Ohm's law. Mathematically, the quantity is expressed as:

$$R = \frac{E}{I}$$

The ammeter-voltmeter method of measuring resistance is an indirect method inasmuch as it is a calculated value. Neither meter reading is a direct indication of resistance value.

By setting either voltage or current constant and monitoring the variable value, we can scale the meter so that it does the resistance calculation for us. There are two types of analog ohmmeters in popular use: the series ohmmeter and the shunt ohmmeter.

SERIES OHMMETER. The *series ohmmeter* provides a constant voltage source and derives its resistance measurement from the variable current value. Figure 1-14 illustrates the circuit of the basic series ohmmeter.

A battery (E) is connected in series with a variable resistor (R1), a fixed meter resistance (R_m), and a 0- to 1-mA meter movement. The two test points P1 and P2 represent test probes that are connected across the unknown

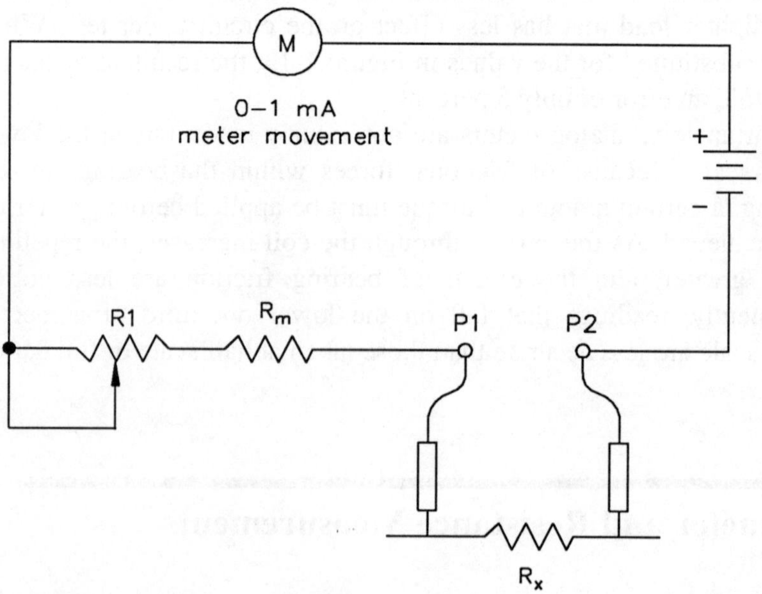

Figure 1-14 Series ohmmeter

resistance (R_x) to be measured. Since the meter reads 1 mA at full scale, the value of R1 and R_m must be such that:

$$\frac{E}{R1 + R_m} = 1 \text{ mA}$$

so that full-scale deflection is achieved when the test probes are shorted together. This condition indicates an R_x value of zero ohms, and the meter scale can be marked accordingly.

When a resistance of any value greater than zero is inserted between P1 and P2, the meter current decreases proportionally. The meter current indicated by the instrument is equal to:

$$I_m = \frac{E}{R_x + R1 + R_m}$$

An R_x value equal to R1 + R_m, for example, would yield a meter reading of:

$$I_m = \frac{E}{(R1 + R_m) + R1 + R_m}$$

$$= \frac{E}{2R1 + 2R_m}$$

$$2I_m = \frac{E}{R1 + R_m}$$

In other words, the meter would indicate half scale, and this point could be marked with the appropriate resistance value. If the total internal meter resistance was 100 ohms, for example, then this point would be marked with a value of 100. By proceeding with this type of calibration using a wide range of resistance values, one is able to construct a meter scale reflective of the unknown resistance. Such a meter scale is illustrated in Figure 1-15(a).

Notice that the scale is nonlinear and tends to be overcrowded near the high-resistance end of the scale. As the value of R_x increases, the internal resistance of the instrument becomes less of a factor and has less influence on the current and meter reading. As R_x approaches infinity, the instrument's current rapidly decreases to near zero, virtually swamping the meter's internal resistance by its shear magnitude, thus creating the crowding conditions noted on the scale.

SHUNT OHMMETER. When very low values of resistance are to be measured, the shunt ohmmeter is superior to the series ohmmeter just described. The shunt ohmmeter relies on the internal resistance of the meter

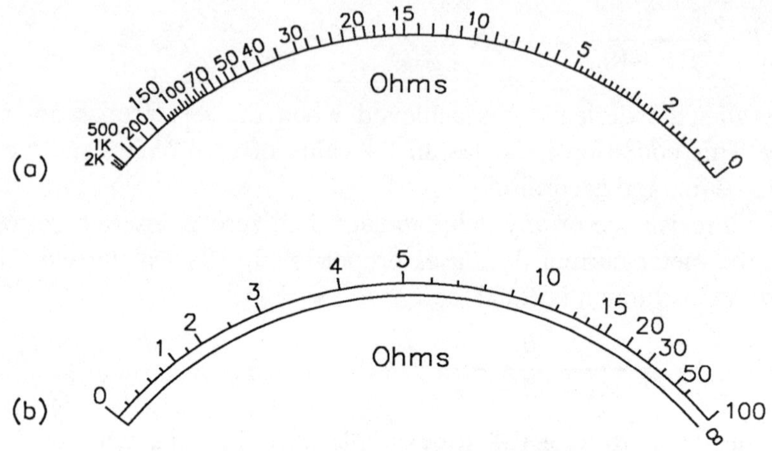

Figure 1-15 Series (a) and shunt (b) ohmmeter scales

movement (R_m) to derive its resistance measurements. The circuit of a shunt ohmmeter is shown in Figure 1-16.

In operation, the current in the ohmmeter is first adjusted by R1 to provide full-scale deflection on the meter with the test probes separated. This reading represents an unknown resistance R_x of infinity, and the scale is marked accordingly.

When an unknown resistance is connected between the probes, R_x acts like a current shunt and passes a portion of the current around the meter, decreasing the meter reading proportionally. When $R_x = R_m$, the amount of current shunted by the unknown resistor is equal to the amount of current recorded by the meter, and the meter pointer indicates half scale. If the internal meter resistance is 100 ohms, then R_x is equal to 100 ohms. This value is marked on the scale, and the calibration process continues. When finished, the scale looks like the drawing in Figure 1-15(b).

Notice that the shunt ohmmeter scale is calibrated from left to right, as opposed to the series-type ohmmeter, which is calibrated from right to left. Again, we have the typical crowding of the scale as R_x approaches infinity. An inherent disadvantage of this circuit is the fact that the battery is always in the circuit, so that unless the instrument is turned off when not in use, the battery runs down very rapidly.

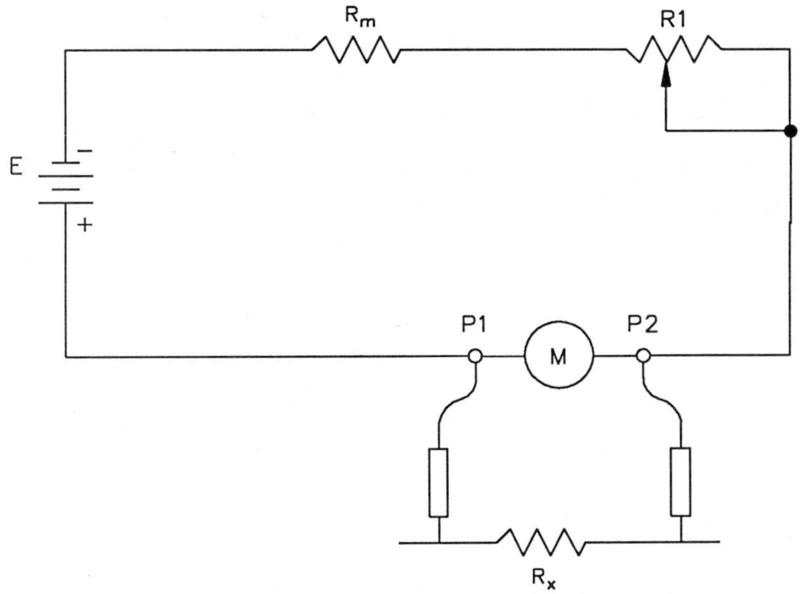

Figure 1-16 Shunt ohmmeter

OHMMETER ACCURACY. Inaccuracies in the measurement of resistance using analog meters can occur in a number of ways. The nonlinearity of the scale puts an immediate stress on the precision of the measurement made near the infinity region. A small displacement in needle position within this region can translate into a resistance difference of up to a magnitude.

Another problem occurs in the way the factory rates the accuracy of the meter itself. The factory-rated accuracy of a meter is based on a full-scale dc voltage indication. The amount of error is determined by an area of indecision brought about by mechanical limitations inherent to the meter, described as the *arc of error*. Frictional losses in the bearings of the coil pivot or taut-band suspension create a hysteresis effect that the meter must overcome before the pointer comes to final rest. Depending on the direction of pointer movement up or down scale, the needle can fall to either side of the arc of error, or anywhere in between.

Suppose that the rated accuracy of the meter is specified as ±1% of full-scale voltage indication. This means that the voltage read at the center of the scale (one-half full deflection) is ±1% the full-scale value, or ±2% of the indicated voltage.

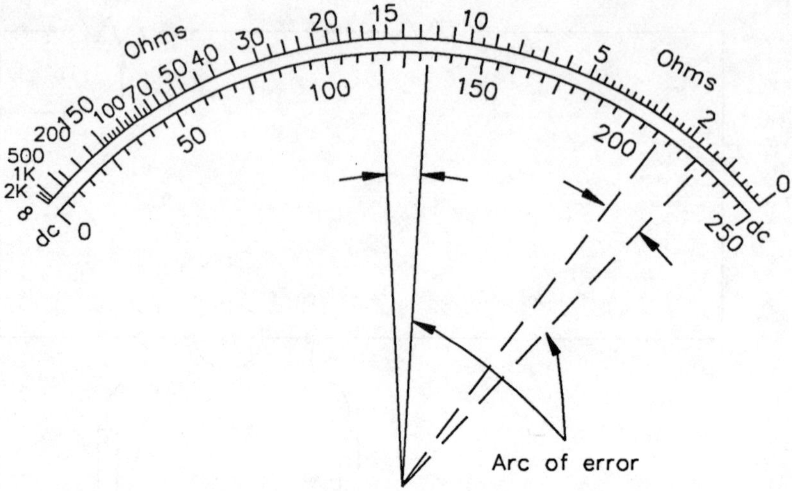

Figure 1-17 Arc of error measurement

Because of the nonlinearity of the ohmmeter scale, the arc of error can cause a relatively greater error in resistance measurements than it does in voltage or current measurements. The relationship between the arc of error for dc voltages and the accuracy of the ohmmeter scale is demonstrated in Figure 1-17. While a series-type scale is illustrated, the same reasoning applies to a shunt-type scale.

When one considers the arc of error in light of scale nonlinearity, it is evident that the greatest accuracy can be achieved when the meter gives an indication as close to half scale as possible. This occurs when R_X is equal in value to the effective internal resistance that exists across points P1 and P2, whether the instrument be a series or shunt type.

Volt-Ohm-Milliammeters

Seldom are analog ohmmeters found as stand-alone instruments. An ohmmeter is normally part of a *volt-ohm-milliammeter* (VOM) or *multifunction meter*. The VOM is a multifunction instrument capable of operating as an ac or dc voltmeter, an ac or dc ammeter, or a multirange ohmmeter. A function switch is provided to permit selection of any function, and a range switch facilitates range changing. Additional functions, such as decibel measurements, are available with the use of special adapters or probes.

In essence, the VOM is many separate instruments in one, with all using the same meter movement as a readout. The circuitry for each function is basically the same voltmeter, ammeter, or ohmmeter circuit described above. The precision resistors used in the instrument normally serve in as many functions of the instrument as possible, to reduce instrument complexity and overall cost. The circuitry is simply arranged so that only two selector switches are required to configure the instrument: function and range.

To use the VOM, one simply chooses the function and range desired, and proceeds to measure accordingly. A typical VOM multimeter is shown in Figure 1-18.

The *function switch* facilitates selection of +dc volts and current, –dc volts and current, or ac volts and current. The *range switch*, located in the center of the instrument below the meter, selects the scale for: voltage (2.5-V through 500-V), current (1 mA through 500 mA), and resistance (1 through 10k). Additional functions are available by using the labeled probe jack located on the front of the VOM.

The terminals marked COMMON (–) and (+) are those normally employed for all voltage, current, and resistance measurements. The +10-A and –10-A

Figure 1-18 Simpson Model 260 (courtesy of Simpson Electric)

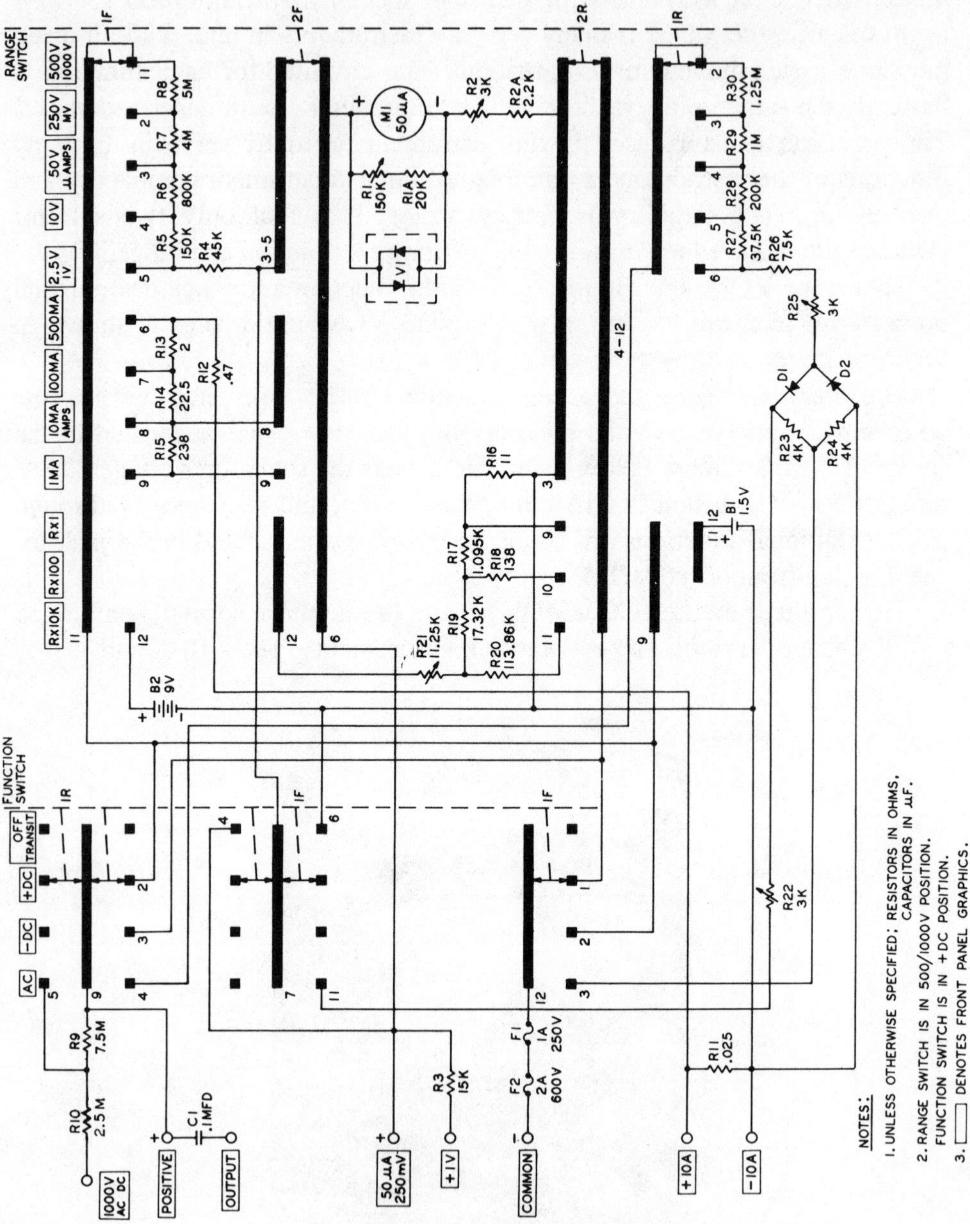

Figure 1-19 Schematic of Simpson Model 260 (courtesy of Simpson Electric)

terminals are used to measure dc current levels ranging from 0-10 A when the range switch is in the 10-A position. These jacks permit direct access to the meter movement and use a meter shunt for proper ranging. Direct access to the meter movement with no current shunt is through the +50 AMPS/250 MV jack, with the range switch set to 50 V and COMMON used as the return line. Maximum input voltage in this mode is 250 millivolts.

To measure voltages, both ac and dc, greater than 500 volts, the test probe is removed from the + input terminal and plugged into the 1000-V terminal. The range switch is set to the 500-V position, effectively doubling the range of the instrument. For voltages less than 2.5 volts but greater than 250 millivolts, one can plug into the +1 V terminal and twist the range switch to the 2.5-V position while reading from a 0-10 scale.

Over the years, the VOM has become more than a versatile instrument, with a wide range of features. For some it is an institution. Consider the plight of the field engineer or technician, whose job would be made very inconvenient, if not impossible, if he had to carry around an arsenal of instruments for service.

Electronic Voltmeters

Voltmeters constructed of moving-coil meters and multiplier resistors have some important limitations. As illustrated above, the most severe is an inherently low internal resistance, which leads to significant loading of many circuits under test and subsequent measurement errors. The instrument is also incapable of measuring very low voltage levels. These limitations can be overcome using an instrument that can amplify and buffer the input signal from the mechanical meter movement. Examples of such devices are the *electronic voltmeter* and the *electronic multimeter.*

The electronic multimeter was originally called the *vacuum tube voltmeter* (VTVM) because it was designed using vacuum tubes. With the advent of the transistor and later the integrated circuit, vacuum tubes were dropped in favor of more stable semiconductor devices and the instrument became known simply as the electronic voltmeter.

TRANSISTORIZED ELECTRONIC VOLTMETERS. The concept of the transistorized electronic voltmeter is basically one of placing a metering device within a differential amplifier circuit. The differential amplifier

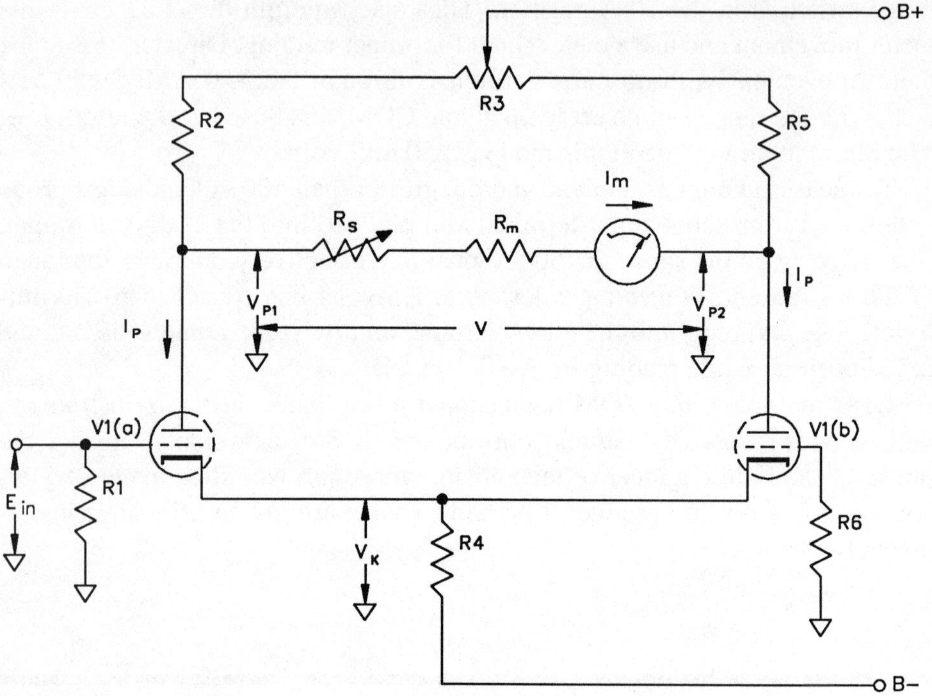

Figure 1-20 VTVM schematic

provides both amplification and isolation of the input signal from the meter movement. Refer to Figure 1-21 for the following analysis.

Transistors Q1 and Q2, together with R2, R5, and R4, constitute a differential, emitter-coupled amplifier. The amplifier gets its name from the fact that when the voltage at the base of Q2 is zero and a signal voltage is applied to the base of Q1, the difference between the two base voltages is amplified and reflected in the collector voltages.

Differential amplifiers require both a positive and a negative power supply to work properly. The common junction of the two power sources is the virtual ground. With the bases of both Q1 and Q2 held at ground (zero volts), the voltage drop across emitter resistor R4 is:

$$V_{R4} = 0 - V_{BE} - (-V_{EE})$$

or approximately V_{EE} less 0.7 volt. The current through R4 is:

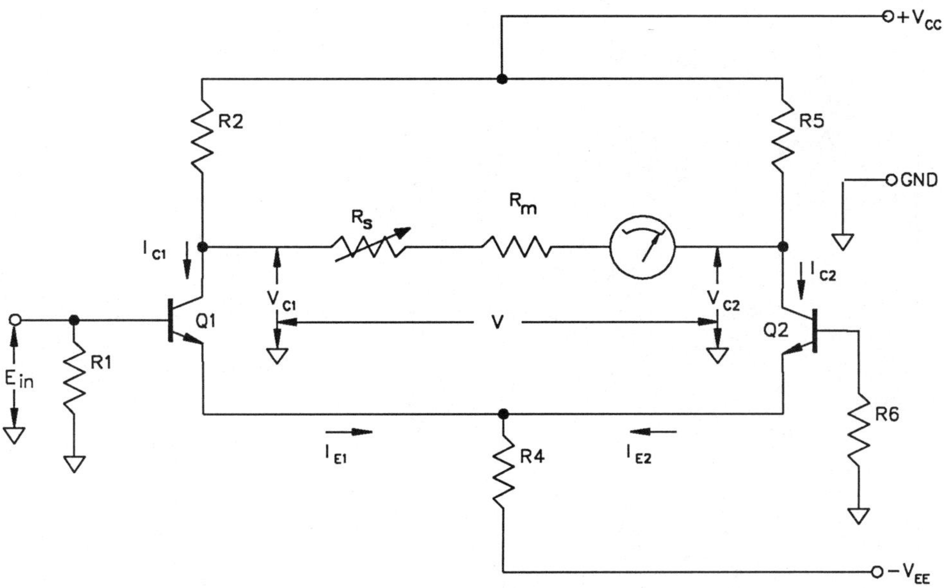

Figure 1-21 Differential amplifier stage

$$I_{E1} + I_{E2} = \frac{V_{R4}}{R4}$$

When $E_{in} = 0$, then $I_{E1} = I_{E2}$. If the transistors are sufficiently matched in characteristics, then $I_{C1} = I_{C2}$, and $V_{R2} = V_{R5}$.

This condition forces the voltage at the collector of Q1 equal to the voltage at the collector of Q2. If a meter or other voltage monitoring device is now placed between the collectors of the two transistors, the meter would register no difference in voltage, and remain at rest.

Because the voltage drop across R4 must be maintained to satisfy the V_{BE} requirements of transistor Q2, a positive voltage applied to the base of Q1 causes the current through Q1 to increase and the current through Q2 to decrease. This in turn produces a larger voltage drop across R2 and a smaller voltage drop across R5, with respectively lower and higher voltages at the collectors of Q1 and Q2.

The consequence of this is a differential voltage between the collectors, which the meter sees as a positive potential at the collector of Q2 and a negative potential at the collector of Q1. The meter voltage is directly proportional to the input voltage E_{in}, resulting in a meter reading that is representative of the signal voltage at E_{in}.

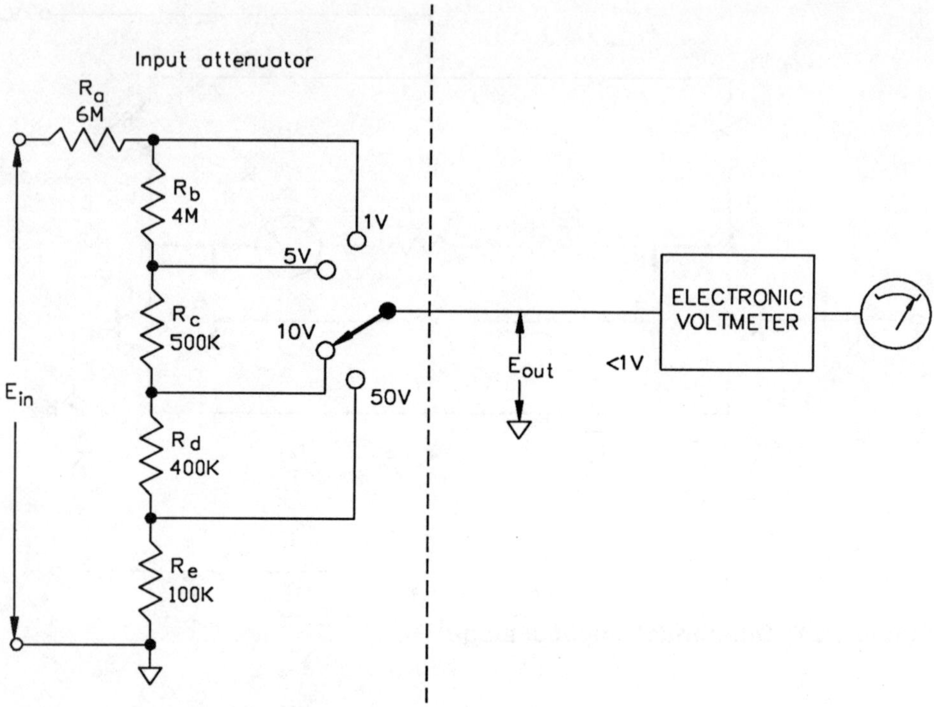

Figure 1-22 Input divider network

A voltage divider made of a resistor ladder limits the voltage to the input of the differential amplifier and essentially sets the instrument's range. The voltage divider, which consists of resistors R_a through R_e, accurately divides the voltage to be measured before it is applied to the input transistor, as illustrated in Figure 1-22. This arrangement limits the input to the input transistor to less than 1 volt when the selector switch is properly set for the voltage under test.

This attenuator configuration has a distinct advantage over other divider arrangements in that it presents a constant input resistance to the circuit under test, regardless of the voltage range setting. The term sensitivity no longer has meaning with an electronic voltmeter. The term generally applied to electronic voltmeters is *input impedance.*

When using the attenuator described above with a bipolar transistor, however, the transistor input resistance has an effect on the instrument's input impedance. Because the base resistance is in parallel with the divider, some error is introduced into the attenuator in terms of amplifier loading. Assuming that R4 in Figure 1-21 is 10 kΩ, the resistance seen at the base of Q1 is approximately 20 kΩ to 40 kΩ, depending on the characteristics of the transistor. When this parallel load is placed across the divider resistors, it lowers the total resistance by an amount that can be determined by simple parallel resistor laws. For the sake of argument, assume that the base input resistance is 40 kΩ. When the selector switch is in the 50-volt position, this 40 kΩ is in parallel with R_e, leading to an effective resistance of:

$$R_e' = \frac{R_{ib}R_e}{R_{ib} + R_e}$$

$$R_e' = \frac{(40k)(10k)}{40k + 10k}$$

$$R_e' = 28.6k$$

This is a far cry from the intended 100k value for R_e, and is totally unacceptable. To resolve the problem, the input impedance of the differential amplifier must be increased.

FET VOLTMETERS. A device capable of the task is the *field-effect transistor*, or FET. When an FET is connected at the input of the differential amplifier, the input resistance of the electronic voltmeter is significantly increased. Because the FET gate current is typically less than 50 nA, it has an equivalent input impedance of 2×10^{10}. MOSFET devices exhibit even larger input impedances. The resulting circuit is illustrated in Figure 1-23. In this configuration, the source terminal voltage changes by the same amount as the voltage at the FET gate terminal, and is called a *source follower*.

Although a definite value of h_{FE} has been assumed in Figure 1-21 for transistors Q1 and Q2, h_{FE} can vary enough from one transistor to another to create an imbalance within the differential stage, even when the transistors are closely matched. For this reason, an electrical zero control (R3) is provided to balance the currents flowing through Q1 and Q2 for proper meter operation. With the test probes of the instrument shorted together, this control is adjusted so that the meter registers zero volts.

OPERATIONAL AMPLIFIER ELECTRONIC VOLTMETERS. Recent advances in integrated circuits have yielded operational amplifiers with characteristics stable enough for use in electronic voltmeters. The input resistance of IC op amps, as they are called, is very high. Those with bipolar transistor input stages have typical input currents on the order of 0.2 A, while FET input op amps measure their input currents at 50 pA and less. The output resistance of an op amp is ideally zero ohms, and is quite capable of supplying the current necessary to drive a deflection instrument.

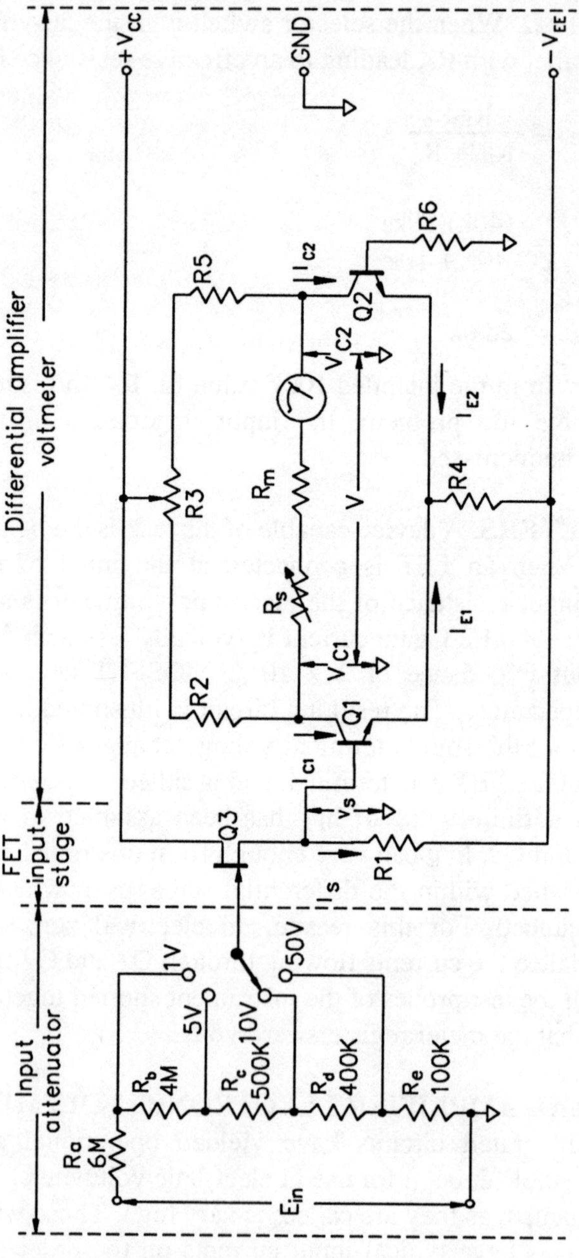

Figure 1-23 FET electronic voltmeter

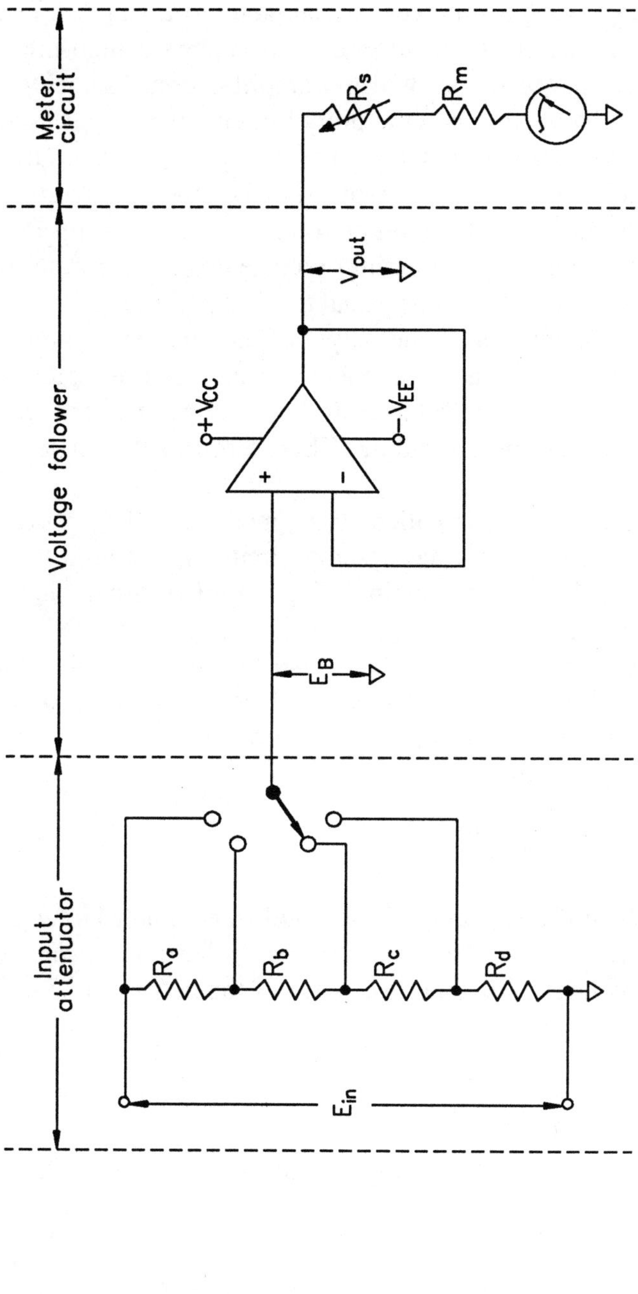

Figure 1-24 Voltage follower op amp voltmeter

Operational amplifier electronic voltmeters come in three circuit configurations, depending on application. They are voltage follower, noninverting, and voltage-to-current converters.

The op-amp voltage follower is analogous to a single transistor emitter follower circuit, whose sole purpose is to change a high-impedance input onto a low-impedance output with no amplification. Referring to Figure 1-24, the gain of the op amp is set at unity by tying the output of the op amp to its inverting input. By setting the gain of the op amp at unity, the voltage follower accurately reproduces the input signal at the noninverting input at its output. Unlike the transistor emitter follower, the op-amp voltage follower does not suffer from a base-to-emitter voltage drop as it mimics the input signal. The output voltage is always equal to the input voltage.

The meter is connected to the output of the op amp through a network of series resistances. Resistance R_m is the internal meter resistance and is an inherent characteristic of the meter itself. R_s is a series current-limiting resistor used to calibrate the meter reading. The attenuator circuit feeds the noninverting input.

When the instrument is called on to measure voltages that are smaller than the meter can display, the op-amp circuit can assume the role of both buffer and amplifier, using a noninverting amplifier circuit similar to the one in Figure 1-25.

A noninverting amplifier is very easily designed. A feedback network, consisting of R4 and R3, returns current from the op-amp output to the inverting input, and thus sets the amplification factor. The gain of the circuit is calculated using the equation:

$$Av = \frac{R3 + R4}{R3}$$

The potential divider current I is selected very much larger than than the input bias current (I_B) of the op amp so that I_B has no significant effect upon the feedback voltage. The total resistance of the network (R3 + R4) is calculated as:

$$R3 + R4 = \frac{V_{out}}{I}$$

The voltage across R3 is always equal to E_{in}, and its value is calculated using:

$$R3 = \frac{E_{in}}{I}$$

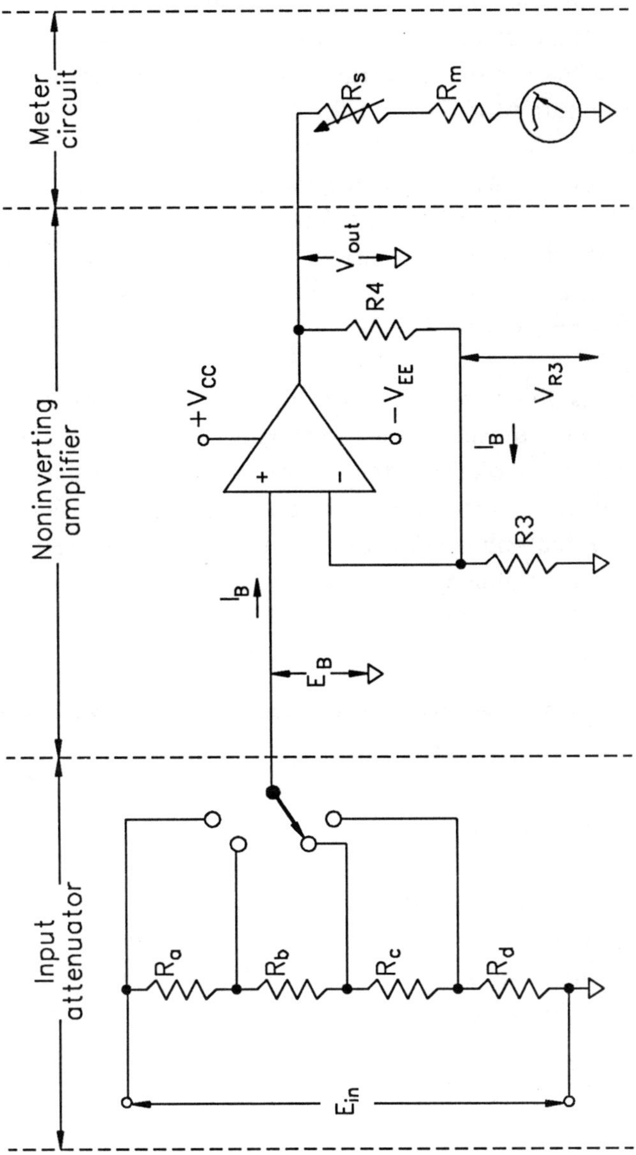

Figure 1-25 Noninverting op amp voltmeter

Apart from any error due to the deflection instrument or inaccuracies in the resistors, there is a small error introduced by the internal gain of the operational amplifier. In both circuits, however, the error is negligible compared to the normal deflection meter error.

The voltage-to-current converter is a variation on the noninverting amplifier configuration of Figure 1-25. In the voltage-to-current converter, a meter is substituted in place of resistor R4, and R3 is made partially adjustable. As with the noninverting configuration, the voltage drop across R3 is always equal to input voltage E_{in}. If the input signal increases or decreases, V_{R3} follows it precisely. So, the current through the meter (I) is equal to:

$$I = \frac{E_{in}}{R3}$$

In other words, the meter reading is a direct function of the input voltage. Part of R3 is made adjustable for instrument calibration. Calibration is performed by inputting a voltage to the noninverting input of the op amp and adjusting R3 until the meter pointer indicates the applied voltage. For proper circuit operation, the voltage developed across R_m + R3 must always be less than the V_{CC} supply voltage.

Electronic Current Measurements

The electronic multimeter is basically a voltage measuring device. Consequently, when using an electronic voltmeter to measure current, the current must first be converted to a voltage. The network in Figure 1-26 is used for current measurement.

This circuit represents a selectable resistance shunt, across which is generated a voltage. When the voltage is measured and displayed on an analog meter with the proper scale markings, it is representative of a current value according to Ohm's law:

$$I = \frac{E_R}{V_m}$$

where E_R is the voltage across the current shunt and V_m is the meter reading.

For each range starting at 1 amp, an additional resistor is added in series until a total of 450 ohms is inserted into the circuit under test. Note that the resistance values are selected so that the shunt voltage never exceeds 0.455 volt at each range setting. The lower the insertion loss of the shunt resistor,

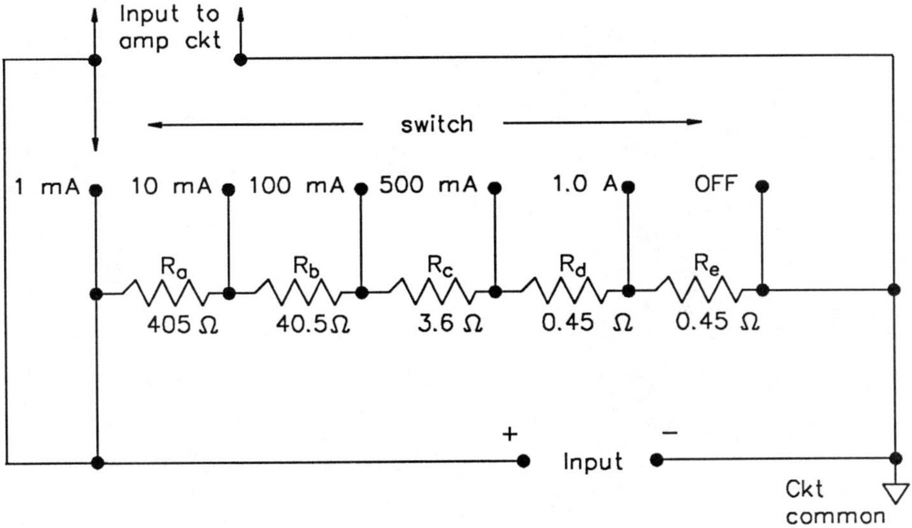

Figure 1-26 Electronic current measurement

the less error the instrument introduces to the circuit under test and the more accurate the measurement.

Unfortunately, the current function of most electronic multimeters inserts a resistance value into the circuit under test that is considerably greater than that presented by a laboratory current shunt, which typically displays a 50-mV voltage drop at the same current value. While the induced error caused by the multimeter is unacceptable for serious laboratory studies, the convenience of an all-in-one instrument for field service and bench work far outweighs the disadvantage in these applications.

Resistance Measurements

The electronic voltmeter can also be used to measure resistance by providing a fixed current through an unknown resistance, then measuring the voltage drop across the unknown resistance and correlating the voltage to a resistance value. All that is required is a battery, several standard resistors, and a suitable switching arrangement. There are three methods normally used by electronic voltmeters to measure resistance.

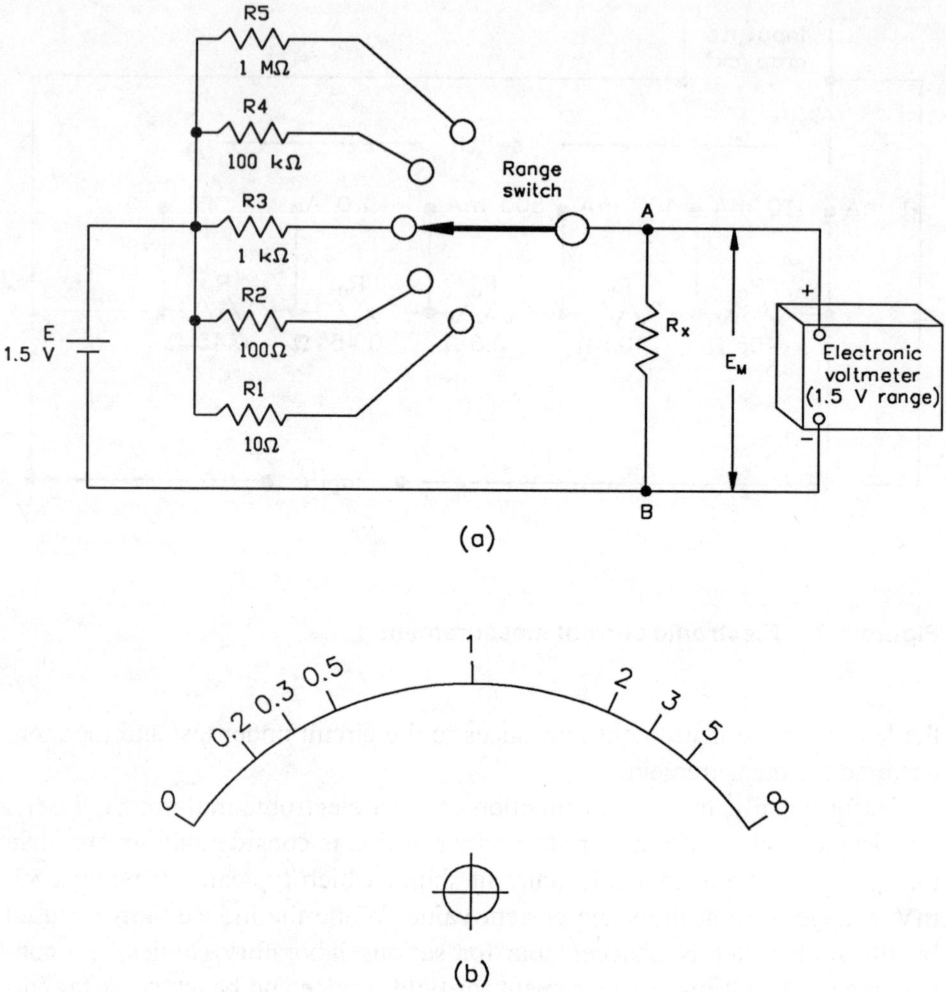

Figure 1-27 Series ohmmeter circuit for electronic instrument

SERIES ELECTRONIC OHMMETER. The circuit illustrated in Figure 1-27(a) is of the *series ohmmeter* type. In this example, a 1.5-volt battery is used in conjunction with the 1.5-volt range of the electronic voltmeter. Resistors R1 through R5 establish a selectable range of currents for use with the ohmmeter function. The unknown resistor connects the current-limiting resistor selected and the negative terminal of the 1.5-volt battery. The electronic voltmeter then measures the voltage drop across the unknown resistance, R_X.

Suppose the range switch is set, as shown, so that R3 (1k) is the current-limiting element. With no resistance connected between terminals A and B, representing a resistance of infinite ohms, the voltmeter will indicate 1.5 volts, or full-scale deflection. If terminals A and B are shorted together, representing zero ohms, no voltage is measured by the voltmeter and the pointer remains at zero, thus displaying 0 ohms.

Now suppose that an unknown resistance R_X greater than zero but less than infinity is connected to A and B. The unknown resistance forms a series voltage divider with R3. The battery voltage E_B is now potentially divided across R3 and R_X, giving us the expression:

$$E_m = E_B \frac{R_X}{R3 + R_X}$$

When $R_X = R3 = 1k$, the voltmeter indicates:

$$E_m = 1.5\ V \times \frac{1k}{1k + 1k}$$

$$= 0.75 \text{ volt}$$

The pointer now rests exactly center on the meter's scale, and the scale is marked 1 at this point. The meter will always indicate half scale when R_X equals the current-limiting resistor (R1 through R5) on whatever scale is selected.

Further calculations demonstrate that the scale is logarithmic, and that the readings become progressively cramped at the full-deflection end, as indicated in Figure 1-27(b), making accurate judgment of the value within this region virtually impossible. Inaccuracies within the meter movement, on the other hand, limit the accuracy of readings on the lower one-third portion of the scale. Consequently, this type of instrument is most accurate when indicating values close to half-scale deflection.

It should be noted that the electronic series ohmmeter requires two adjustments before use. First, the voltmeter must be zeroed electrically when the test probes are shorted together in the zero resistance mode. Next, the test probes are separated in the open-circuit mode and the meter is calibrated for full-scale deflection. The unknown resistance can be measured only after both adjustments have been made. As the 1.5-volt battery ages, its terminal voltage drops, and a time will come when full-scale (open-circuit) calibration is impossible. The cure is battery replacement.

SHUNT ELECTRONIC OHMMETER. A *shunt-type ohmmeter* uses a regulated power supply to provide a stable 6-volt power source, instead of employing a battery. By stabilizing the voltage source, one adjustment can be eliminated. Unlike a battery, the regulated power supply voltage does not fluctuate with use, and a one-time calibration for full-scale deflection is sufficient.

A circuit for a shunt-type ohmmeter is shown in Figure 1-28. With terminals A and B open circuited, $R_X = \infty$ and:

$$E_m = E_{source} \frac{R2}{R1 + R2}$$

For the values shown in Figure 1-28:

$$E_m = 6V \frac{1.33k}{4k + 1.33k}$$

$$= 1.5 \text{ volts}$$

With the test probes shorted together $R_X = 0$ ohms, thus forcing $E_m = 0$ volts. For any value of R_X:

$$E_m = E_{source} \frac{R_T}{R1 + R_T}$$

where:

$$R_T = \frac{R2R_X}{R2 + R_X}$$

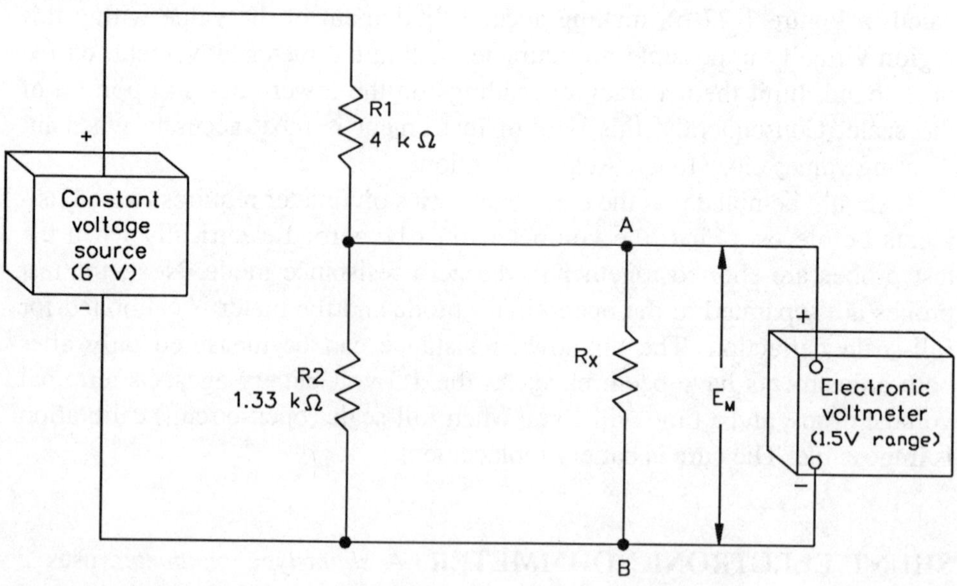

Figure 1-28 Shunt ohmmeter circuit for electronic instrument

For $R_X = 1k$:

$$R_T = \frac{(1.33k)(1k)}{1.33k + 1k}$$

$$R_T = 570$$

$$E_m = 6V \frac{570}{4k + 570}$$

$$E_m = 0.75 \text{ volts}$$

This is one-half full-scale deflection, and the scale is marked with a one at this point. Notice that this is the same E_m range used for the series electronic ohmmeter described above. Consequently, the ohm scale shown in Figure 1-27(b) is the same for the shunt meter.

Also notice that the calculated combination of R1 and R2 in parallel ($4k \times 1.33k/4k + 1.33k$) equals 1k. In shunt ohmmeter circuits of this type:

$$R_X = \frac{(R1)(R2)}{R1 + R2}$$

when the meter indicates half-scale deflection.

The major advantage of the shunt electronic ohmmeter over the series electronic ohmmeter is ease of calibration. Only one adjustment is required before resistance measurements are made, that of zeroing the meter with the test probes shorted together. Because of their regulated power supply requirements, shunt electronic ohmmeters are generally of the laboratory type and are intended for relatively low resistance measurements.

LINEAR-SCALE OHMMETERS. A major limitation of either the series or shunt ohmmeter is the lack of accuracy. Scale crowding at full-scale deflection, coupled with inherent mechanical limitations of the meter movement at the low end, restricts the useful range of the instrument to a very small area within the center of the scale.

Linear-scale electronic ohmmeters increase instrument accuracy by providing a linear scale similar in performance to a voltmeter. The linear scale permits the instrument to use a larger portion of the meter range for measurements, thus improving accuracy.

Linear-scale electronic ohmmeters use a constant current source to provide scale linearity. The voltage E_{RX} is measured across the unknown resistor, thus giving a resistance value equal to:

$$R_X = \frac{E_{RX}}{I_{RX}}$$

Because I_{RX} is a fixed quantity, and does not change with changing resistance values, then:

$$R_X = \frac{E_{RX}}{k}$$

In this format, R_X is a direct function of voltage and can be displayed on a conventional linear-scale voltmeter.

Consider the circuit in Figure 1-29. Transistor Q1, together with resistors R1, R2, and R_E, operates as a constant current source. Resistors R1 and R2 divide V_{CC} so that 5.7 volts appears on the base of Q1. This forces the voltage across R_E equal to 5 volts (5.7 V $-V_{BE}$ = 5 V). The current through R_E is now equal to:

$$I_E = \frac{5\text{ V}}{R_E}$$

and is a constant quantity. With R_E fixed at 5k, I_E is:

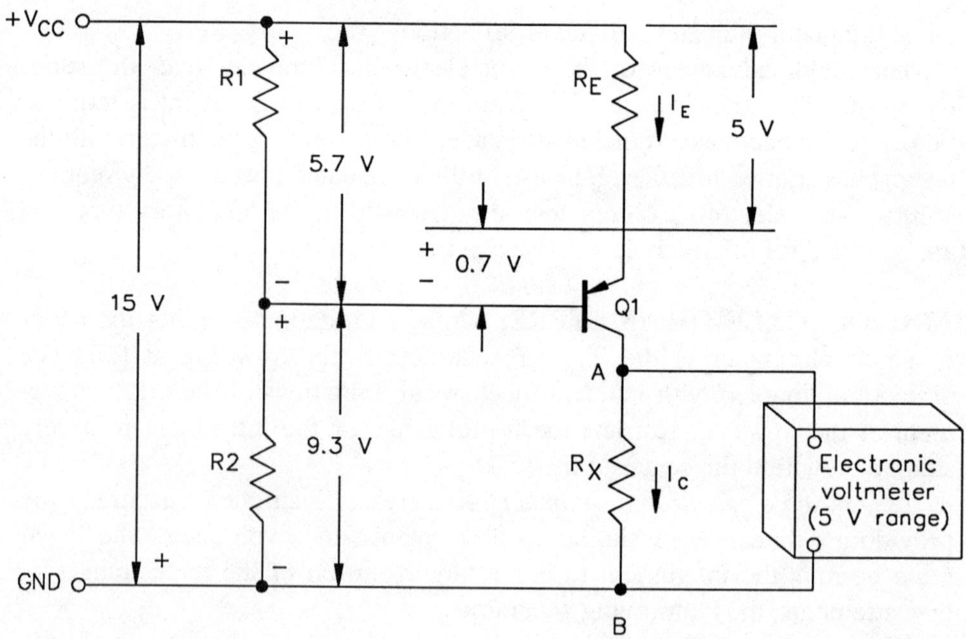

Figure 1-29 Linear-scale ohmmeter

$$I_E = \frac{5}{5000}$$

$$= 1\text{ mA}$$

Assuming that I_{CBO} is negligible compared to I_E, then $I_E = I_C$ and is likewise a constant quantity.

When the current passes through resistance R_X, a voltage is developed across the unknown quantity equal to:

$$E_{RX} = R_X \times 0.001$$

Transposing the equation reveals that:

$$R_X = E_{RX} \times 10^3$$

If the voltmeter indicates that E_{RX} is 2.5 volts, then the unknown resistance R_X is:

$$R_X = 2.5 \times 10^3 = 2.5\text{k}$$

Similarly, if $E_{RX} = 10$ volts, then $R_X = 10\text{k}$, etc.

For multirange operation, a switching arrangement is provided for selection of R_E, from which several resistance ranges may be displayed. Substituting a 50k resistor for R_E yields a meter scale that reads in 10k multiples. With each R_E substitution, the value of I_E changes, as does the range of the instrument.

At very high resistance values, however, the voltmeter has a significant effect on the accuracy of the measurement. The effect can be easiest described by considering the voltmeter as a bulk resistance of finite value in parallel with the unknown resistance:

$$E_{RX} = \frac{R_X R_m}{R_X + R_m} I_C$$

At low resistance values, the effect of the voltmeter impedance on the measurement is negligible and the equation essentially reduces to R_X times I_C. As R_X increases, the voltmeter's effect on the parallel resistance value becomes more pronounced, as exhibited in the following example using a 2-megohm resistor for R_X and a typical impedance value of 11 megohms for the electronic voltmeter.

$$E_{RX} = \frac{(2\text{M})(11\text{M})}{2\text{M} + 11\text{M}} I_C$$

Assuming $I_C = 10\ \mu\text{A}$, then:

$$E_{Rx} = 1.69\,\mathrm{M} \times 10^{-5}$$

$$= 1.69 \text{ volt}$$

In other words, the meter indicates 1.69 volts when it should read 2.00 volts. The error in measurement caused by the voltmeter is a full 8.5%.

AC Electronic Voltmeters

The measurement of ac voltages presents some unique problems not present in dc measurements. The problems stem from the fact that ac voltage has additional parameters, besides magnitude, that affect the measurement. Let's begin with an analysis of the most common ac waveform, the sine wave.

Sine Wave Analysis

Figure 1-30 illustrates the parameters of a sinusoidal waveform, or sine wave. The voltage of the waveform goes from zero to a positive maximum and back to zero, and then goes from zero to maximum in the negative direction and back to zero. This compete alternation is called one *cycle*. The time duration for one cycle is called the *period* (T). The number of periodic cycle repetitions in a one-second interval is called the *frequency*.

The instantaneous value of the voltage of the sine wave is given by the mathematical expression:

$$e = E_m \sin(\omega t)$$

where

e = instantaneous value (volts)

E_m = maximum amplitude (volts, peak value)

ω = radian frequency (radians per second)

t = time (seconds)

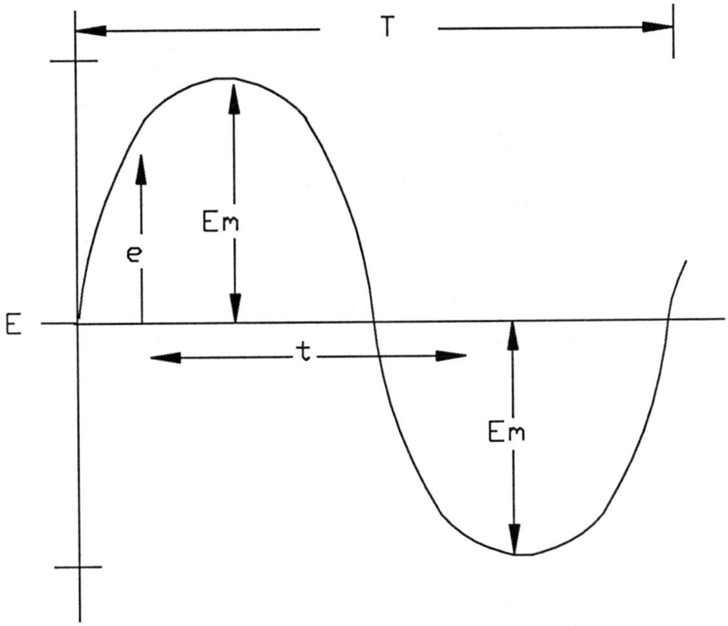

Figure 1-30 Sine wave parameters

The following expressions are used to further define the properties of a sine wave:

Peak-to-peak value ($E_{p\text{-}p}$): The peak-to-peak voltage is twice the maximum (E_m) value. $E_{p\text{-}p} = 2E_m$

Average value (E_{avg}): The average value of any voltage is defined algebraically as the total area under the voltage curve divided by the period in radians. In the case of the sine wave, the average value is zero because the positive portion is exactly equal to the negative portion; hence, the total area is zero.

Root-mean-square (E_{rms}): The root-mean-square, or effective value, of the ac is defined as that value of ac voltage that has the same heating effect as a dc voltage. For example, it requires 14.14 volts of ac across a 1-ohm resistor to deliver the same heating effect as 10 volts of dc. Therefore, the E_{rms} value can be expressed as:

$$\frac{E_m}{E_{rms}} = \frac{14.14}{10} = 1.414 = \sqrt{2}$$

therefore:

$$E_m = \sqrt{2}\, E_{rms}$$

and

$$E_{rms} = \frac{E_m}{\sqrt{2}}$$

$$= 0.707\, E_m$$

Of the three defined values, the rms is by far the most important quantity, since it is the only quantity that permits a direct and accurate comparison between the effects of dc and ac signals, regardless of the wave shape.

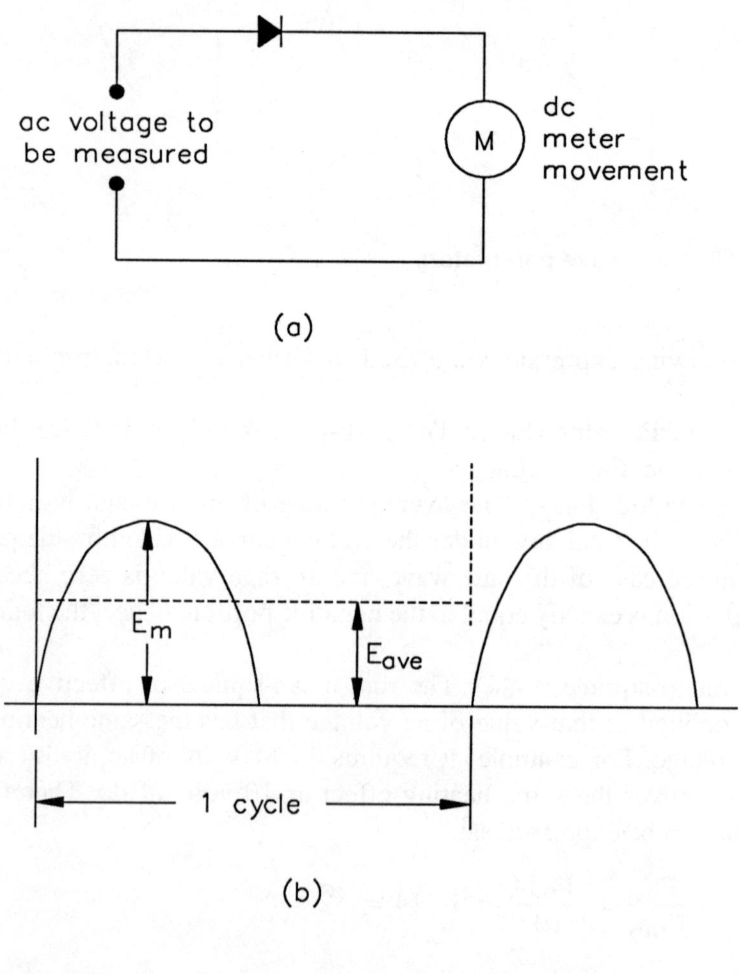

Figure 1-31 Half-wave rectifier

All electronic voltmeters use a dc meter movement for their readout. Consequently, the ac signal input must be converted from ac to dc before it can be displayed. The conversion must be such that the true rms value of the waveform is retained. There are a number of ways to do the conversion.

HALF-WAVE RECTIFIER METERS. Basically, the conversion of ac to dc is done using rectifier diodes. The rectifier diode is a one-way electronic valve that essentially shears off the negative peak of the waveform when used alone, thus converting the wave shape to a pulsating dc waveform, as illustrated in Figure 1-31(b). The process is called *half-wave rectification* because only half the original sine wave exists after it takes place.

The integrated circuit operational amplifier is an ideal candidate for ac to dc conversions using diode rectification. Connecting a rectifier in series with the meter circuit of an op-amp voltage-follower voltmeter, as illustrated in Figure 1-32, converts the instrument into a half-wave rectifier ac voltmeter.

In this design, the ac signal is input to the attenuator stage through coupling capacitor C1. The coupling capacitor is provided at the input of the ac voltmeter to block unwanted dc voltages from affecting the measurement. As with the dc electronic voltmeter, the attenuator stage provides the instrument with a wide selection of input voltage ranges using a simple switching ar-

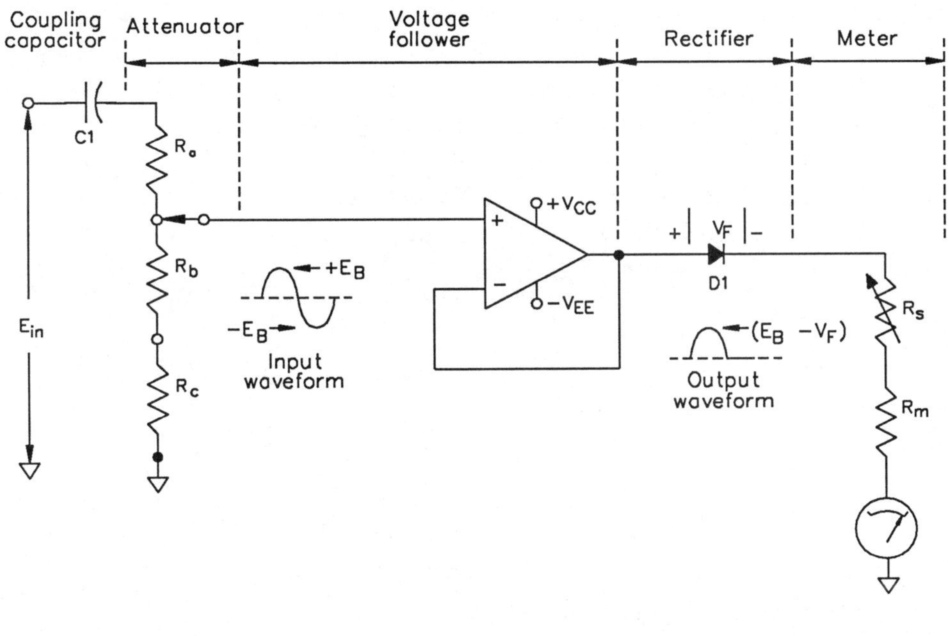

Figure 1-32 Half-wave rectifier voltmeter

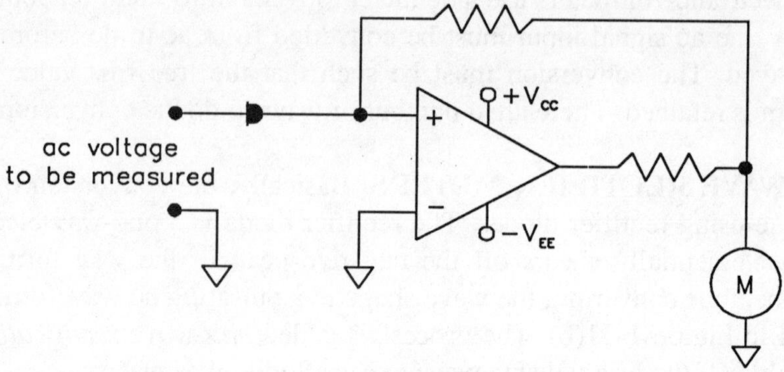

Figure 1-33 Backward diode voltmeter places the rectifier ahead of the op amp.

rangement. The attenuated waveform is input directly to the noninverting input of the op amp. The output of the amplifier is returned to the inverting input, thus setting the gain of the amplifier at unity.

A rectifier may also be connected to the input of the op amp, where it converts the ac to dc before it is impressed across the meter. By placing the rectifier before the op amp, the amplifier deals only with dc values and the bandwidth of the amplifier is not a critical factor in instrument design. This design is not as desirable as that of Figure 1-32 because of the nonlinearity properties inherent to solid-state rectifier diodes. Backward diodes, which are variations of tunnel diodes, are an exception and work well with this configuration for measuring low-value, high-frequency ac voltages.

Analog meters are average-voltage-responding instruments. The average value of the half-wave-rectified output is given by:

$$E_{avg} = \frac{E_m}{\pi} = 0.318\ E_m$$

Since

$$E_m = 1.41\ E_{rms}$$

then

$$E_{avg} = 0.318(1.41)E_{rms} = 0.45\ E_{rms}$$

or

$$E_{rms} = \frac{E_{avg}}{0.45}$$

$$E_{rms} = 2.22\ E_{avg}$$

The coefficient 2.22 is called the *form factor*, which is the ratio of $E_{rms}/E_{avg} = 2.22$. This means that when a meter reads an E_{rms} value of 100 volts, it is actually measuring:

$$\frac{E_{rms}}{2.22} = 45 \text{ volt}$$

The form factor allows us to calibrate the meter scale in E_{rms} while actually measuring average values.

Unfortunately, the forward voltage drop V_F across the rectifier in the circuit of Figure 1-32 is a source of error. When correctly forward-biased, a silicon diode typically has a forward voltage drop V_F of 0.7 volt. V_F increases slightly with increase in I_F, and if I_F falls to a very low level (below the knee of the characteristic curve) there can be a substantial drop in V_F. V_F also varies with temperature changes. While V_F can be taken into account in design calculations, its variations cannot.

To avoid the errors, the voltage follower feedback loop to the inverting terminal is taken from the cathode of the rectifier instead of from the amplifier output, as demonstrated in Figure 1-34. The result is that the half-wave rectified output precisely follows the positive half cycle of the input voltage. There is no V_F voltage drop from input to output because the invert-

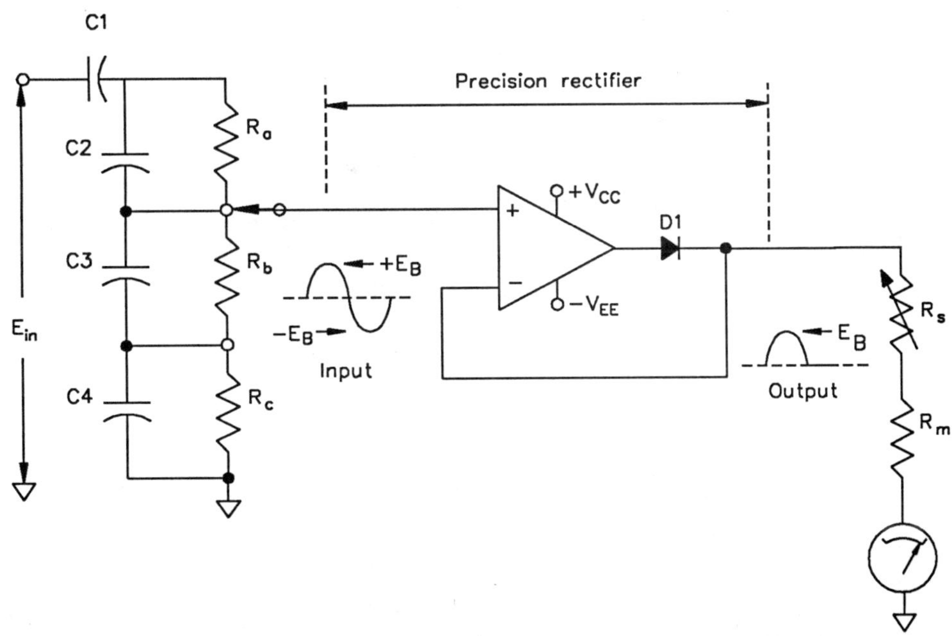

Figure 1-34 Precision rectifier

ing terminal follows the input to the noninverting terminal. With the inverting input referenced to the rectifier cathode, the voltage to the meter always equals the voltage at the amplifier input. Such a circuit configuration is commonly referred to as a *precision rectifier circuit.*

Note that capacitors C2, C3, and C4 are connected across the attenuator resistors in Figure 1-34. These capacitors are normally employed in ac voltmeter attenuators to compensate for the input capacitance of the operational amplifier. Input capacitance problems also occur in oscilloscopes, and the issue is discussed at length in Chapter 5.

Low-level ac voltages too small for normal display must be accurately amplified before being rectified and applied to the meter. The circuit shown in Figure 1-35 combines amplification with half-wave rectification. Ignoring the rectifier diode for the moment, we see that the circuit is a noninverting amplifier, as described on page 30, with a gain factor of:

$$Av = \frac{R2 + R3}{R3}$$

Inclusion of the diode causes the positive half cycles to be amplified and the negative half cycles ignored. Again, the configuration of the feedback loop

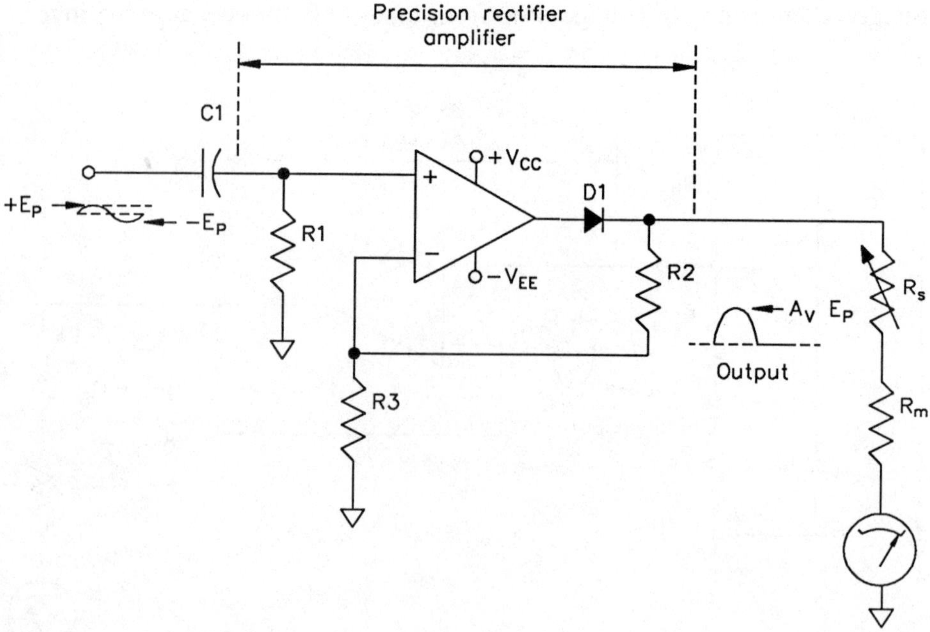

Figure 1-35 Precision rectifier amplifier

guarantees that the amplification is precise and that no rectifier volt drop is involved in the measurement.

FULL-WAVE RECTIFIER METERS. When *full-wave rectification* is used, both the positive and negative peaks of the waveform are converted into dc signals, as shown in Figure 1-36. Generally, the full-wave rectifier is of the bridge type. A bridge rectifier passes the positive half cycles of the sinusoidal input waveform and inverts the negative half cycles. Referring to Figure 1-36, when the input signal is positive, diodes D1 and D4 conduct, causing current to flow through the meter from the negative meter terminal to the positive. When the input goes negative, diodes D1 and D4 are reverse-biased, and diodes D2 and D3 conduct current through the meter. Because of their arrangement, D2 and D3 force the current to again flow through the

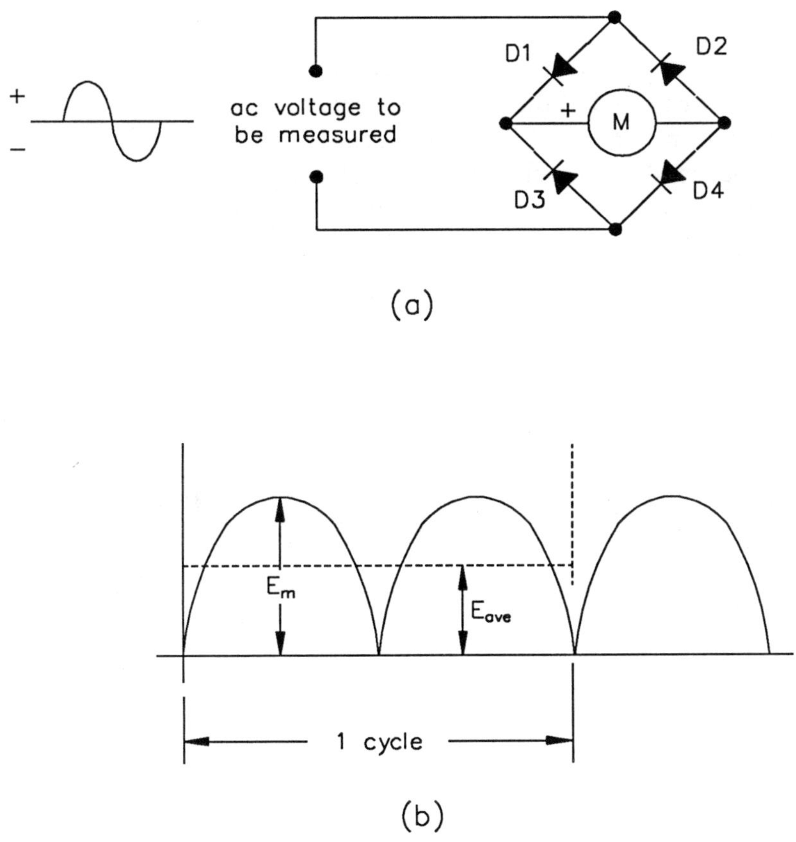

Figure 1-36 Full-wave bridge rectifier

meter from the negative meter terminal to the positive. The resulting current flow is a series of positive half cycles without any intervening spaces.

When a full-wave rectifier is used, the form factor is only 1.11 because:

$$E_{avg} = \frac{2E_m}{\pi} = 0.636\ E_m$$

$$E_m = 1.414\ E_{rms}$$

Therefore

$$E_{avg} = 0.636(1.414)E_{rms} = 0.9\ E_{rms}$$

and

$$E_{rms} = \frac{E_{avg}}{0.9}$$

$$= 1.11\ E_{avg}$$

As indicated earlier, the rectifier-type ac voltmeter is calibrated to read the effective, or rms, value of the measured average voltage.

The circuit in Figure 1-37 employs a full-wave bridge rectifier to convert the ac signal to dc. This circuit is basically a voltage-to-current converter with rectification. In this design, the voltage drop across R3 is always equal to the input voltage.

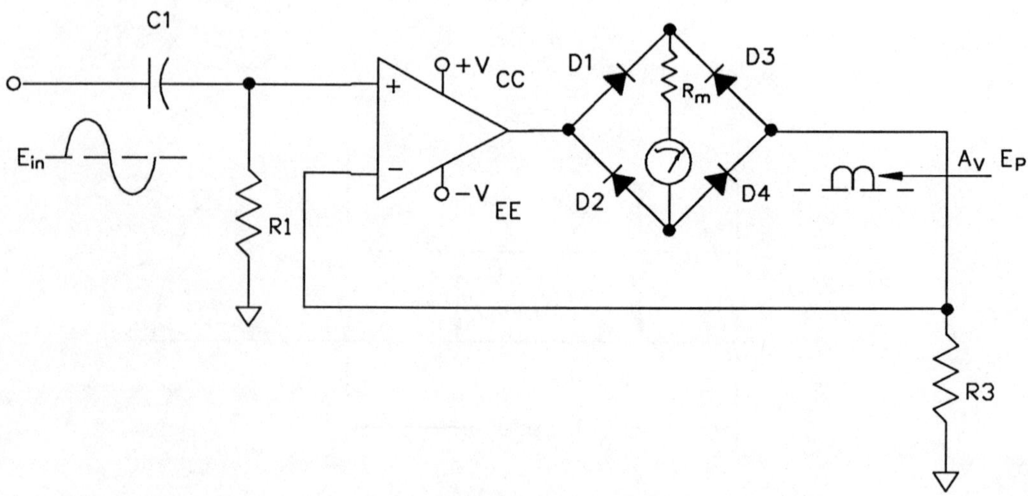

Figure 1-37 Full-wave voltage-to-current converter

If the input increases or decreases, V_{R3} follows it precisely, thus eliminating variations in the diode's characteristics and the accumulated (V_{FD1} + V_{FD4}) forward voltage drop. During positive half cycles, diodes D1 and D4 are forward-biased so that current flows through the meter from top (+) to bottom (–). When the input is negative, D2 and D3 are forward-biased, causing current to flow from (+) to (–).

True-responding RMS Voltmeters

Rectifier voltmeters operate satisfactorily only with measuring inputs that have purely sinusoidal waveforms. This is because the instrument is not actually measuring rms voltage but is deriving it from another parameter, namely average voltage. Each meter scale is calibrated to indicate rms values on the assumption that the input is a pure sine wave.

When the waveform is not sinusoidal, the instrument does not indicate the true rms value of the measured voltage, but instead a value related to the average voltage of the waveform involved. The table in Figure 1-38 shows how the correlation between rms and average values changes with waveform. Unless the voltmeter scale is calibrated for each waveform specifically, the measurements are meaningless.

True rms measurements can be made using an *rms converter*. Basically, an rms converter is a mechanical or electronic measuring device that samples the true rms heating power of the quantity under test and converts it into a voltage or current that is suitable for conventional display. Examples of two common rms converters are shown in Figures 1-40 and 1-41.

THERMOPILE CONVERTER. In Figure 1-40 a *thermopile converter* using matched thermopile units is used in a balancing circuit to measure rms voltage. The thermopile relies on the characteristics of a *thermocouple* monitoring device to convert rms values into heat and back again into electrical signals.

A thermocouple is nothing more than a junction consisting of two dissimilar metal wires welded together. A number of alloys, such as copper and Constantan, have the unique property of generating a thermovoltage when placed in contact with each other. The voltage is brought about by the exchange of electrons between the dissimilar metals, and is proportional to the temperature of the thermocouple junction. The hotter the junction temperature,

	WAVEFORM	$V_{average}$	V_{rms}
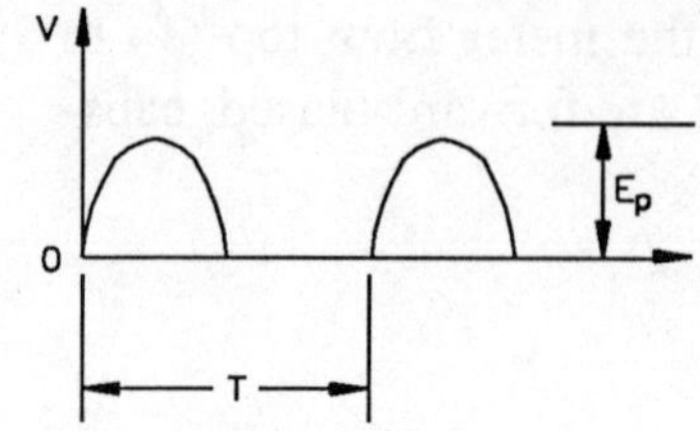	Half-wave rectified ac	0.318 E_p	0.354 E_p
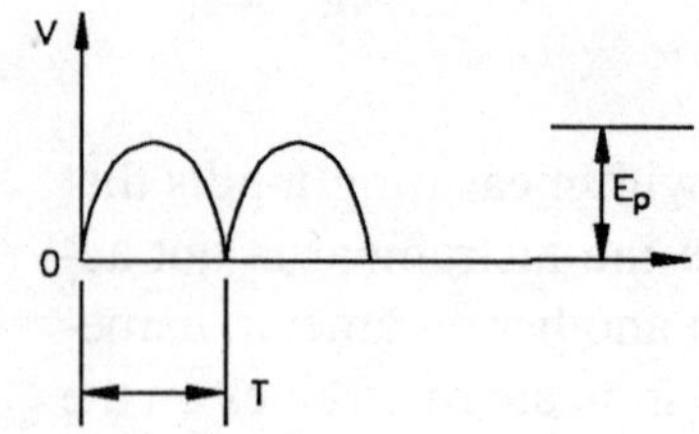	Full-wave rectified ac	0.636 E_p	0.707 E_p
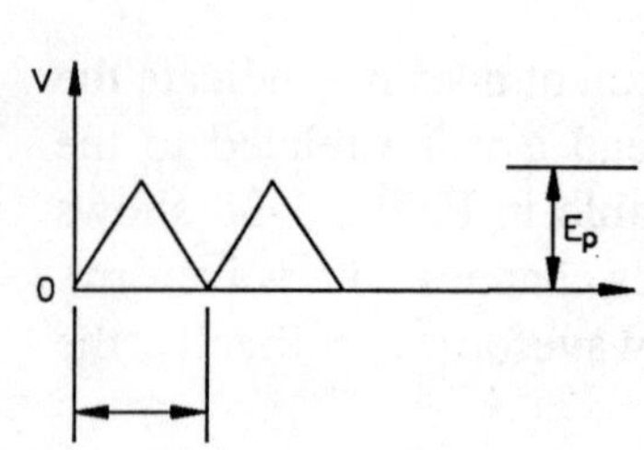	Triangular wave	0.5 E_p	0.576 E_p
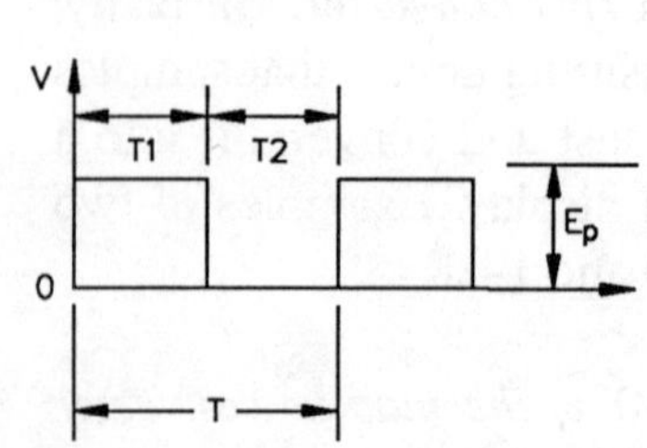	Square wave $T_1 = T_2$	0.5 E_p	0.5 E_p
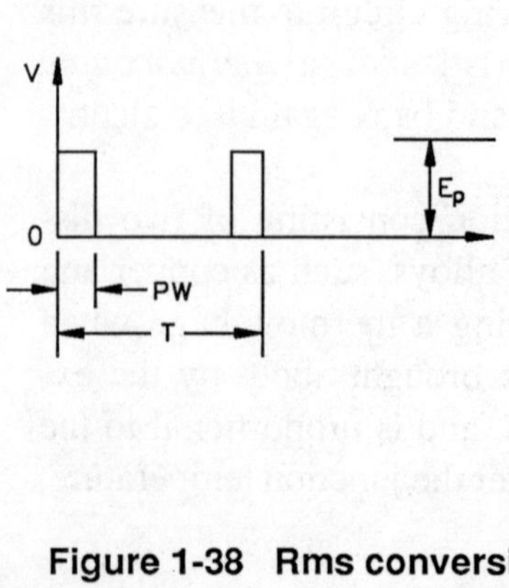	Pulse wave	$E_p \times$ duty cycle where duty cycle $= \frac{PW}{T}$	$E_p \sqrt{\text{duty cycle}}$

Figure 1-38 Rms conversion table

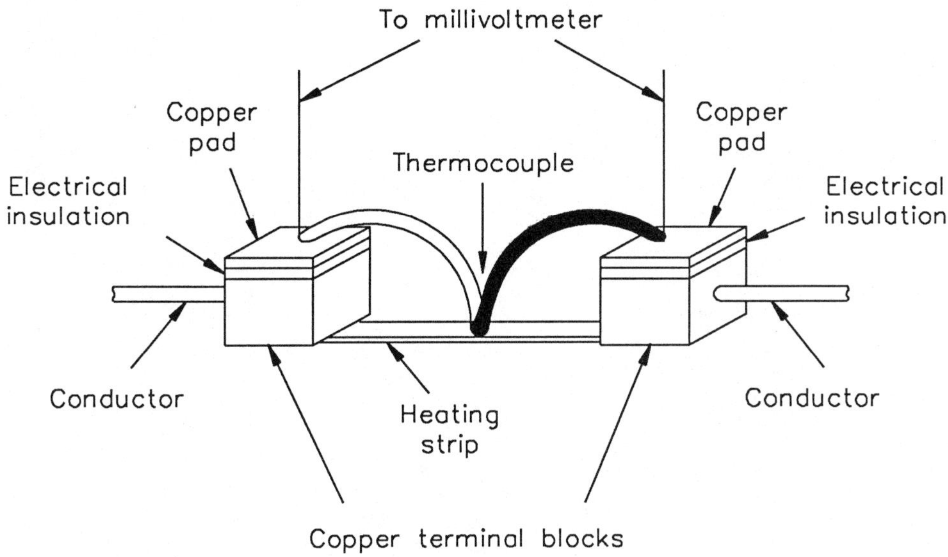

Figure 1-39 Compensated thermopile converter

the higher the junction voltage. This voltage can be read by a millivoltmeter and compared to a standard chart to derive the junction temperature.

In a thermopile, the thermocouple is attached to a flat heating strip, as shown in Figure 1-39. The junction may be directly welded to the heater, or thermally (but not electrically) attached to the heater element by a bead of ceramic material. When a current is passed through the heating strip, its temperature rises. This temperature rise is duly noted by the thermocouple junction and expressed as a voltage across its output leads.

In Figure 1-39, the ends of the thermocouple wires are connected to copper pads that are electrically insulated from the heater, yet remain in thermal contact with the large copper terminal blocks. This has the effect of keeping the ends of the thermocouple wires at the same ambient temperature as the terminal blocks. This establishes what is called a *cold junction*. For a thermocouple to generate voltage, there must be a difference in temperature between the cold junction and the thermocouple junction. Because the thermocouple wires are maintained at the same ambient temperature as the heater strip when no current is flowing, the emf generated at the heated junction results only from the heating effects of current passing through the heater strip. This thermopile device is called a *compensated thermocouple*,

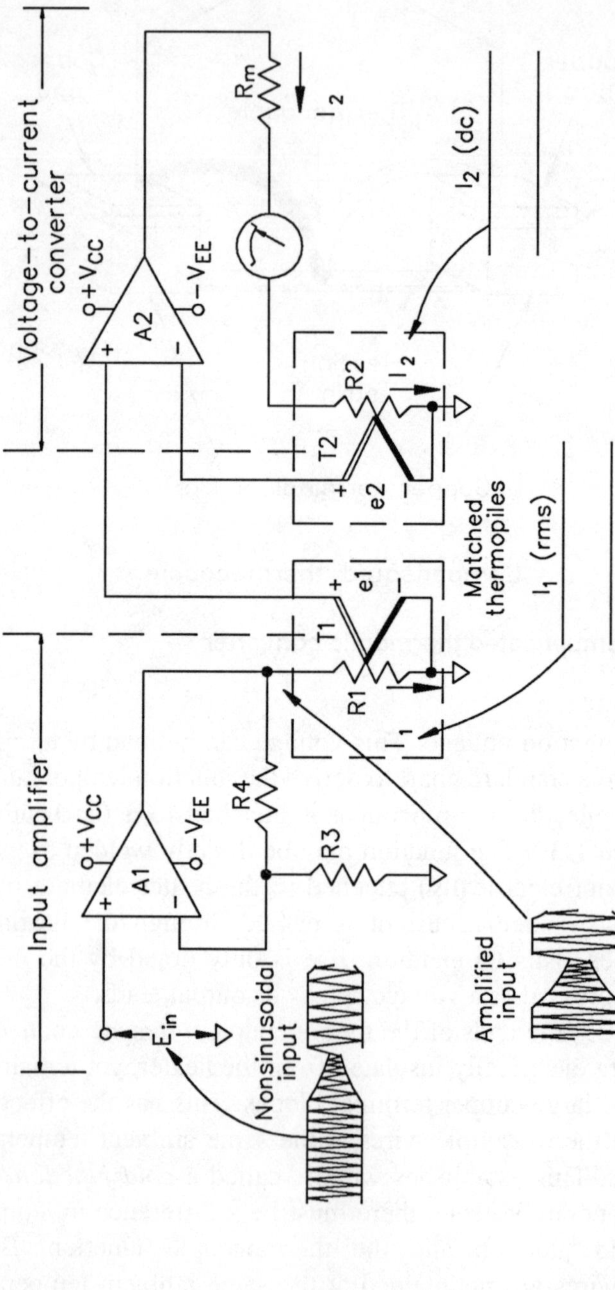

Figure 1-40 Matched thermopile rms converter

meaning it is compensated against the effects of ambient temperature changes. A typical maximum thermopile output voltage is approximately 12 mV; typical heater strip currents are in the range of 2 mA to 50 mA.

Because the heating strip of the thermopile is insensitive to waveform type, it represents a true picture of the rms qualities of the input voltage. Regardless of waveshape, the rise in thermopile output voltage is consistent with the rms heating properties of the input signal.

In the rms converter circuit of Figure 1-40, a matched pair of thermopile units are used in a balancing circuit in such a way that the dc voltage delivered from the circuit is proportional to the rms value of the input ac waveform. This is a direct implementation from the very nature of rms value, in which rms characteristics are based upon the heating capability of an ac waveform.

Operational amplifier A1 is a buffer amp that provides isolation between the input signal and the thermopile. The current passed from A1 to resistor R1 has the same waveform as the input voltage. Consequently, the heat generated in R1 is directly proportional to the rms value of the voltage to be measured. This in turn makes the output voltage of thermocouple T1 directly proportional to the rms value of input E_{in}.

Amplifier A2 functions as a voltage-to-current converter, with the output current passing through a deflection meter and R2, the heating element for the second thermopile. The monitoring thermocouple for thermopile T2 is connected to the inverting input of A2. This arrangement treats the input of A2 as a differential amplifier. Only when the output voltages of T1 and T2 thermocouples are matched is the circuit at equilibrium. If e_2 is less than e_1, the output at A2 increases, increasing I_2 until its heating effect raises e_2 to the same level as e_1. Thus, if the thermopiles are matched devices, the heating effect of current I_2 must be equal to the heating effect of I_1. This means that the direct current indication on the meter is equal to the rms value of the input voltage E.

An obvious disadvantage of the thermocouple or thermopile technique is the need for very expensive, closely matched thermopile units. Another limitation is the thermal response time; it often takes several seconds for the thermopile units to stabilize to a change in input signal.

THE OPERATIONAL RMS CONVERTER. Figure 1-41 shows the second common approach to rms conversion, the *analog operational technique*. This approach is actually a computational method that relies on the actual computation of the rms value of an input voltage using analog circuits that perform the various mathematical operations. In other words, it is a mathematical description of the rms value said in electronics.

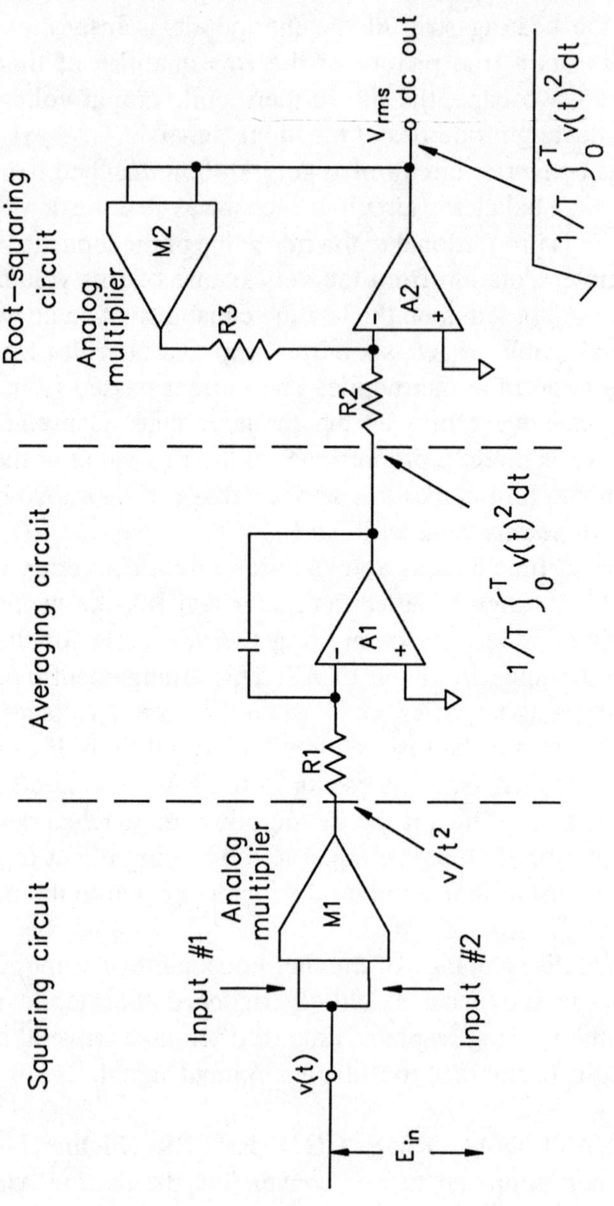

Figure 1-41 Analog operational rms converter

For any repetitive waveform, the root-mean-square value can be calculated using the relationship:

$$E_{rms} = \sqrt{\frac{1}{T}\int_0^T v(t)^2\,dt}$$

where v(t) represents the ac waveform, with E the instantaneous amplitude as a function of time (t).

The input stage of the operational converter is a squaring circuit that takes the input signal, breaks it into two equal factors (x and y), and multiplies the two values to derive a square value of the input voltage. This can be expressed as $v(t)^2$ in our equation.

The second stage is an averaging or integrator circuit. In this stage a feedback capacitor is used to derive the function:

$$\frac{1}{T}\int_0^T v(t)^2\,dt$$

from our input stage.

The final stage is a square-root compiler circuit composed of a precision rectifier circuit and an analog multiplier feedback network. The results are an output voltage proportional to the true root-mean-square (three functions, count them) input voltage.

Peak-reading Voltmeters

In some instances, it is more important to be able to measure the peak E_m voltage of a waveform than its true rms or average value. Measurements of this type are critical to power supply design (where rectifier breakdown voltage V_R may be exceeded) and digital pulse circuits.

Figure 1-42 shows the circuit for a typical peak-reading electronic voltmeter. Generally, this circuit is placed ahead of the voltmeter inside a test probe, much in the same way that oscilloscopes place much of their input circuitry in a scope probe. The reasons are two-fold. First, it reduces component count inside the instrument itself. Second, and more importantly, it improves the high frequency response of the instrument by limiting the length of cable the signal must travel before being processed. (An in-depth discussion of test probe and cable losses is presented in Chapter 5.)

In the circuit pictured in Figure 1-42, capacitor C1 is in series with the input terminal of the dc electronic voltmeter. Capacitor C1 prevents dc input

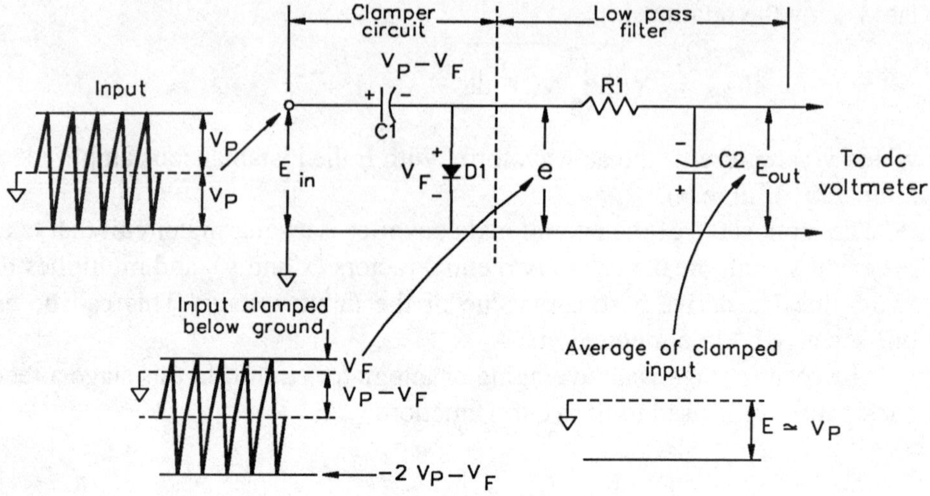

Figure 1-42 Peak-reading test probe

voltages from affecting the meter reading and allows only the ac component to pass. As the input signal goes positive, D1 becomes forward-biased and C1 charges up to $E_m - V_F$ with the polarity shown. When the input signal starts on its negative swing, D1 becomes reverse-biased and C1 retains its previous peak voltage value.

Neglecting the diode forward-voltage drop V_F, the voltmeter sees this input as an average value of approximately one-half the peak-to-peak ($V_{p\text{-}p}$) voltage. Resistor R1 and capacitor C2 function as a low-pass filter to remove the alternating component of the input voltage and to pass the dc peak value. This value can be used to calibrate the meter scale in rms, peak (E_m), or peak-to-peak ($V_{p\text{-}p}$) measurements.

Frequency Response

One very important characteristic that affects the ac measuring process is frequency. The various types of instruments used have internal resistance, capacitance, and inductance. Because the latter two parameters relate to lumped impedance in a frequency-sensitive manner, the frequency response

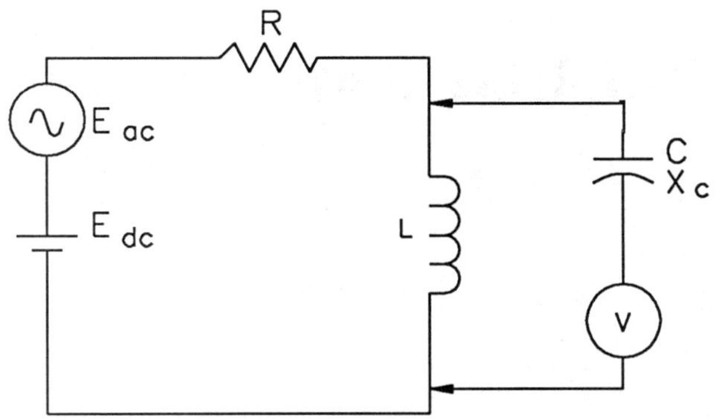

Figure 1-43 Ac meter loading effect

and bandwidth of an instrument become major influences in the measurement of an ac signal.

As in the case of dc measurements, the ac meter is inserted into the circuit under test — voltmeters in parallel and ammeters in series. The net effect of the meter's affect is detailed in Figure 1-43.

As the frequency varies, certain elements of each component come into play. At very low frequencies, the capacitive reactance of the input-blocking capacitor must be such that the input impedance of the meter is 100 times greater than the reactance of the capacitor at the lowest frequency present. At higher frequencies, the shunt capacitance of the instrument should be isolated enough by the test probe that its effect on the waveform is minimized (see page 161 for further explanation).

Because of the proliferation of oscilloscopes as general-purpose measuring instruments, we have become accustomed to viewing on the screen a nonsinusoidal ac waveform that appears to be comprised only of what we see occurring at a certain repetitive rate. From Fourier analysis, however, we know this is far from the truth. The harmonic content of many waveforms — particularly those, like square waves, with sharp corners — extend far beyond the fundamental frequency and are intricate parts of the waveform energy spectrum. As a rule of thumb, the usable passband of the test analog test instrument should be at least ten times the repetition rate of any nonsinusoidal waveform under test.

2. DIGITAL INSTRUMENTS

While it can be argued that analog instruments have earned their place in the laboratory, they are still limited by finite mechanical accuracy. Analog measurements are largely subjective, influenced by visual and psychological perception errors which may cause viewers to report slightly different readings for the same event. While much of the error can be attributed to parallax, which is caused by viewing the meter pointer and scale at an angle other than straight on, our psychological inner self tends to make us "fudge" on measurements even when we have interpreted them properly in the physical sense. If a measurement is slightly off the mark, we have the tendency to compensate for the error by allowing the recorded reading to creep closer to our perceived impression of the true value. Aggravating the situation is the fact that the meter itself does not always register identical values the same way twice, depending on which way the pointer happens to be moving. Mechanical errors brought about by bearing friction and heat prevent the mechanical movement from being totally predictable from measurement to measurement.

The digital readout, on the other hand, is a truly objective one, with all users perceiving the same indication. The direct numerical readout reduces the human error and the tedium involved in long periods of measurement. "Cheating" is less prevalent because the numbers are before you, in black and white (or red on black, as the case may be). Digital electronics also brings with it a wide range of peripheral benefits not available in analog instruments, including auto-ranging, waveform computation, and high common-mode rejection qualities.

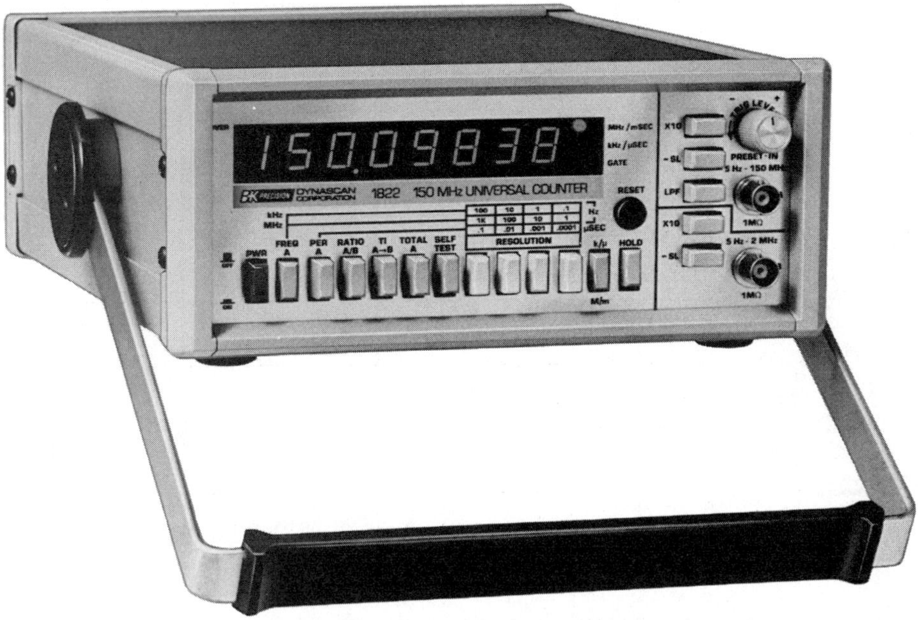

Figure 2-1 175 MHz universal counter (courtesy of Dynascan Corporation)

Digital Displays

The first digital display devices were mechanical, made up of numbered wheels and driven by servo-controlled motors. Next on the scene were various formats consisting of miniature incandescent lamps. These devices evolved to a surprisingly satisfactory state of the art and gave rise to control techniques that now enable the popular light emitting diodes (LED) and liquid-crystal displays (LCD) to dominate instrumentation readouts.

The weak link in the incandescent lamp display is the fragile filament. Even when operated at low current, the incandescent bulb has a life expectancy that is measured in hours. Short life notwithstanding, the thermal response time of the filament also poses a problem in instrument readouts with a relatively long lag time. Although lamps are little used now, they established the popular seven-segment display format that is used by all digital readouts today.

Gas-discharge plasma readouts emit light via the ionization of neon, argon, and other noble gases. A small amount of mercury is often used to manipulate the ionization potential and to modify the spectrum. These displays have been made into many forms, some of which have external

Display format of segmented module

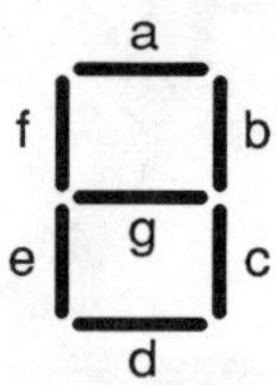

Truth table of segmented module

	1	2	3	4	5	6	7	8	9
a		X	X		X	X	X	X	X
b	X	X	X	X			X	X	X
c	X		X	X	X	X	X	X	X
d		X	X		X	X		X	X
e		X				X		X	
f				X	X	X		X	X
g		X	X	X	X	X		X	X

Figure 2-2 Seven-segment display

electrodes. Light output is very good and the definite on-off characteristics of plasma devices allow for convenient multiplexing. A long-enduring and very satisfactory gas-discharge readout was the Nixie tube made by Burroughs. These neon display tubes had nine internal cathodes, each shaped as a digit. The cathodes were stacked one behind the other in a line perpendicular to the line of sight. When a cathode was selected for excitement, it would glow in the shape of the number it represented. The fact that the nine digits were each in a different plane was somewhat disconcerting, but acceptable. A disadvantage of plasma displays is the high ionization potential required to ignite the lamp. It is on the order of 150 to 200 volts, which creates interfacing problems with today's low-voltage, solid-state circuitry. While no longer the dominant readout, plasma displays continue to find applications.

Vacuum fluorescent readouts were also a product of vacuum-tube technology. The fluorescent readout had a multiplicity of anodes that were coated with a light-emitting phosphor similar to the material found inside fluorescent lamps. While blue-green was the most popular color, others existed. The anode would fluoresce when bombarded by electrons from the cathode emitter. In most cases, the digit was formed using the seven-segment format described above. While these readout devices had the shortcomings inherent to vacuum tubes, including their fragile, short-lived filaments and high voltage requirements, they nevertheless had considerable aesthetic appeal and still find their way into modern applications, such as automotive and computer displays.

Electroluminescent displays also make use of a light-emitting phosphor. Instead of the electron tube structure, though, these displays are essentially capacitors. A phosphor-rich film is sandwiched between two transparent, conductive plates. When ac voltage is applied to the plates, the capacitance displacement current excites the phosphor and causes it to glow.

Light emitting diode devices, developed as a result of continued advancement with semiconductor materials and fabrication, are the workhorses of digital display. They are bright, cost-effective, compatible with solid-state voltage and current levels, and long-lasting. Originally available only in red, LEDs now come in a wide range of colors, including red, orange, yellow, green, and blue.

Liquid crystal displays (LCDs) do not emit light, but instead alter the characteristics of the crystal material to reflect incident light. Optically, LCDs shine where LEDs fail; the brighter the ambient light, the brighter the LCD display. Conversely, LEDs wash out with increasing ambient light levels, and can scarcely be seen. On the other hand, the LCD display becomes shadowy at the low light levels at which the LED offers maximum readability. Because they are not light emitters, LCD displays offer the lowest power consumption of any electrically activated display. It is for this reason that LCD displays are very popular with portable and battery-operated equipment.

Electronic Digital Counter

Common to all digital instruments is the *electronic digital counter*. The electronic counter, most often thought of as a device that totalizes input events, is the basic building block for all digital test and measurement applications.

A simplified block diagram of an electronic digital counter for frequency measurements is shown in Figure 2-3. A simple frequency counter consists of seven major functions: input conditioning, timebase oscillator, timebase dividers, main-gate flip-flop, main gate, counting register, and readout display. Central to the operation of the digital counter is the main gate.

The main gate is nothing more than a logic AND circuit. When both the input conditioning and main-gate flip-flop are logically true, the main gate

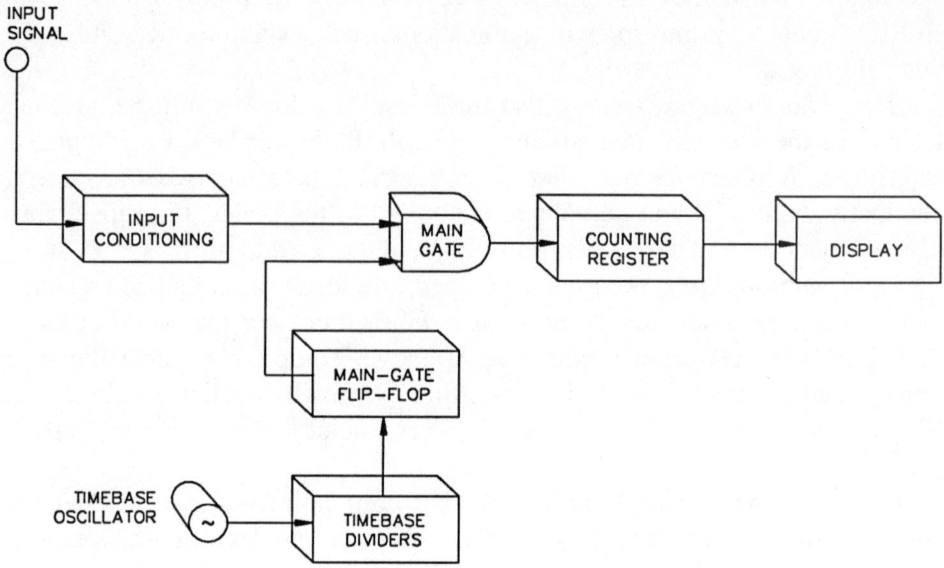

Figure 2-3 Electronic digital counter block diagram

opens for a period of time that is determined by the timebase divider. While the main gate is open, the conditioned input signal pulses are passed through to the counting register, where they are tallied and then scaled for output by the display circuitry. At the end of the counting period, the main gate is closed and the counter reset for the next sampling period.

INPUT SIGNAL CONDITIONING. First, the input signal is conditioned to remove unwanted noise and make the signal level compatible with the internal digital processing circuitry. Because the frequency counter must accommodate a wide range of input signals, with varying degrees of voltage, noise component, and dc offset, the input must pass through many conditioning circuits before it is fully conditioned. The individual components of the input conditioning function are shown in Figure 2-4.

The input signal to be measured is first passed through an attenuator to reduce the amplitude of the signal, if necessary. The attenuator is a voltage-dividing RC network that typically provides switch-selectable input division by 1, 10, or 100. In some instruments, the RC network may be replaced with a potentiometer to provide continuously variable input attenuation.

An ac signal with a large dc component can shift the level of the signal outside the usable range of the instrument. Consequently, input coupling to the attenuator may be either dc or ac, depending on the signal input charac-

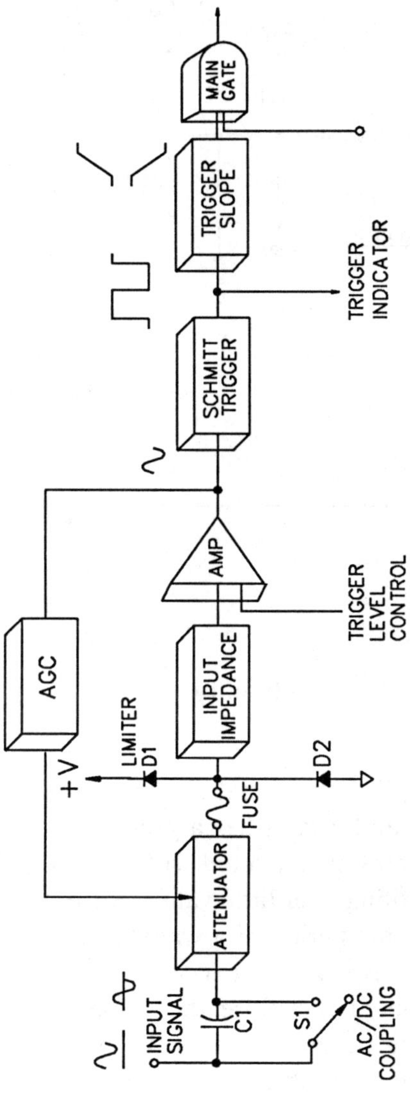

Figure 2-4 Signal conditioner block diagram

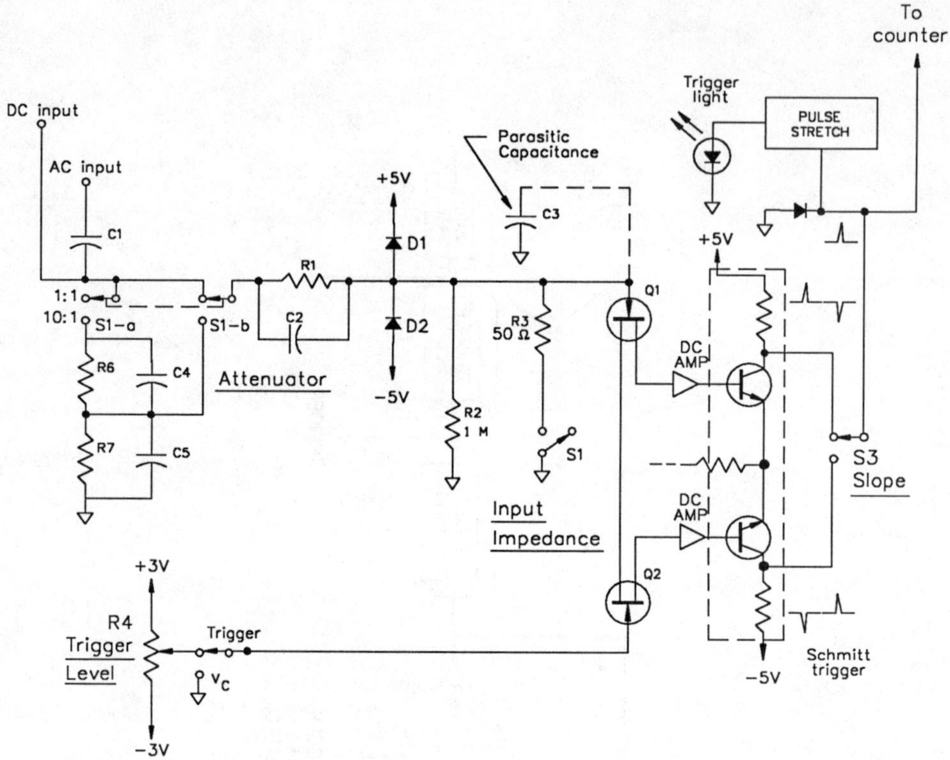

Figure 2-5 Input conditioning circuitry

teristics. With ac coupling, the input signal passes through a capacitor that blocks the dc voltage and establishes a ground reference for the waveform, thus eliminating problems encountered with input signals that contain a large offset voltage. Ac coupling is of little value, however, if the duty cycle — the ratio of the on time to the period of a waveform — is low, because the time constant of the input network may exceed the width of the input pulse. The result is an integration of the pulse that appears as an averaging voltage rather than a countable pulse. For the same reason, ac coupling should be avoided when measuring variable duty cycle signals, such as the kind found in switching power supplies and power controllers.

To prevent instrument damage from an accidental overload, two diodes (D1 and D2) are used to limit the input voltage in the event the input signal should exceed the range of the input attenuator setting. The input signal is then routed through an impedance-matching network into a wide-band amplifier.

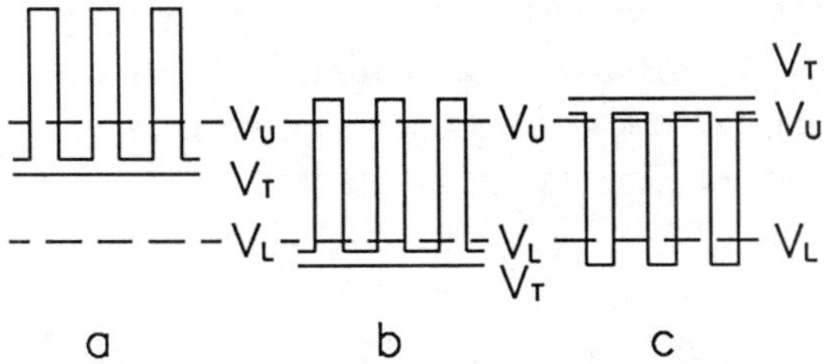

Figure 2-6 Trigger level

The input to the amplifier is generally high impedance (one megohm), but some frequency counters permit a choice between either a high-impedance, 1-megohm input or low-impedance, 50-ohm input. The lower impedance gives the instrument better sensitivity, by reducing the amount of noise picked up by the input cable. A sensitivity of 20 mV to 25 mV is possible with a 50-ohm input, as opposed to the typical 250-mV sensitivity of the high-impedance input. The lower-impedance input also reduces the effect of pulse integration when using ac input coupling. Instrument loading, as discussed in Chapter 1, must be taken into consideration when choosing an input impedance.

The *trigger level* establishes a reference voltage, above which the input signal will trigger the frequency counter. If the trigger level (V_T) is set too high, as shown in Figure 2-6(a), the input signal will never fall below the lower hysteresis voltage, and no pulses will be counted. By lowering the trigger threshold, as shown in Figure 2-6(b), the input signal traverses the complete hysteresis range, and each input pulse will be counted. For negative input voltages, the trigger level (V_T) is set as shown in Figure 2-6(c).

The AGC circuit adjusts the sensitivity of the amplifier to accommodate inputs of different magnitudes. The AGC smooths out any rough spots that may be encountered in everyday input signals, should they vary somewhat in amplitude, thus providing a constant output voltage to the Schmitt trigger. A trade-off exists between the response of the AGC and the minimum frequency signal that can be counted, which is usually about 50 Hz with common AGC circuits. Furthermore, common AGC circuits are also limited to their ability to respond to high-frequency AM (amplitude modulated) signals. The AGC circuit tends to respond to the peak voltage levels and ignore the valleys. In an attempt to track a high-frequency AM input, the response time may not be fast enough to catch some low-level modulations, which leads to erroneous counting.

The output of the amplifier goes to the input of a Schmitt trigger. The Schmitt trigger provides the hysteresis necessary for proper triggering of the input signal and conditions any signal with a slow rise time, a slow fall time, or both (as in the case of a low-frequency sinewave), to provide a rectangular output with sharp edges. The signal is then presented to one input of the main gate.

TIMEBASE OSCILLATOR. The output pulse from the Schmitt trigger passes to the counting circuits only when output Q from the main-gate flip-flop is at a logic 1. The flip-flop changes state each time a negative-going output is received from the timebase dividers. Therefore, if the timebase divider is configured to supply an output signal once every second, the flip-flop output is alternately at level 1 for one second and at level 0 for one second. Consequently, the main gate is alternately open for one second and closed for one second. In other words, the main gate alternately passes the Schmitt trigger output pulses to the counting registers for one second, and then blocks them for one second. The timing of the main-gate flip-flop is controlled by the timebase dividers, which are driven by the timebase oscillator.

The timebase oscillator is an accurate crystal oscillator that drives the timebase dividers. The timebase divider scales the output of the timebase oscillator and provides a pulse train whose frequency is switch-selected, generally in steps of ten (10, 100, 100, etc.), for scale ranging.

Because the timebase oscillator controls the period of time the main gate is open, its importance in determining accuracy cannot be overstated. There are basically three types of crystal oscillators: room temperature crystal oscillator (RTXO), temperature-compensated crystal oscillator (TCXO), and oven-controlled crystal oscillator.

Temperature is the natural enemy of any oscillator. The frequency stability of an oscillator is directly dependent on the temperature stability of the tuned-circuit components. In a crystal oscillator, this is the crystal itself.

TYPE	FREQUENCY	STABILITY	TEMP. RANGE
Double oven	2.5 MHz	$+1 \times 10^{-10}$	0 to +40°C
Single oven	5 MHz	$\pm 1 \times 10^{-8}$	–20 to +65°C
TCXO	2 – 20 MHz	$\pm 1 \times 10^{-7}$	–30 to +50°C
RXCO	4 – 21 MHz	$\pm 2.5 \times 10^{-6}$	–10 to +60°C

Table 2-1 Crystal oscillator stability

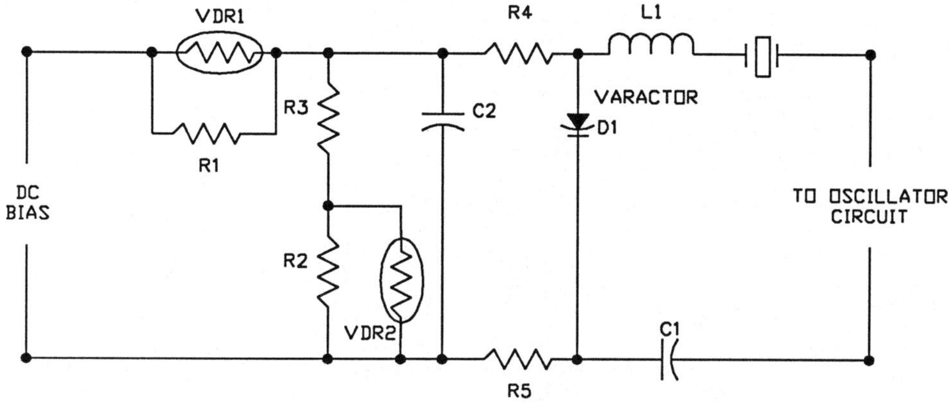

Figure 2-7 TCXO compensation

Temperature effects can be minimized by thermostatic control, temperature compensation, or choice of an appropriate crystal type. The merits of these methods are summarized in Table 2-1.

Thermostatically controlled ovens with proportional control are the most stable of all oscillators. Single-oven types are able to control the oven temperature to ±0.001°C, for an ambient variation of 0 to 70°C. Double ovens (an oven within an oven) can control to ±0.0001°C over a similar temperature range. When the oven temperature coincides with the crystal turnover temperature, a very low overall temperature coefficient is achieved. It generally takes 24 hours after the application of power to the instrument for an oven to attain its maximum specified accuracy, but an oven can usually attain an accuracy of 7 parts in 10^{-9} within 20 minutes.

When an oven is not practical, a crystal can be temperature-compensated (TCXO) with a series reactance whose temperature coefficient cancels the temperature coefficient of the crystal. Another alternative is to use a voltage-variable capacitor (varactor) with a temperature-dependent bias, as shown in Figure 2-7. The value of inductor L1 is selected so that:

$$\omega_S < \sqrt{\frac{1}{(L1)(C1)}} < \omega_O$$

where ω_O is the desired oscillation frequency and ω_S is the spurious frequency. Although this technique does not give as low a temperature coefficient as a thermostatically controlled oven, it is useful because of its lower power consumption and absence of any appreciable warm-up time.

The RTXO oscillator is the simplest of the three types and the least expensive to build. With the RTXO, no attempt is made to compensate for fre-

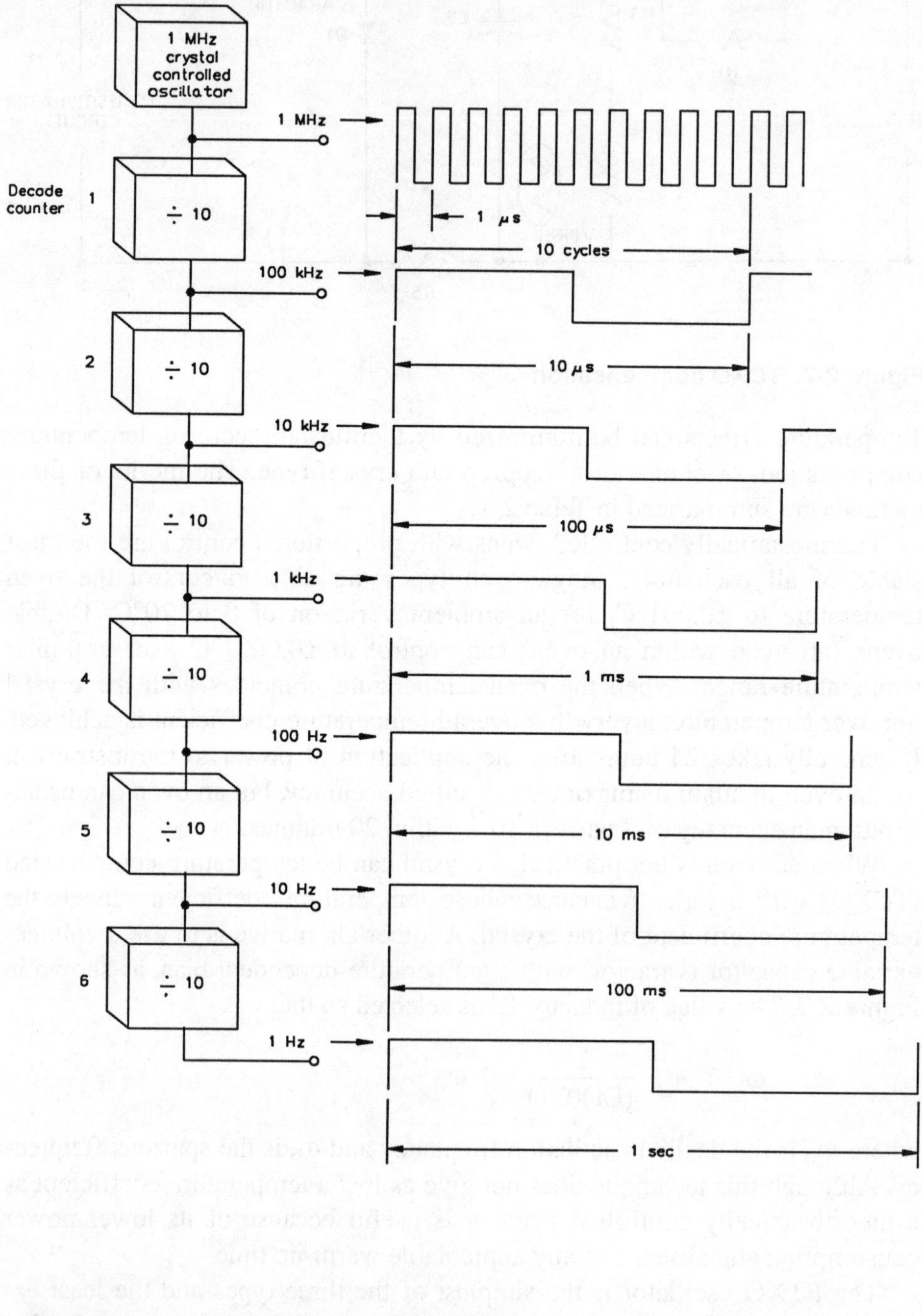

Figure 2-8 Timebase oscillator and dividers

quency drift caused by the effects of temperature. Choice of the proper cutting angle, however, will produce a crystal with a low temperature coefficient over the desired temperature range.

TIMEBASE DIVIDERS. The main-gate flip-flop clock frequency is obtained by dividing down the timebase oscillator frequency in the timebase dividers. The timebase dividers contains a bank of decade counters, each of which divides the input frequency by 10.

The output frequency from the final flip-flop of a decade counter is exactly one-tenth of the input triggering frequency. This means that the time period of the output waveform is exactly ten times the time period of the input waveform. The 1-MHz output from the crystal oscillator in Figure 2-8 has a time period of 1 μs, and the output waveform from the first decade counter, which is driven by the 1-MHz timebase oscillator, has a time period of 10 μs. The time period of the output from the second decade counter (triggered by the output of the first decade counter) is 100 μs, and that from the third decade counter is 1 ms, etc. With all six decade counters, the available time periods range from 1 μs to 1 second.

COUNTING REGISTER. When the main-gate flip-flop opens the main gate, digital pulses from the Schmitt trigger stream into the counting register. The counting register is an array of decade counters that count and tally the digital pulses. Each display digit has its own decade counter, which consists of a latch circuit and a BCD to seven-segment decoder.

The decade counter is the pulse accumulator and totalizer. It is within this circuit that the pulses are counted and the count converted to decimal format. The counters are cascaded to provide carryover as the addition proceeds. The output of the first decade (units) sends a pulse to the second decade counter (tens) each time the count rolls over. When the second counter has accumulated ten pulses from the first counter, it rolls over (zero) and outputs a pulse to the third (hundreds) decade counter, and so forth. The number of display digits determines the digital counter's resolution.

The latch circuit is used to provide isolation between the decade counter and readout display while counting is in progress. During counting, the display is disconnected from the decade counter output. At the end of the counting period, a signal to the latch loads the results of the tally into the latch register, causing the display to change to the decimal equivalent of the final condition of the counting circuits. The BCD to seven-segment decoders convert the binary format of the decade counter output to seven-segment format for the display.

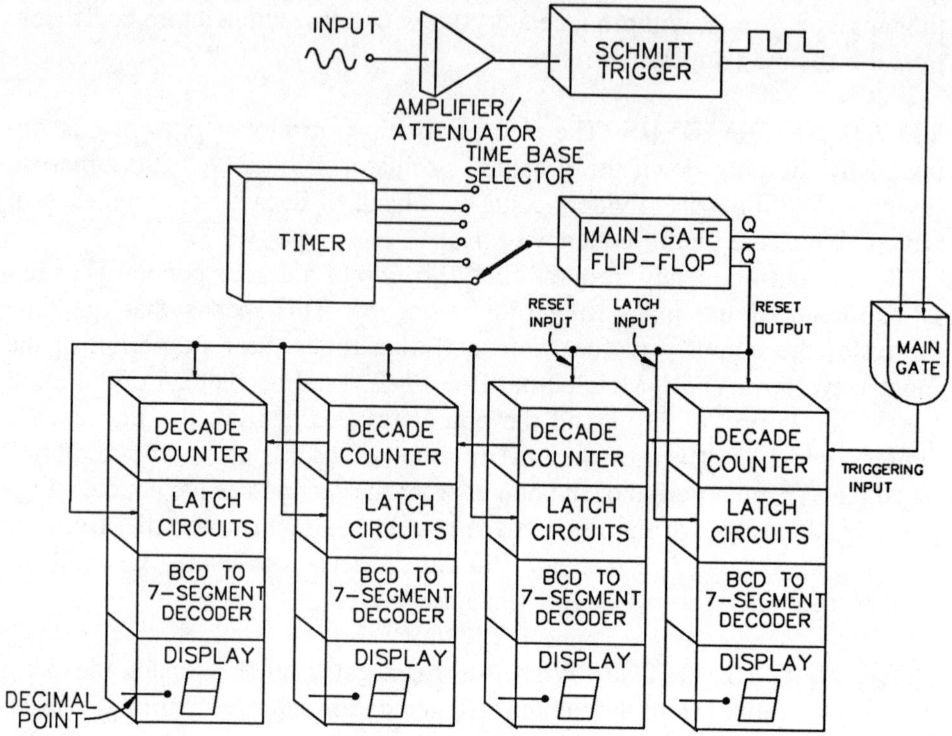

Figure 2-9 Counting register

Digital Frequency Counters

The digital counter has come a long way since it first appeared on the commercial market. Even in its original form it was recognized as perhaps the most useful measuring instrument to emerge from the laboratory since the oscilloscope. The original application of the digital counter was that of a *frequency counter*.

When the main gate is controlled by an accurate time interval, the counter is in the *frequency mode*. The frequency mode is similar to the totalize mode described above, with the difference being in the way the gate control is operated.

For frequency measurements, the counting time interval is precisely controlled to be multiples of a second. If, for example, 45,500 pulses pass through to the counting register while the main gate remains open for one

Type of measurement	Counter mode	Book reference
Period	Period	Figure 2-10
Ratio	Ratio	Figure 2-11
Time Interval	Interval	Figure 2-12
Phase	Interval	Figure 2-14
Risetime	Interval	Figure 2-14
Time Interval	Reciprocal	Figure 2-16
Rf Burst	Reciprocal	Figure 2-20

Table 2-2 Counter modes

second, then the input signal frequency is 45.5 kHz. Depending on the number of digits in the readout, this number could be displayed directly or expressed as powers of ten.

To measure frequencies that would otherwise exceed the maximum count of the counting register, the time interval of the main gate can be reduced, with an appropriate change made to the scale factor. A similar number of pulses (45,500) accumulating in the counting register within a 0.1 second period, for example, would represent a frequency of 455 kHz; 25,000 pulses counted within 10 μs is 2.5 MHz, and so forth. Scaling can be done automatically by the positioning of the decimal point on the digital readout, as the timebase period is switched from range to range.

Instead of manually switching to the appropriate timer range, some instruments have an automatic range selection system called *auto ranging*. The auto ranger usually consists of a circuit that generates a voltage approximately proportional to the input frequency. Depending on the actual voltage level, one of several transistor switches is turned on to select the correct time base and decimal point position.

Using the frequency counter as a frequency meter, however, is exploiting but one of its several electronic counter functions. A counter can also indicate the period of an input signal, compare two signals in the ratio mode, indicate the time between two points on a waveform, or do reciprocal counting. The mode of operation is a simple reconfiguration of the basic counter building blocks shown in Figure 2-3.

PERIOD MODE. The *period mode* of a signal is the reciprocal of its frequency. Period measurements are made by determining the amount of time a signal takes to complete one cycle of oscillation. In this mode, the counting register counts pulses from the timebase oscillator, and the input signal is used to enable the main-gate flip-flop. The circuit configuration is shown in

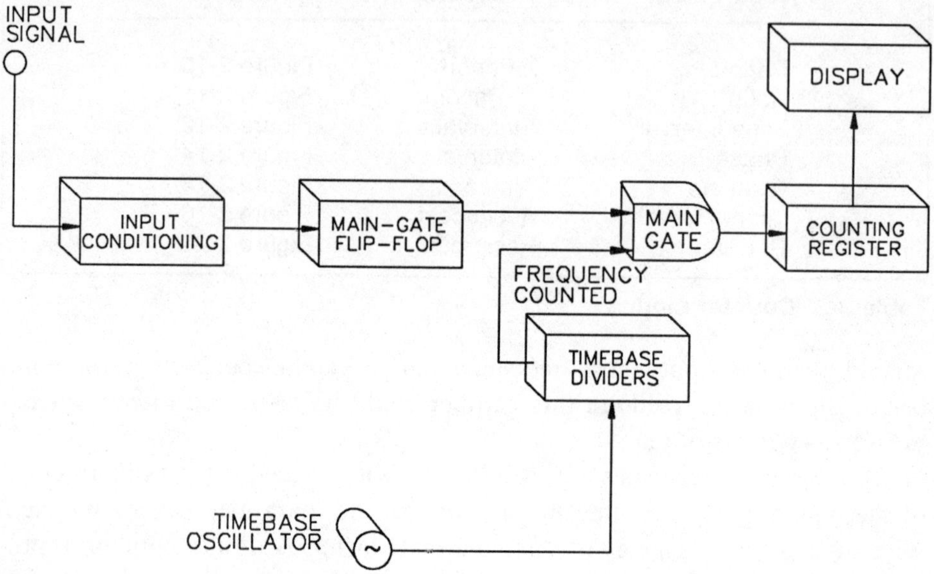

Figure 2-10 Period mode

Figure 2-10. The number of timebase pulses counted is directly proportional to the time period of the input signal.

For example, let's say the signal to be measured is 100 kHz (10 μs) and the timebase reference oscillator is 1 MHz (1 μs). The main gate triggers on the input pulse and opens for a period of 100 microseconds, and during this time 10 pulses from the timebase oscillator pass through the main gate. These 10 pulses are then processed and scaled by the counter to provide a readout indication of 100 microseconds.

RATIO MODE. A third mode of operation is the *ratio mode.* In the ratio mode, the ratio of two input signals is displayed on the counter readout. Frequency ratio may be measured, as shown in Figure 2-11, by disconnecting the timebase oscillator from the gating circuitry and allowing the lower-frequency signal to control the main gate.

In effect, what happens is that the lower-frequency signal determines how long the gate is open and thus how long the gate passes the higher-frequency signal from the upper input. Therefore, the gate output is the ratio of the two signals. If the lower-frequency signal allows but one higher-frequency signal

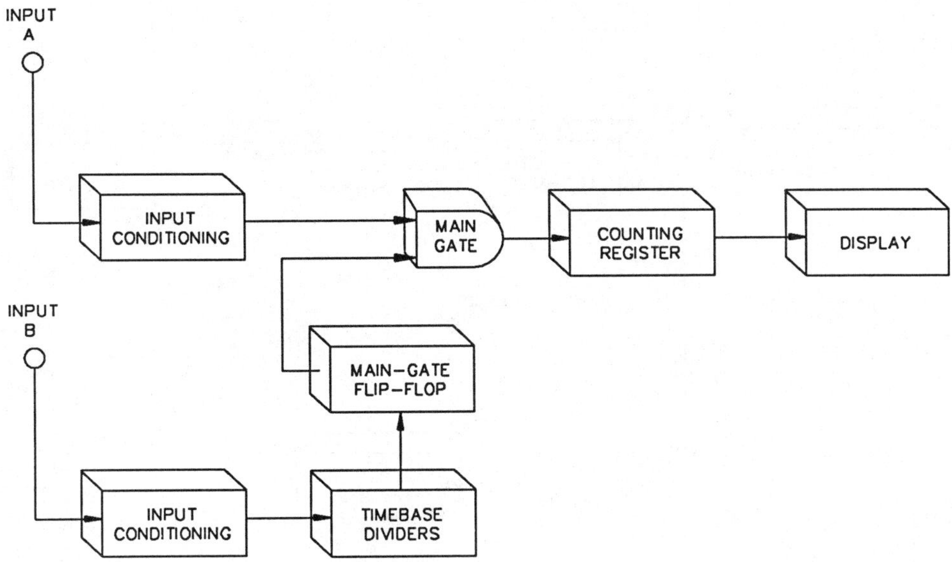

Figure 2-11 Ratio mode

to pass, then the ratio is 1:1. Should the lower-frequency signal allow 10 higher-frequency cycles to pass, then the ration is 10:1. This scheme will not work with reference frequencies that are lower than the signal to be referenced (i.e., when the frequency to Input A is less than the frequency at Input B), because the main gate would never be open long enough to pass even a single pulse to the counting register.

Time Interval Mode

The *time interval* mode permits the counting of any number of events occurring between any two points in time. In effect, it is the equivalent of an electronic stopwatch, and uses start and stop inputs to control the opening and closing of the main gate. The equivalent circuit diagram is shown in Figure 2-12. During the counting period, the counting register counts pulses from the timebase oscillator. Selection of the timebase divider may alter the timebase frequency in most instruments.

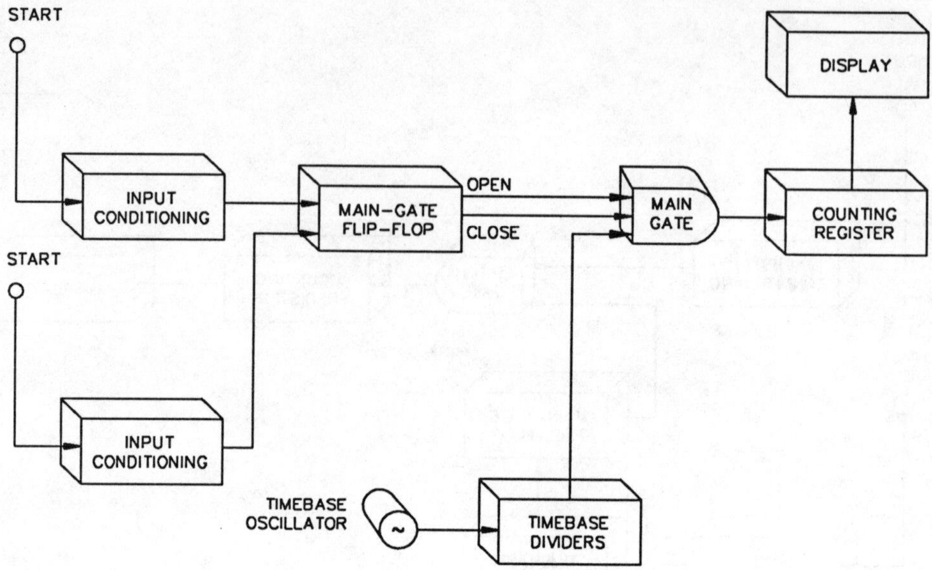

Figure 2-12 Time interval mode

Phase Measurements

A special application of the time interval mode is *phase measurement*. A frequency counter can be used to measure the phase difference between two equal-frequency signals by taking a time interval measurement between identical points on the waveform and calculating the phase from the measured time.

Consider the timing diagram in Figure 2-13. In this illustration, the values T1 and T2 are identical points on the reference-phase and unknown-phase waveforms. $T\phi$ is the time difference between successive positive-going zero crossings of each signal. To measure $T\phi$, the frequency counter is set to operate in the **START/STOP** time interval mode, with the count starting at T1 and ending at T2.

The phase interval of time measurement is possible only with a more elaborate gate control than those associated with other modes previously discussed. As shown in Figure 2-14, the main gate has three inputs: one from the main-gate flip-flop and two from the signal conditioner stages, **START** and **STOP** inputs.

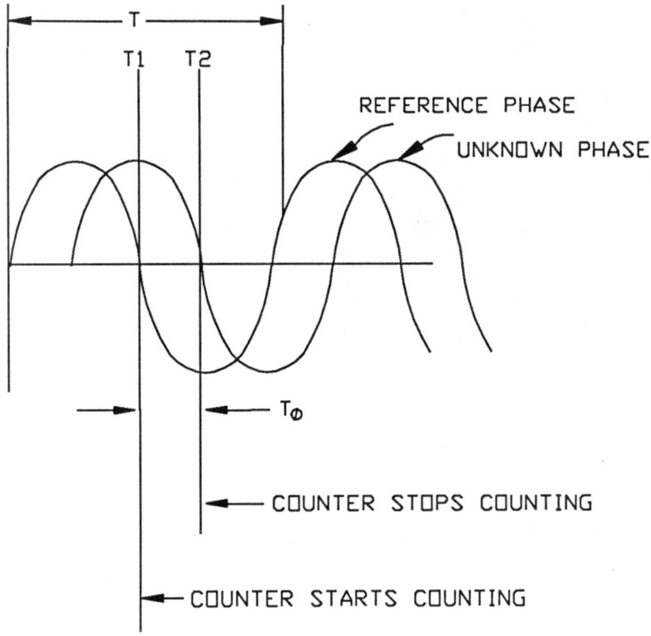

Figure 2-13 Phase measurement mode

The main-gate flip-flop feeds an accurate oscillator frequency from the timebase oscillator to the counting register when the **START** signal triggers the main gate on. A similar trigger from the stop input halts the counting process. The phase angle between the two signals is then calculated from the equation:

$$f(\text{degrees}) = (T\phi/T) \times 360$$

The **START** and **STOP** points are selected by triggering levels on the input waveforms.

RISE-TIME MEASUREMENTS. *Rise-time measurements* can also be made using the modified circuit configuration of Figure 2-14. When measuring rise time, the start and stop trigger levels are set at 10 percent and 90 percent, respectively, of the maximum voltage V_M (see Figure 2-15).

Standard TTL gates, for example, have a nominal high output voltage of 2.4 volts. In this example, the rise time is measured by setting the **START** control to trigger at 0.24 volt and the **STOP** control to trigger at 2.16 volts. The intervening pulse count from the timebase oscillator/timebase divider is the rise time.

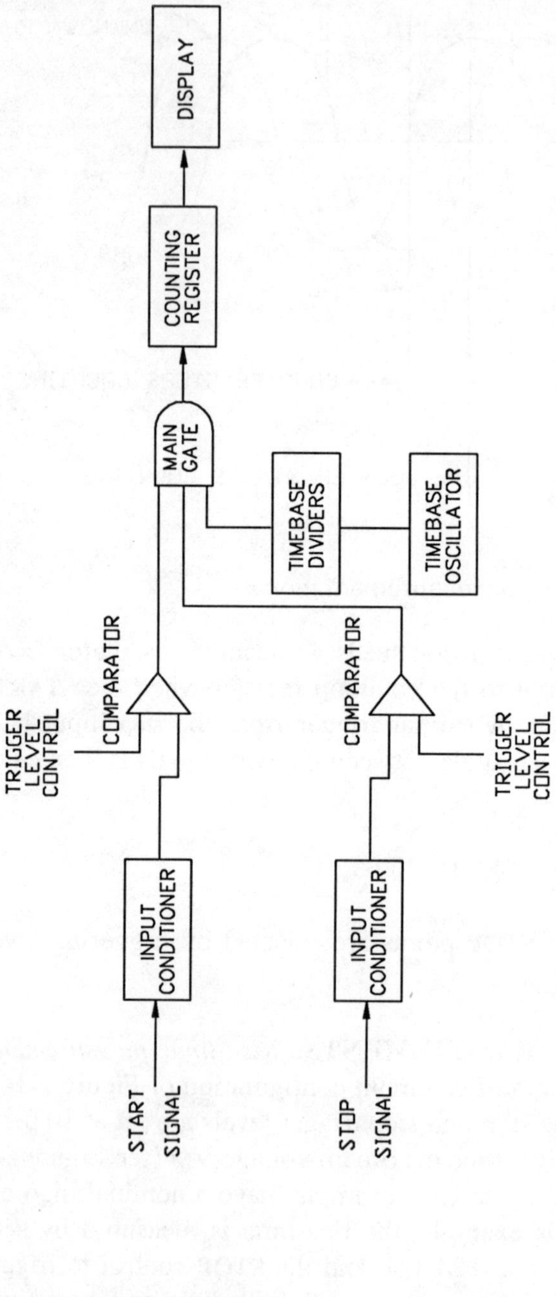

Figure 2-14 Phase mode timer control

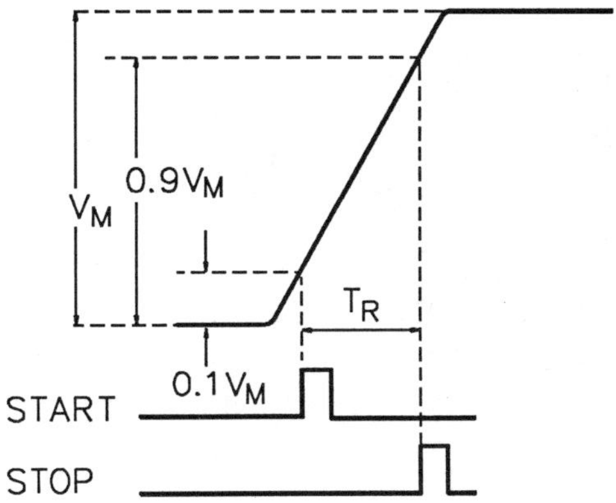

Figure 2-15 Rise-time measurement

Reciprocal Frequency Counters

The *reciprocal frequency counter* differs from the ordinary frequency counter in that it uses separate registers to accumulate time- and event-counts, as noted in Figure 2-16. The contents of these registers are then processed to provide outputs that represent either period or frequency. It works like this: as long as the main gate is open, the event counter accumulates pulses from the external source, while the time counter accumulates pulses from an internal clock. The two counts may then be displayed or manipulated to achieve a wide range of measurements.

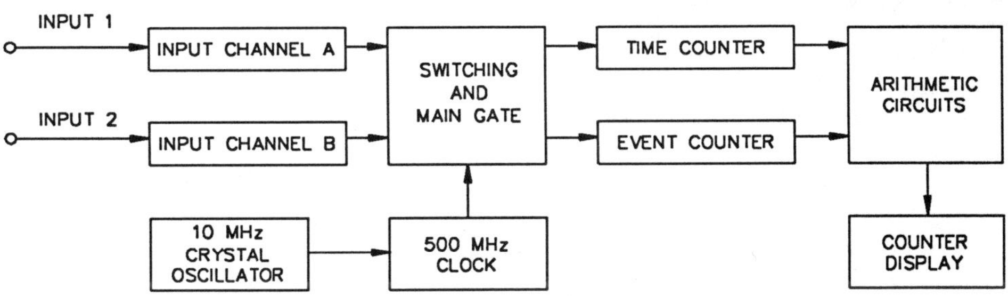

Figure 2-16 Reciprocal counter

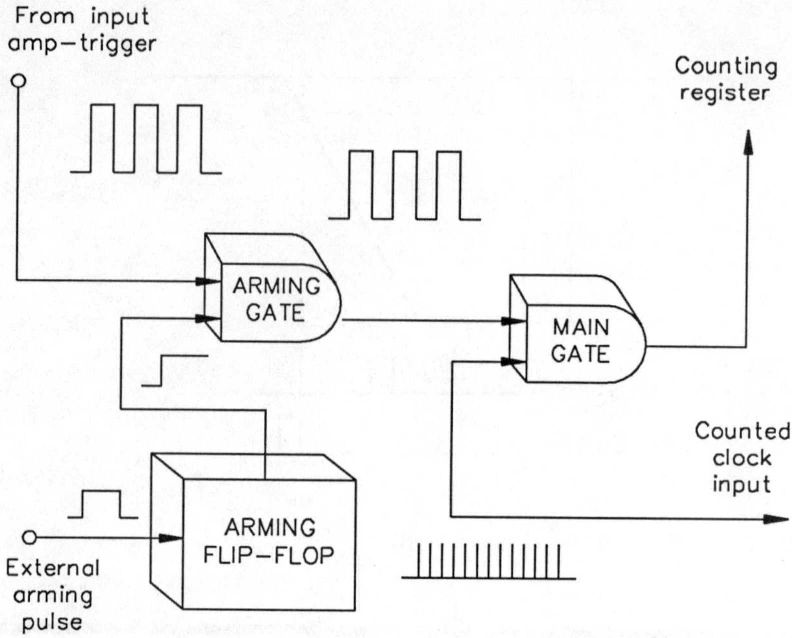

Figure 2-17 Armed input for reciprocal counter

Some reciprocal counters allow counting to be externally armed, as shown in Figure 2-17. External arming could be useful, for example, in measuring the frequency of a pulsed-RF signal, as shown in Figure 2-18. By

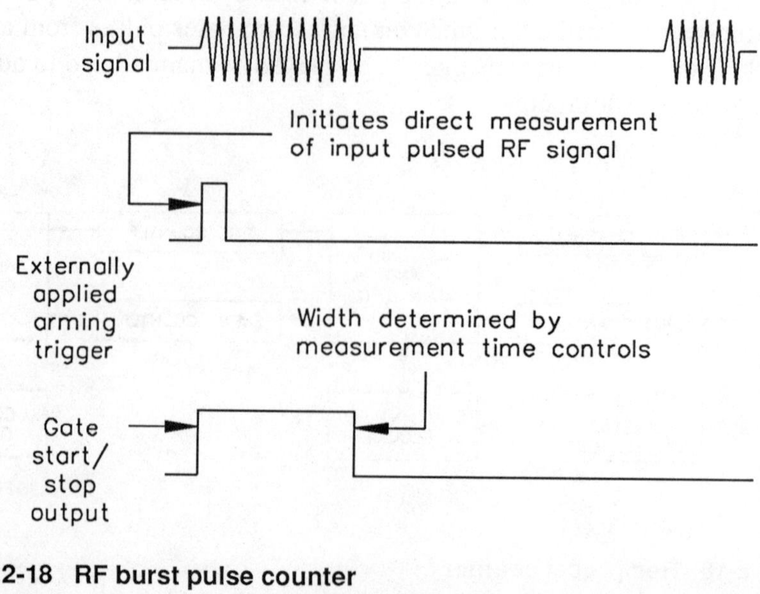

Figure 2-18 RF burst pulse counter

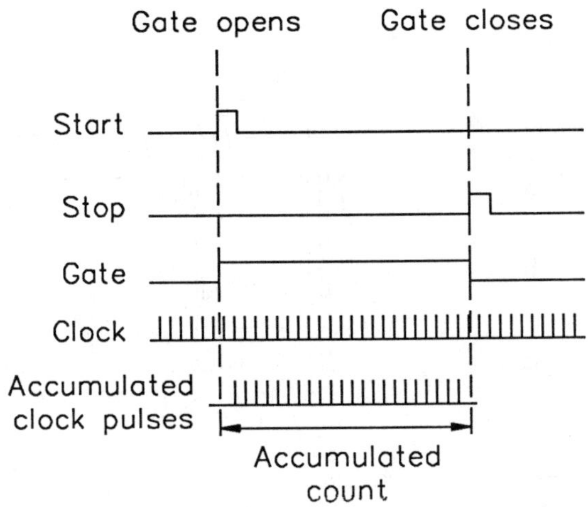

Figure 2-19 Time-interval measurements with a reciprocal counter

setting the measurement-time control to the width of the RF burst, an accurate frequency measurement of the burst can be made.

In a reciprocal counter, time-interval measurements are made by counting the number of pulses between independent **START** and **STOP** inputs, as shown in Figure 2-19. The resolution is determined by the frequency of the timebase oscillator. A 10-MHz clock, for example, could provide a 1×10^{-7}, or 100-nanosecond, resolution.

Counting Errors

Although the notion of connecting a frequency counter to a frequency source to get an instantaneous digital readout down to the last hertz and beyond is intriguing, care must be exercised in the measurement. Everything is fine when the frequency counter deals with nothing but smooth, repetitive waveforms with high amplitude. However, erroneous readings can result if the wave has certain irregularities, involves transients, or is accompanied by RFI noise or other disturbances.

The input stage of a counter contains a Schmitt trigger circuit, which has an upper trigger point and a lower trigger point. The dead band separating the two is called the *hysteresis*. The effect of hysteresis is demonstrated in Figure

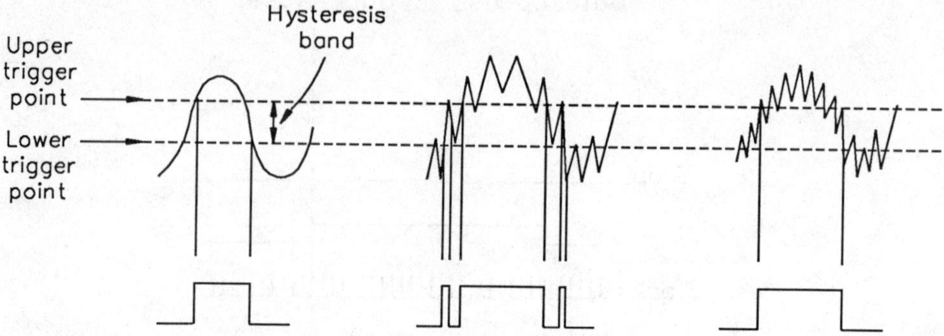

(a) "Clean" sine wave crosses the hysteresis band twice during each cycle — triggers counting circuits once per cycle.

(b) Noisy sine wave crosses hysteresis band more than twice during each cycle — produces counting errors.

(c) Attentuated noisy waveform crosses hysteresis band twice in each cycle.

Figure 2-20 Hysteresis and counting errors

2-20. Assuming that the amplifier/attenuator stage is set for unity gain (1), the Schmitt output goes positive when the input sine wave passes the upper trigger point. When the sine wave goes below the lower trigger point, the Schmitt trigger output returns to its previous level.

Now look at the effect of noise on the input signal, as illustrated in Figure 2-20. The noise spikes cause the signal to cross the hysteresis band more than twice in each cycle. Thus unwanted additional output pulses are generated that introduce errors in the frequency measurement. Such errors can be overcome by expanding the hysteresis band, but at the expense of not being able to detect low-level signals.

In typical frequency counters, the input circuitry is optimized for frequency counting by detecting zero crossings. This means that every time the waveform crosses over the zero reference line, the counter is triggered. That makes it difficult to measure rise times, propagation delays, and low-periodic-rate pulses because of the limited trigger-level range used (typically one volt or less). At best, the level control varies the center of the hysteresis band; at worst it is offset by several millivolts.

An alternative is to compensate for the ambiguity in the trigger point by using *hysteresis compensation*, in which a dc voltage equal to 1/2 the hysteresis band is added to or subtracted from the selected trigger level or reference voltage.

This method is most effective when used in conjunction with an oscilloscope. The main function of the oscilloscope is to reveal the true nature of the signal presented to the digital counter. Armed with the knowledge of the waveform, certain techniques, such as rectification, can be used to eliminate sources of interference. Wherever feasible, connections to the digital counter should be made with coaxial cable. In such instances, it is also preferable to use the low-impedance (50-ohm) input mode.

Rather than use hysteresis compensation, the time-interval probe solves the problem of trigger ambiguity by using an automatic calibration scheme. The user grounds the probe to be calibrated and then presses a front panel switch. That causes the reference voltage V_R to decrease in a stair-step fashion in 1-mV steps until the device triggers. For negative slope calibration, the voltage increases in 1-mV steps. The system then uses the value of V_R to adjust itself so that the actual trigger voltage corresponds to the trigger level selected by the user. Recalibration, when slopes or probes change, assures constant triggering accuracy. Some instruments allow the trigger voltage to be set from −9.99 volts to +9.99 volts, in 10-mV steps, by adjusting front panel thumbwheel switches.

The time-interval probe has other advantages, including high input impedance (1 MΩ) and low input capacitance (10 pF). Ordinary high-impedance probes have about 40 pF of capacitance, and, at high frequencies, delays through the probe, cable, and circuitry can cause an inaccurate determination of time interval.

Counting errors are also introduced by the main gate itself. In the system described in Figure 2-3, the timebase could switch the main gate on or off while an input pulse from the Schmitt trigger is being applied. The partial pulse that gets through the main gate during the switching period may or may not succeed in triggering the counting circuits. Thus there is always a possible error of ±1 cycle in the count of input cycles during the timing period. The accuracy of a digital frequency meter is usually stated as 1 count timebase error.

So far in this discussion of digital and frequency counters, it has been assumed that the counting period is a one-time event. While this is generally true, most digital counters also provide for counting averaging. In the *averaging mode*, a counter can average the measurement reading over a number of periods or time intervals to give better resolution. Using the averaging mode, one can eliminate the uncertainty of the ±1 count and arrive at a more precise measurement, particularly with waveforms that tend to generate erroneous displays, as described above.

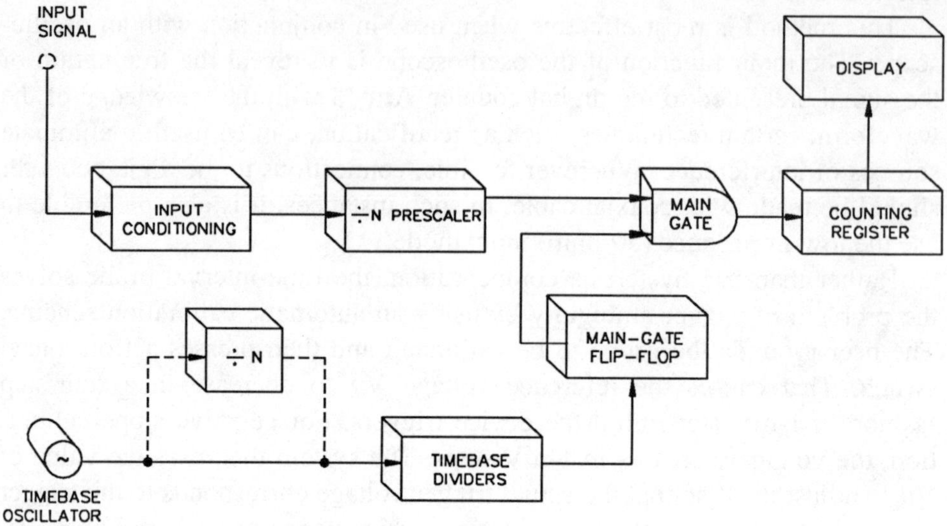

Figure 2-21 Prescaling electronic counter block diagram

Prescaling Counter

To extend the frequency range of a frequency counter, a *prescaler* can be used. The prescaler counter (shown in Figure 2-21) divides the input signal before it goes to the main gate for subsequent counting. The design is economical because it allows a low-cost frequency counter to do the work of an expensive microwave frequency counter. Essentially, there are two techniques employed for frequency counter prescaling: precounting and heterodyning.

Using the *precounting method*, the signal is first conditioned, then fed to an array of prescaler decade dividers. These dividers, like the counting register or timebase dividers, output a pulse for every ten pulses counted. This allows a 1-GHz signal, for example, to be divided by 1000, using three decade counters, for display on a conventional 1-MHz frequency counter. To compensate for a division-by-*n* reading, the counter can either leave its gate open *n* times as long, or multiply the counting register's content by *n*.

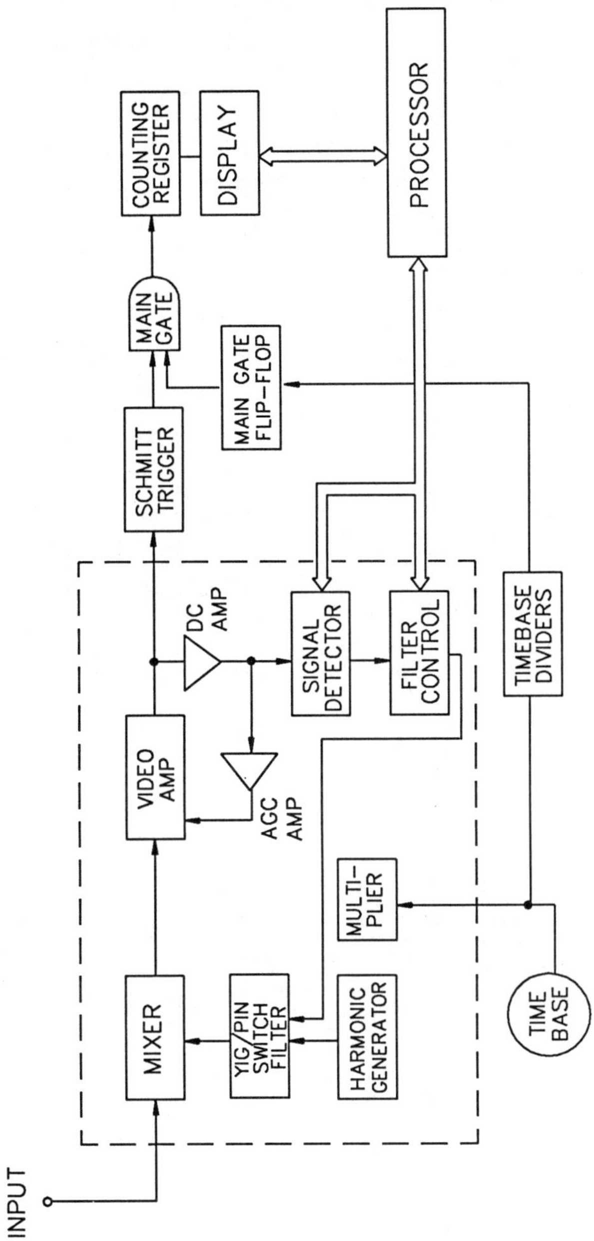

Figure 2-22 Heterodyne converter counter

HETERODYNE CONVERTER COUNTER. The *heterodyne converter* counter mixes the input signal with a second frequency to produce a lower difference frequency. The frequency measurement is made by counting the lower difference frequency. The second frequency is usually derived from the timebase oscillator, which drives a harmonic generator or tuned cavity, as depicted in Figure 2-22.

In heterodyne counting, a beat frequency is produced that the frequency counter can cope with. For example, heterodyning a 1-GHz signal against a 500-MHz signal produces sidebands of 1.5 GHz and 500 MHz, the lower of

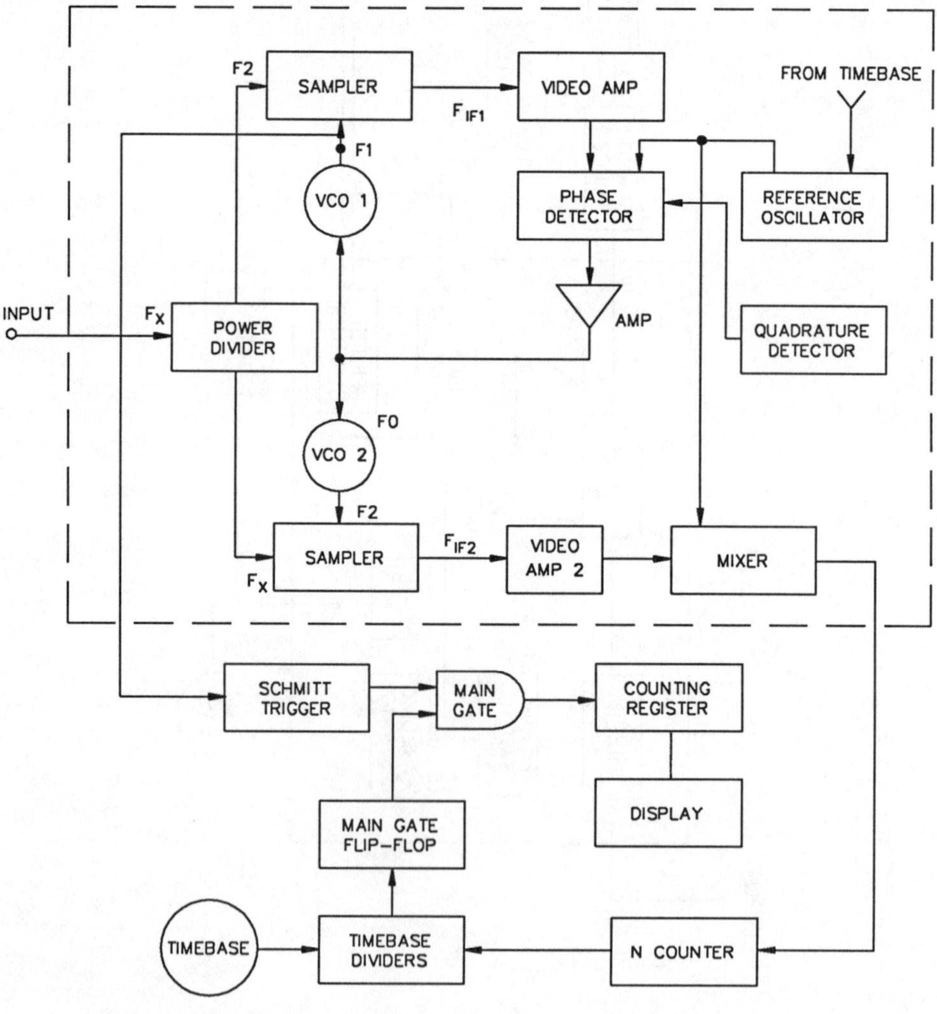

Figure 2-23 Transfer oscillator counter

which is filtered and passed to the frequency counter. Frequency is determined by:

$$f(\text{measured}) = f(\text{actual}) - f(\text{reference})$$

or

$$f(\text{actual}) = f(\text{measured}) + f(\text{reference})$$

Heterodyne counters may provide a resolution of ±1 Hz, and they operate from dc up to 18 GHz. Their drawback comes from the operator having to use a conversion chart to determine the true input frequency from the readout frequency.

TRANSFER-LOCK TECHNIQUE. The *transfer-lock oscillator counter* is somewhat like the heterodyne method converter counter in that two signals are compared. Unlike the heterodyne method with its fixed reference frequency, the transfer oscillator counter uses a variable frequency oscillator (VFO) to track the input signal.

The VFO is at a lower frequency and harmonically related to the input signal being measured. In operation, the VFO frequency is counted, rather than the actual input frequency, thus eliminating errors created by noise and ambiguous transients. Two advantages of the transfer oscillator counter are a

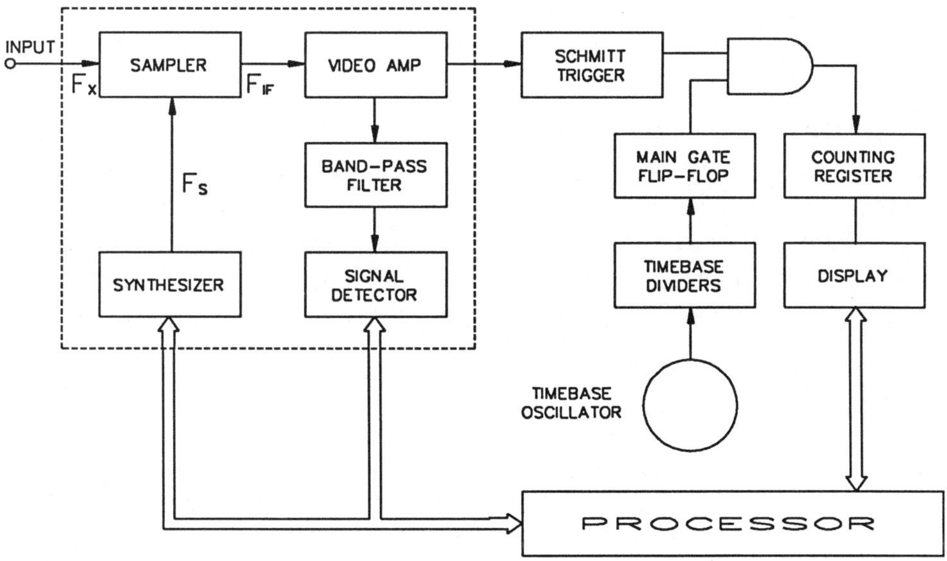

Figure 2-24 Harmonic-heterodyne converter

Conversion Characteristic	Heterodyne Converter	Transfer Oscillator	Harmonic Converter
Frequency Range	20 GHz	23 GHz	40 GHz
Acquisition Time	150 mS	150 mS	350 mS
Sensitivity	–30 dBm	–35 dBm	–30 dBm
Signal-to-Noise Ratio	40 dB	20 dB	20 dB
FM Capture Range	30-40 MHz	1-10 MHz	10-50 MHz
AM Tolerance	2%	>90%	>90%
Amplitude Discrimination	4-30 dB	2-10 dB	2-10 dB

Table 2-3 Microwave conversion performance

very wide frequency range of 20 Hz to 40 GHz and the ability to measure pulsed signals as well as CW signals. Disadvantages include loss of resolution, longer measurement periods, and the need for a skilled operator in the complex computation necessary for the harmonic lock and subsequent frequency calculation. After lock is achieved, the input frequency can be calculated using the equation:

$$F_X = nF_1 - F_{1F1}$$

HARMONIC-HETERODYNE CONVERTER. The *harmonic-heterodyne converter* is a hybrid of the two techniques just described. A block diagram is shown in Figure 2-24.

A harmonic-heterodyne converter acquires input signals in a manner similar to the transfer oscillator, but it measures frequency in a manner similar to the harmonic oscillator. Frequency is determined by the processor's multiplying the synthesizer frequency by *n* and adding the result to the reference frequency. The input frequency can be calculated as:

$$F_X = nF_S + F_{IF}$$

This method is similar to the way the heterodyne converter works, in that the *n*th harmonic of a stable oscillator is mixed with the input to produce a difference frequency.

Most frequency counters can measure frequency-modulated signals, but the heterodyne converter and harmonic-heterodyne frequency counters have a distinct advantage in making that measurement, as shown in Table 2-3. Also noted in Table 2-3 is the fact that the heterodyne converter has a limited tolerance to AM signals because of its AGC circuitry.

Digital Voltmeters

The *digital voltmeter* (DVM) is perhaps the most prolific of all digital test instruments. Basically, the DVM circuitry is similar to the analog voltmeters discussed in Chapter 1, with the expectation that the readout is displayed digitally rather than on a mechanical meter movement.

Although there are numerous techniques used to convert analog values to digital numbers, the basic principles of operation are the same: The voltage to be measured must be converted to a number of pulses proportional to the unknown voltage; the pulses are counted using a digital counter circuit, as described above, and the count is displayed on a seven-segment readout. Five methods of analog-to-digital conversion are commonly used. They are:

voltage to frequency integrating
single ramp
dual-slope integrating
successive approximation potentiometeric
continuous balance potentiometeric

ELECTROMECHANICAL DIGITAL VOLTMETER. The *electromechanical digital voltmeter*, shown in Figure 2-25, uses the continuous balance potentiometeric method. Although this instrument has long been replaced by solid-state versions, it is interesting and instructive to contemplate this early approach to digital instrumentation. The mechanical readout was comprised of four wheels, each bearing numerals from 0 to 9. The readout device was mechanically linked to a servo motor, which also drove a precision potentiometer. The output of a comparator governs the operational direction of the servo motor and the overall action as it hunts for a null (zero voltage) by adjusting the position of the potentiometer wiper arm. At null, the servo motor would cease operation, and the digits displayed on the mechanical wheels could be read as voltage.

Stability of the reference voltage and the linearity of the precision potentiometer were largely responsible for the accuracy of the instrument. The comparator and the amplifier were often discrete circuits, but hybrid modules and other packaged circuitry were also used.

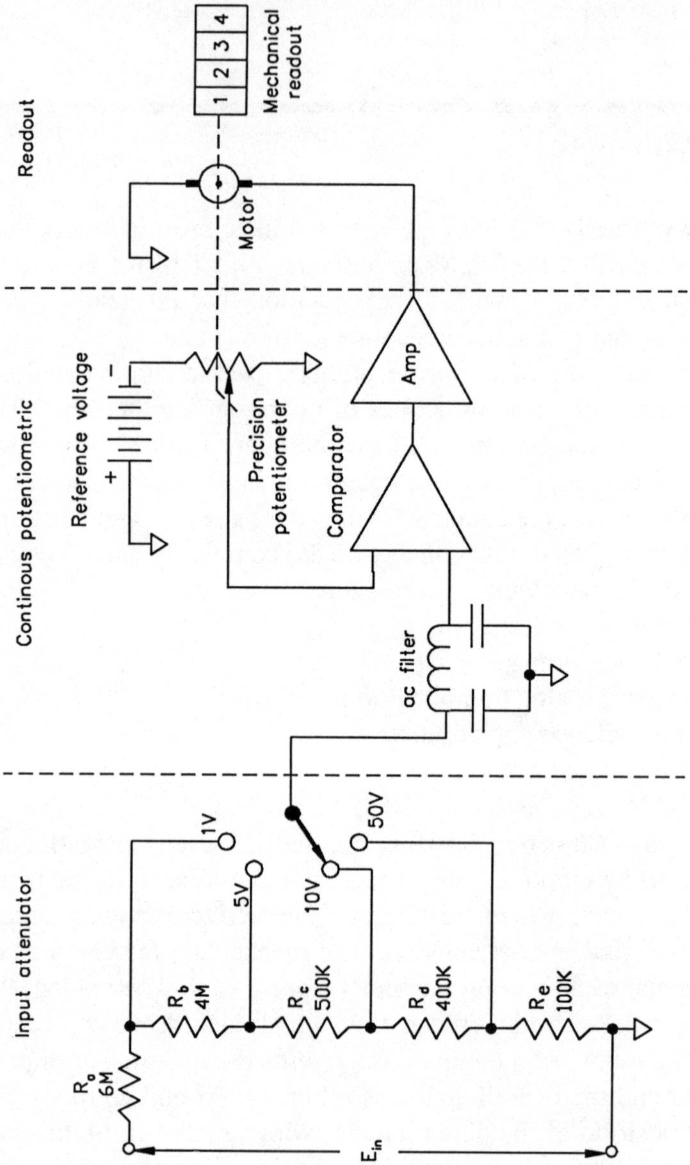

Figure 2-25 Electromechanical digital voltmeter

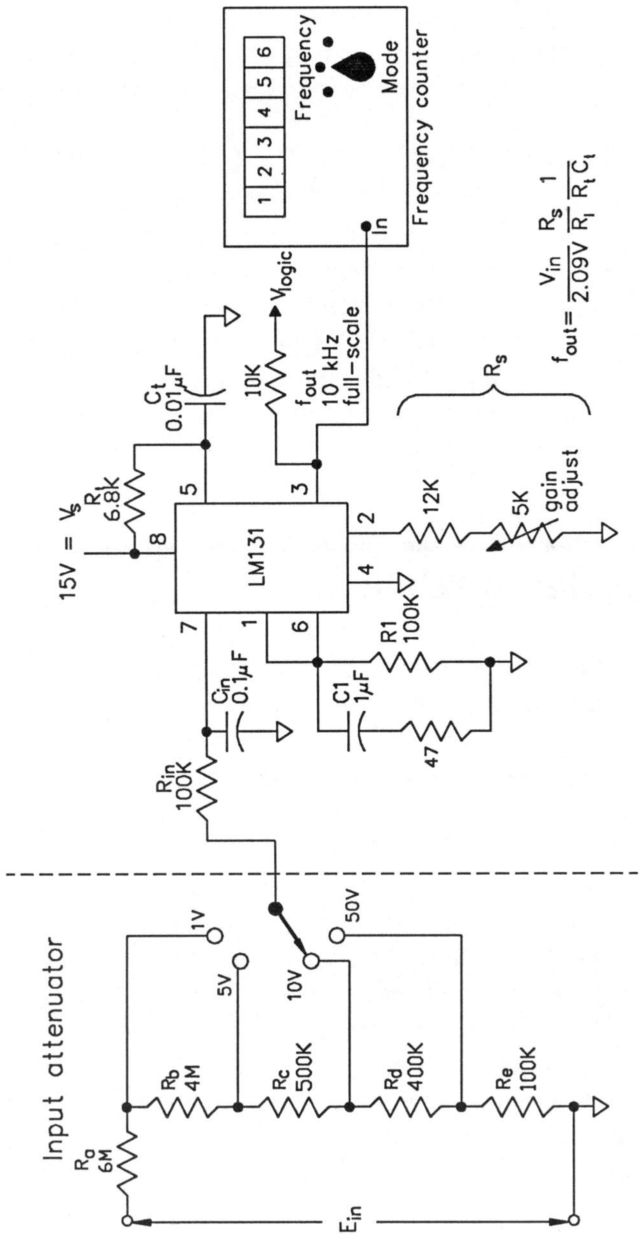

Figure 2-26 Voltage-to-frequency conversion voltmeter

Later versions of the electromechanical DVM replaced the servo motor with relays that would simulate mechanical movement using the *successive approximation potentiometeric method* to display a digital result. The first of these digital meters was introduced by Non-Linear Systems, now better known for their compact oscilloscopes and Kaypro computer.

VOLTAGE-TO-FREQUENCY CONVERSION. The first truly digital voltmeter came as a result of expanded development with integrated circuits, among the most popular of which was a single-chip voltage-to-frequency converter. In the voltage-to-frequency converter, a voltage controlled oscillator (VCO) is used to output a frequency that is proportional to an input voltage. When this circuit is coupled with a frequency counter, the result is a digital display representative of the input voltage. But this design, too, has fallen in favor of the more accurate ramp-type digital voltmeter.

Ramp-type Digital Voltmeters

Ramp-type digital voltmeters are the mainstay of the DVM industry. In ramp-type digital voltmeters, the voltage to be measured is converted into a time period. This time period is then fed to a digital counter, and the display is read as a voltage. The process is shown in Figure 2-27.

SINGLE-RAMP INTEGRATOR. One system that accurately converts voltage into time period uses a circuit known as a *Miller integrator*. The Miller integrator circuit, shown in Figure 2-28, consists of resistor R1, capacitor C1, and an operational amplifier. With no input voltage and C1 discharged, the operational amplifier behaves as a voltage follower with its non-inverting terminal grounded. Consequently, the inverting voltage and the output terminal are at ground level.

When a positive input voltage is applied to the inverting input through R1, the op-amp inverting input is held at ground level by C1, and the input current is equal to:

$$I_i = \frac{V_i}{R1}$$

I_i is a constant current that is much greater than the input bias current of the operational amplifier. Effectively, all of I_i flows to capacitor C1 to linearly

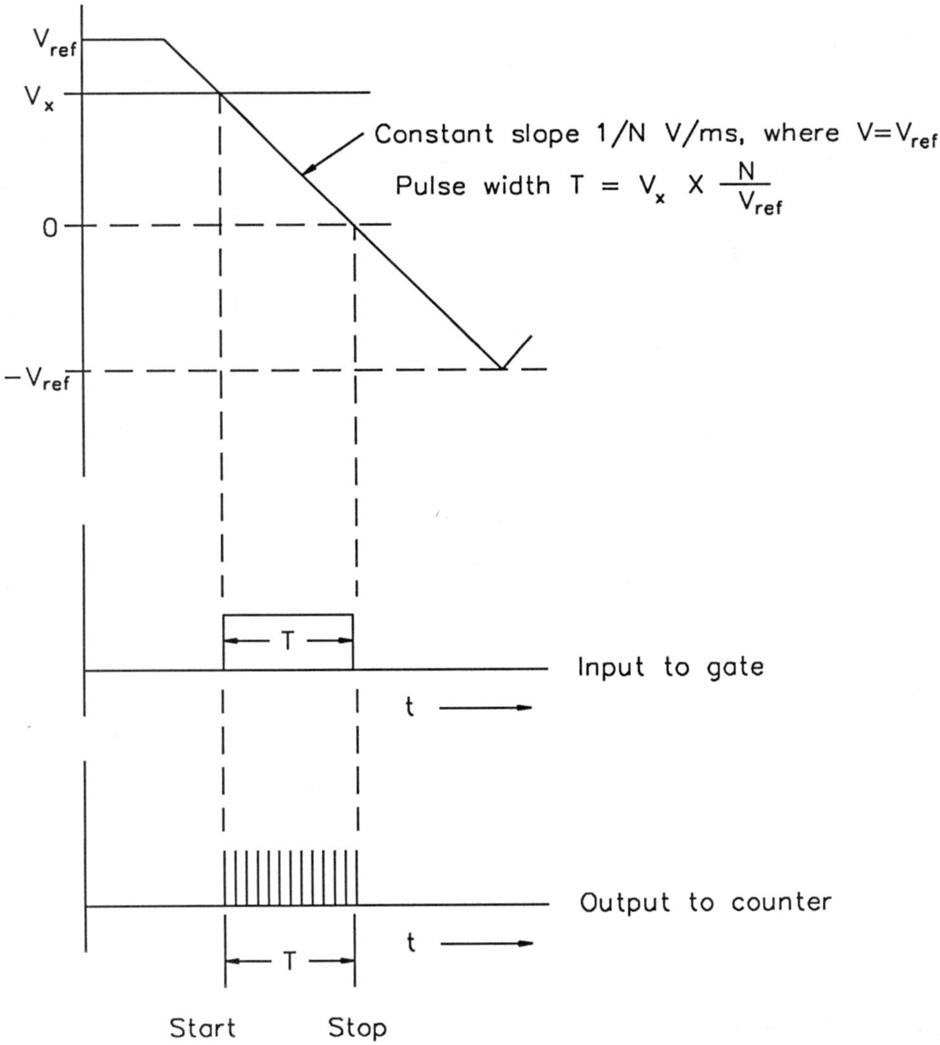

Figure 2-27 Ramp-type DVM

charge the capacitor. This produces a negative-going linear ramp output voltage at Vo.

This waveform is sent to a digital counting circuit configured for the interval mode (see page 73). The **START** and **STOP** signals are derived from the ramp waveform, according to the procedure outlined in Figure 2-27. The ramp pulse is used to gate the timebase oscillator pulses to the counter so that the number of pulses reaching the counter is proportional to the unknown voltage.

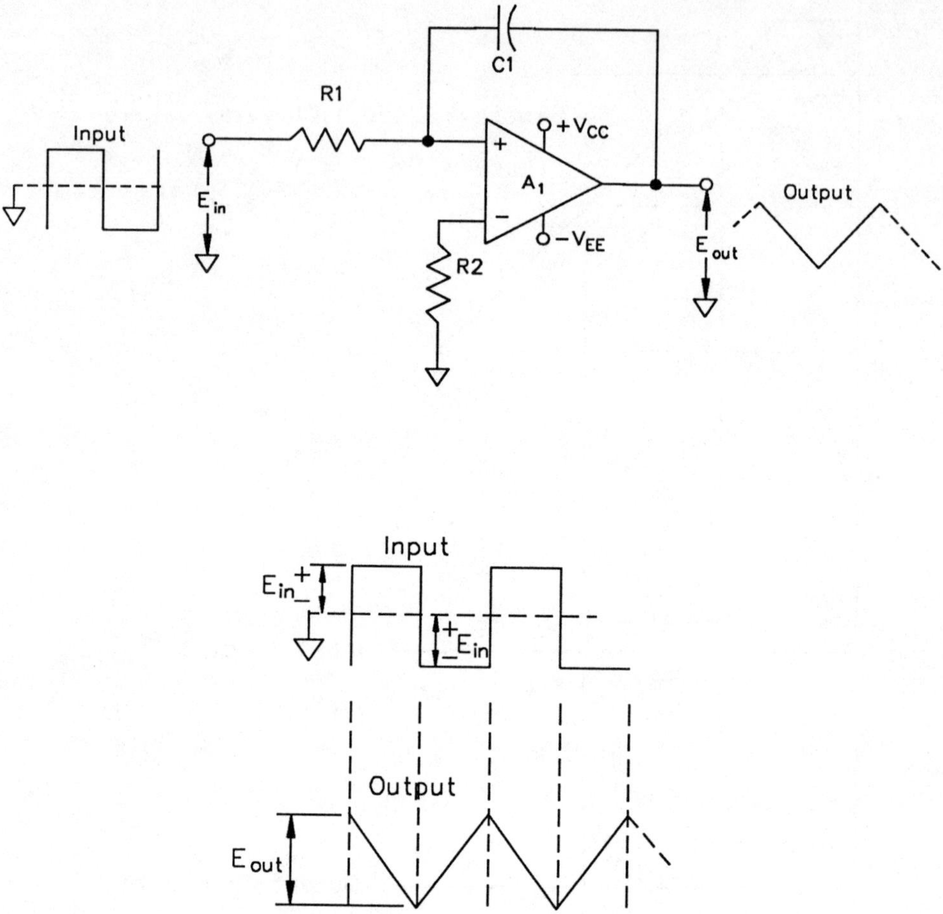

Figure 2-28 Miller integrator

DUAL-SLOPE INTEGRATOR. The *dual-slope integrator*, illustrated in Figure 2-29, is the most preferred form of voltage-to-digital display type of digital voltmeter.

In this design, the voltage to be measured is applied via a voltage follower to the Miller integrator. The voltage follower output is switched by Q1 to the input of the Miller integrator. An accurate current source is also connected to the Miller integrator input, and the integrator output level is monitored by a zero-crossing detector. The zero-crossing detector is basically a high-gain operational amplifier used as a comparator. A large positive output is produced when its input is slightly above ground, and a large negative output

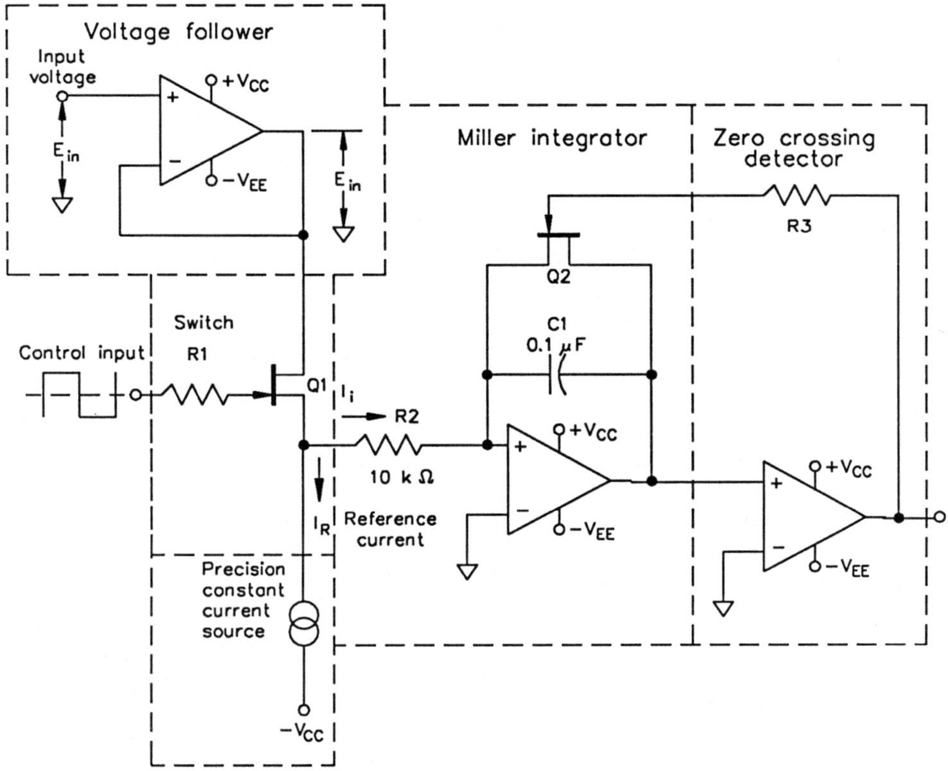

Figure 2-29 Dual-slope integrator

occurs when its input is slightly below ground. The zero-crossing detector controls Q2. When the output of the zero-crossing detector is high, Q2 is on and C1 is short-circuited. When the zero-crossing detector's output is low, Q2 is off and C1 can be charged.

The square-wave input to Q1 controls the time interval during which I_i flows into R3, and the time during which I_R flows out of R3. This square wave is generated by using decade counters to divide the output frequency of a clock source, as explained above. When the control input is negative, Q1 is biased off and V_i is isolated from the rest of the circuit. During this time, the reference current flows through R3. The level of the reference current is equal to:

$$I_R = \frac{-V_Z}{R3 + R4 + R5}$$

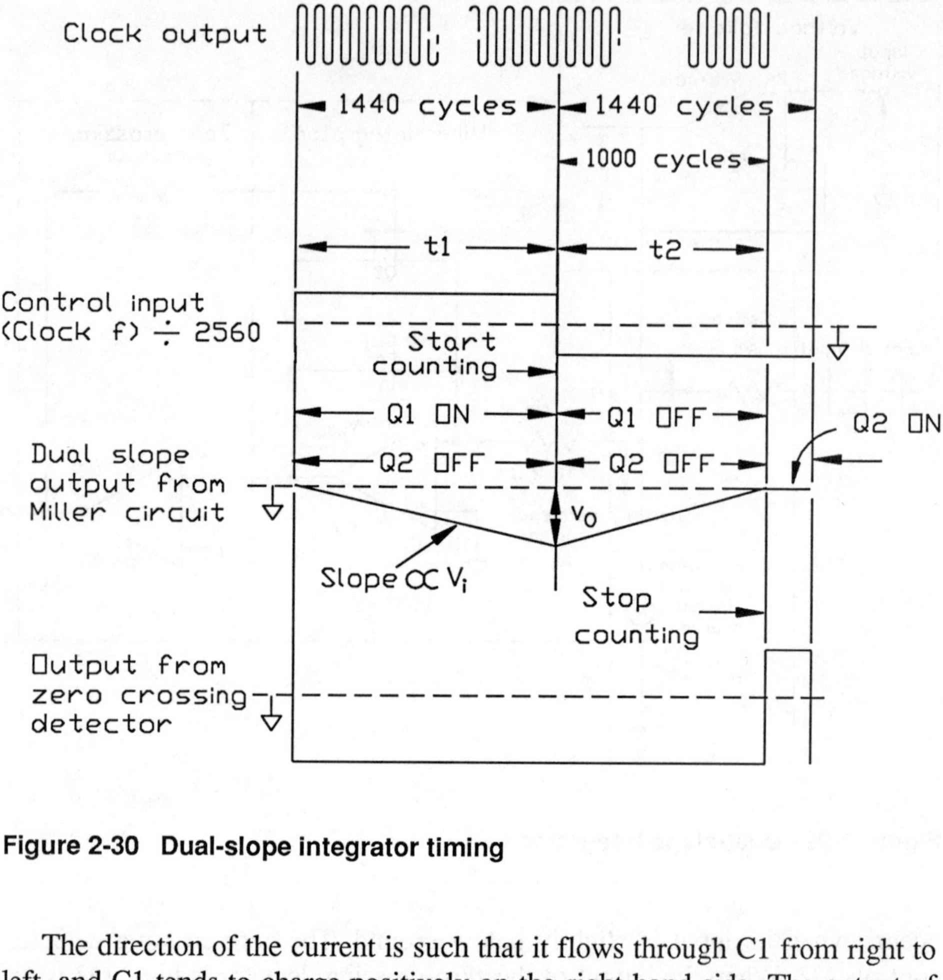

Figure 2-30 Dual-slope integrator timing

The direction of the current is such that it flows through C1 from right to left, and C1 tends to charge positively on the right-hand side. The output of the Miller integrator now rises to ground level, and the zero-crossing detector generates a positive output. This biases Q2 on, which now shorts capacitor C1. Therefore, at the end of the negative half of the square-wave input to Q1, capacitor C1 is short-circuited, and the Miller integrator output is held close to ground.

Transistor Q1 switches on when the square-wave input becomes positive. This action connects the input voltage to resistance R5, and provides the input current to the Miller integrator. Capacitor C1 now charges with negative polarity on the right-hand side, thus producing a negative-going output. Consequently, the zero-crossing detector has a negative output, which biases Q2 off and permits C1 to charge. The output from the Miller integrator is a linear negative ramp voltage (Figure 2-30), which continues during the positive portion of the square-wave input to Q1. Since I_i is directly proportional to the input voltage, the slope of the ramp is also proportional to the input voltage. This means that the ramp amplitude is directly proportional to the input voltage.

When the square-wave input again becomes negative, Q1 switches off and the reference current I_R starts to flow through R5 and C1. I_R now discharges C1 and the Miller circuit output becomes a positive-going ramp. The positive-going ramp continues until it arrives at ground level, at which point the zero-crossing detector switches Q2 on, discharging C1.

The time t2 it takes for the ramp voltage to climb to zero is directly proportional to the input voltage. Time t2 is measured by starting the counting circuits at the negative-going edge of the square-wave input to Q1, and stopping at the positive-going edge of the output from the zero-crossing detector.

An important advantage of the dual-slope integration method is that small drifts in the clock frequency have little or no effect on the accuracy of the measurement. If, for example, the clock frequency is *f*, then the period of one clock equals:

$$T = \frac{1}{f}$$

Assuming that I_i = 100 μA and t1 = 1444, then:

$$V_O = \frac{I_i t1}{C1} = \frac{100\ \mu A \times 1444T}{C1}$$

and

$$t2 = \frac{C1 V_O}{I_R} = \frac{C1}{144\ mA} = 1000T$$

The number of clock pulses during t2 is given by:

$$\frac{t2}{\text{Time period of clock pulses}} = \frac{1000T}{T} = 1000$$

The time period T cancels out of the equation, which makes the clock frequency drift a nondetermining factor.

In the methods described above, it is understood that the processes are repeated a number of times per second, so that if the input voltage is changing, the change is continuously indicated by the readout.

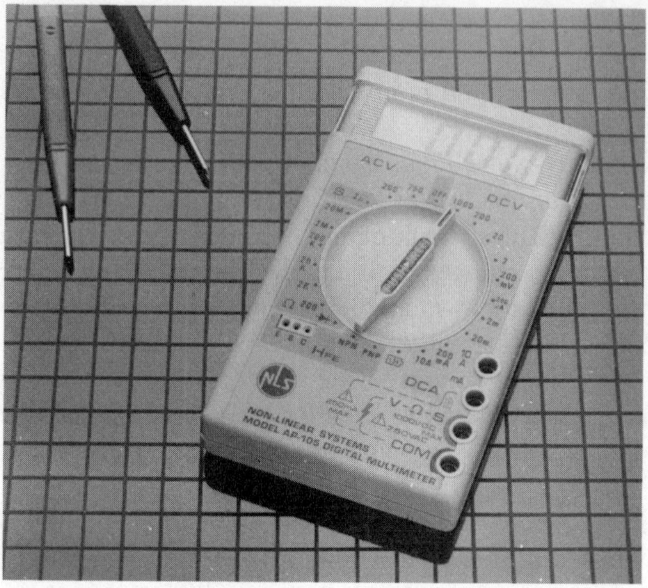

Figure 2-31 Digital multimeter (courtesy of Non-Linear Systems)

Digital Multimeter

As with the analog multimeter, the *digital multimeter* (DMM) is capable of measuring voltage, current, and resistance. Basically, the DMM is an electronic voltmeter that can be configured to express current and resistance measurements through a voltage. Current is derived by measuring the voltage drop across a known resistance and scaling the digital readout accordingly. Resistance is measured using the constant-current technique described in Chapter 1.

3. SIGNAL GENERATORS

The signal generator is an indispensable laboratory instrument used in a wide range of laboratory and field applications. Signal generators are typically classified by type as being low-frequency sine-wave generators, function or waveform generators, radio-frequency sine-wave generators, sweep frequency generators, frequency synthesizers, and pulse generators. Some instruments are capable of generating more than one type of signal, while others are dedicated components designed for very specific applications.

Sine-wave Generators

Low-frequency sine-wave generators are generally dedicated instruments with a narrow frequency range limited to one type of application. While there are several types of sine-wave oscillator circuits that can be used for signal generation, the *Wein bridge* is the most popular for low-frequency signal sources.

The Wein bridge oscillator is a typical RC oscillator composed of three parts: an amplifier, a gain-controlling element, and an RC frequency-sensitive network. The principle disadvantage of RC oscillators is the low Q of the frequency-determining network and the consequently large frequency shifts, which are caused by small changes in the amplifier phase. These disadvantages can be substantially overcome by the use of negative feedback around the amplifier to stabilize its amplification and the phase shift, as emphasized in the Wein bridge design.

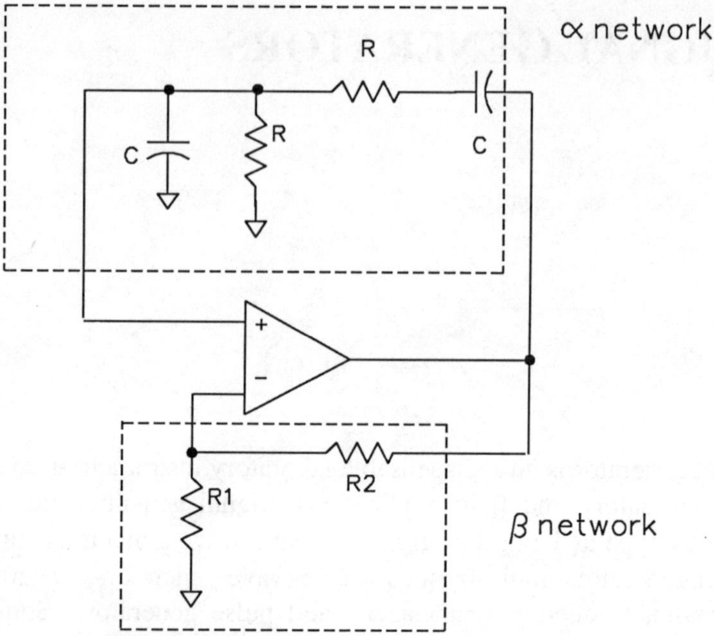

Figure 3-1 Wein bridge oscillator

Referring to the schematic diagram of Figure 3-1, the positive feedback network (α) of the Wein bridge oscillator has maximum transmission at the desired frequency of oscillation when $\omega_o = (RC)^{-1}$. The R and C values in the feedback network are selected to offer minimum impedance at the desired oscillation frequency. It can be shown that the input impedance of the network as seen by the amplifier output is:

$$Z_{in}(\omega_o) = \frac{3R}{\sqrt{2}} \angle 45^\circ$$

for u = v = 1. At the minimum impedance, the amplifier input voltage across R2 and C2 are in phase with the output and reinforce the input to the operational amplifier. The oscillation frequency is equal to:

$$f = \frac{1}{2\pi CR}$$

The attenuation of the positive feedback path is:

$$\beta = \frac{R1}{R1 + R2}$$

For the convenience of using equal values for both R's and C's, the scaling factors of u and v are often chosen to be unity. Continuous tuning over a wide frequency range is achieved by ganging the two capacitors. Note that in Figure 3-1 the stray capacitances are relatively harmless if the series C is located as shown. Furthermore, it cannot be ignored in designs of this type that the positive feedback must be greater than the negative feedback to sustain oscillation.

Function Generators

The *function generator* is the most versatile of the low-frequency signal generators because it provides a wide assortment of wave shapes, including triangular, square, sine, and pulses, in a single instrument. Occasionally, a function generator will also support ramp (sawtooth) waveforms like those used by oscilloscopes for the horizontal sweep (refer to Chapter 4). Because of their diverse selection of wave shapes, function generators are sometimes called *waveform generators*. Waveform generators find a wide range of applications in communications and telemetry equipment, as well as for testing and calibration in the laboratory.

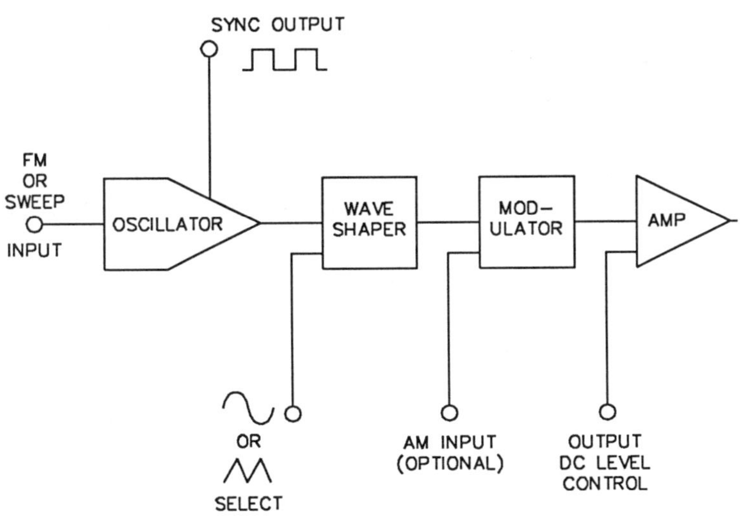

Figure 3-2 Function generator block diagram

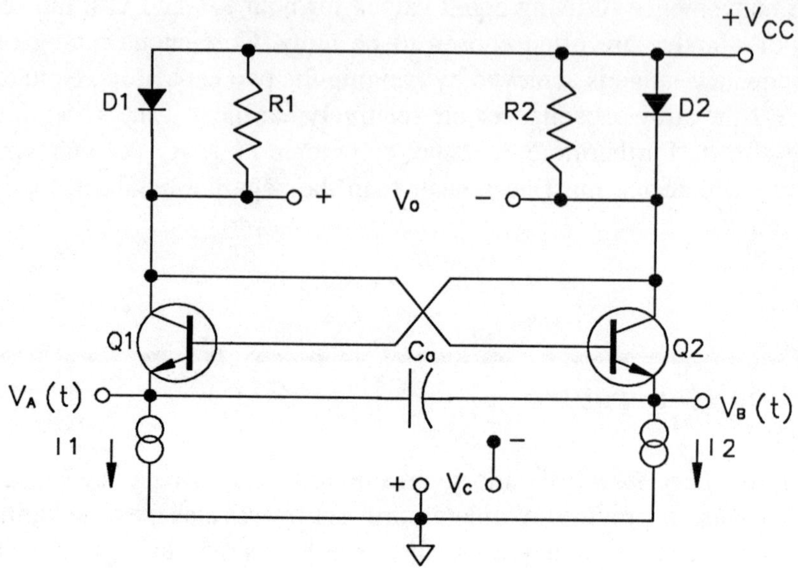

Figure 3-3 Emitter-coupled astable multivibrator

Essentially a function generator is a stable oscillator circuit that outputs well-defined waveforms. A function generator consists of four sections, which includes an oscillator to generate the basic periodic waveform, a wave shaper, an optional modulator to provide AM and FM capabilities, and an output buffer amplifier. A simplified function generator is shown in Figure 3-2.

At the heart of the generator is a voltage-controlled oscillator, or VCO. The oscillator is usually an astable multivibrator similar in operation to the flip-flop. The multivibrator is well-suited to the application because it generates square, triangle, and ramp waveforms within a single circuit. Figure 3-3 shows a simplified emitter-coupled multivibrator of the type associated with waveform generators.

This circuit provides constant-current charging and discharging of an external timing capacitor, C_O, which then generates a square wave as well as a triangle or linear ramp output. At any given time, either Q1 and D1 or Q2 and D2 are conducting in such a way that capacitor C_O is alternately charged and discharged by constant current I_1. Assuming Q1 is conducting, the base of Q2 is biased off by the nearly zero collector voltage of Q1. As the charge through C_O increases, the voltage at the collector of Q1 rises proportionally. At a point determined by the V_{BE} characteristics of Q2, transistor Q2 flips into conduction, forcing the base of Q1 to zero volts and causing Q1 to drop out of con-

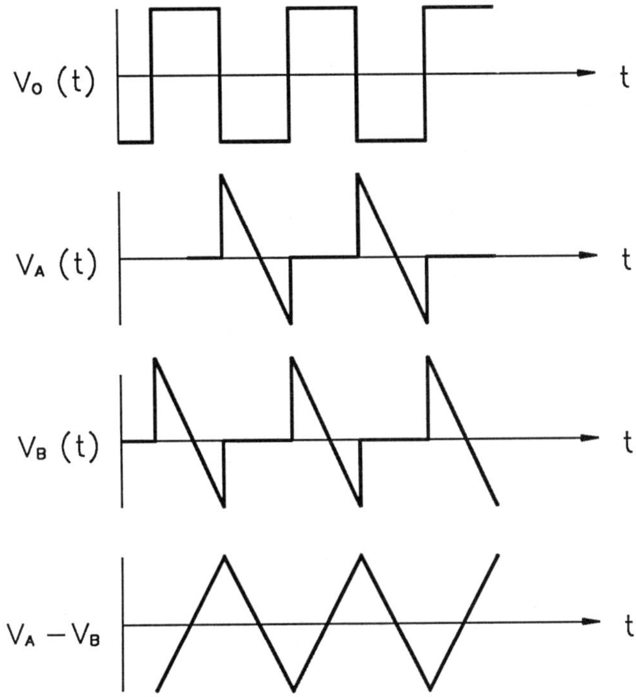

Figure 3-4 Multivibrator waveforms

duction. Capacitor C_O now charges in the opposite direction, until conditions are again right for Q1 to toggle back on and Q2 to cut off.

The output across D1 and D2 corresponds to a symmetrical square wave, having a peak-to-peak amplitude of $2V_{BE}$, or twice the transistor base-emitter voltage drop. Output V_A, which is constant when Q1 is on, becomes a linear ramp with a slope equal to $-I_1/C_O$ when Q1 goes off. Except for half-cycle delay, output $V_B(t)$ is the same as $V_A(t)$. Both linear ramp waveforms, V_A and V_B, have peak-to-peak amplitudes of $2V_{BE}$. Their frequency of oscillation o_f can be determined from the formula:

$$o_f = \frac{1}{4V_{BE}C_O}$$

Frequency o_f can be controlled by variation of charging current I_1 via control voltage V_C.

By subtracting one ramp output voltage from the other using a simple differential amplifier, $V_A - V_B$, a linear triangular waveform can be generated. Symmetry of triangle and square wave outputs may be adjusted by replace-

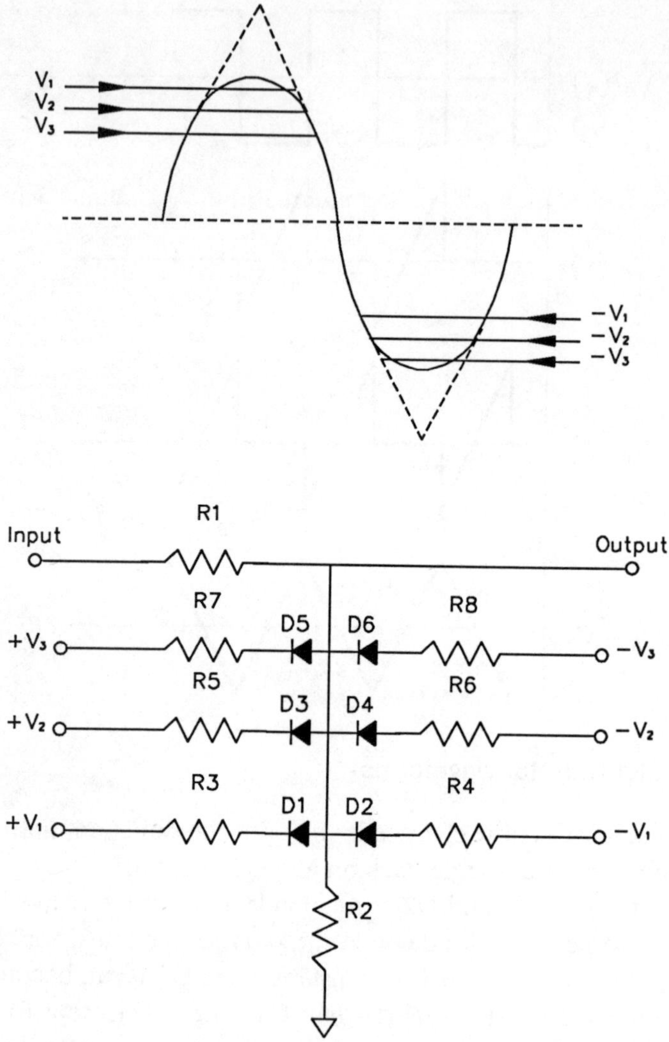

Figure 3-5 Diode wave shaper

ment of one of the two current sources in Figure 3-3, where I_2 does not equal I_1. The duty cycle of the output waveforms now becomes:

$$\text{Duty Cycle} = 50\frac{I_1}{I_2} \text{ percent}$$

The duty cycle of the output may be varied over a wide range by controlling the ratio of currents I_1 and I_2.

In the design of function generators, sinusoidal output is normally obtained by passing a triangular wave through a wave-shaping circuit. A widely used method for converting a triangular wave into an approximate sinusoidal waveform involves a diode-resistor ladder network. Referring to the diagram in Figure 3-5, we see (ignoring diodes D1 through D6, for the moment) that R1 and R2 form a voltage divider that attenuates the triangular waveform by:

$$V_O = V_i \frac{R2}{R1 + R2}$$

The diodes, however, influence the waveform by introducing a finite number of break points at which the output waveform is affected. Consider D1 and R3, which is biased by voltage $+V_1$. At the point at which D1 becomes forward-biased ($V_1 + V_F$), R3 is effectively placed in parallel with R2, changing the attenuation factor to:

$$V_O \approx V_i \frac{R_P}{R1 + R_P}$$

where

$$R_P = \frac{(R2)(R3)}{R2 + R3}$$

The signal levels above $+V_1$ are now attenuated to a greater extent than levels below $+V_1$, leading to a nonlinear effect on the waveshape. With the introduction of D2 through D6, and their respective biasing resistors — each at a different potential — the output can be a good approximation of a sine wave, as illustrated.

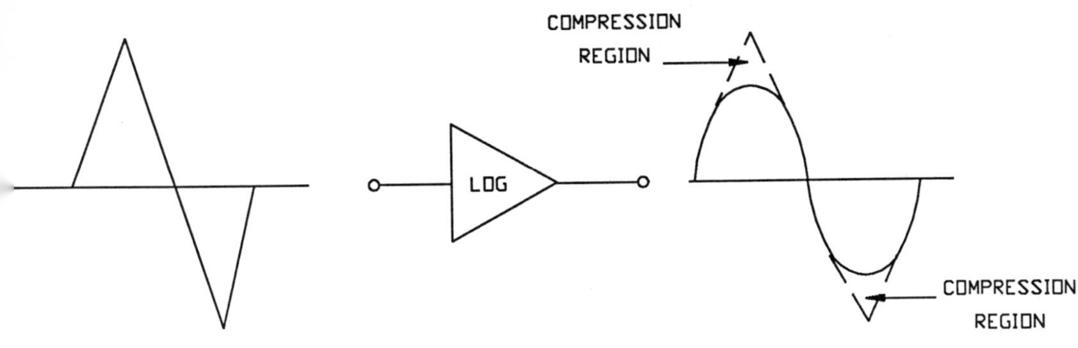

Figure 3-6 Sine-wave shaper

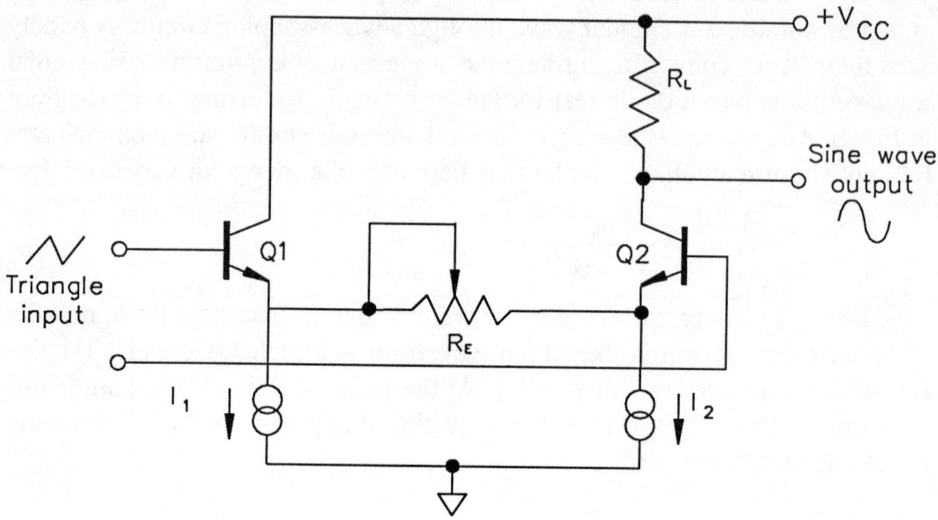

Figure 3-7 Logarithmic compression amplifier

Another method by which triangular waves are reduced to sine waves involves a *logarithmic amplifier*. Basically, a logarithmic amplifier is one in which the amplification factor of the circuit depends on the level of the input signal. As the input voltage to the amplifier increases, its gain decreases. This creates a compression effect (illustrated in Figure 3-6) that rounds off the sharp peaks of the triangular wave and gives the output a sine wave appearance.

A simplified logarithmic amplifier stage is shown in Figure 3-7. Reduction of the emitter-degeneration resistance R_E allows either transistor Q3 or Q4 to be brought near their cutoff point when the input triangle waveform reaches its peaks.. With the proper choice of signal input amplitude and bias-current levels, the transfer characteristics at the peaks of the input triangle waveform are logarithmic rather than linear. Thus the peaks of the triangle become rounded, and the output appears as a low-distortion sine wave. Use of this technique permits output harmonics to be reduced to less than 0.5% with only a single adjustment (R_E).

The typical laboratory quality function generator produces sine, square, and triangle waveforms at frequencies between 0.003 Hz and 200 kHz, and ramp waveforms at frequencies between 0.003 Hz and 30 kHz. The source impedance is generally 600 ohms. The main output waveform is push-button selectable and is available both at a BNC connector and standard, 3/4-inch spaced five-way binding posts. The output level can be varied up to a maxi-

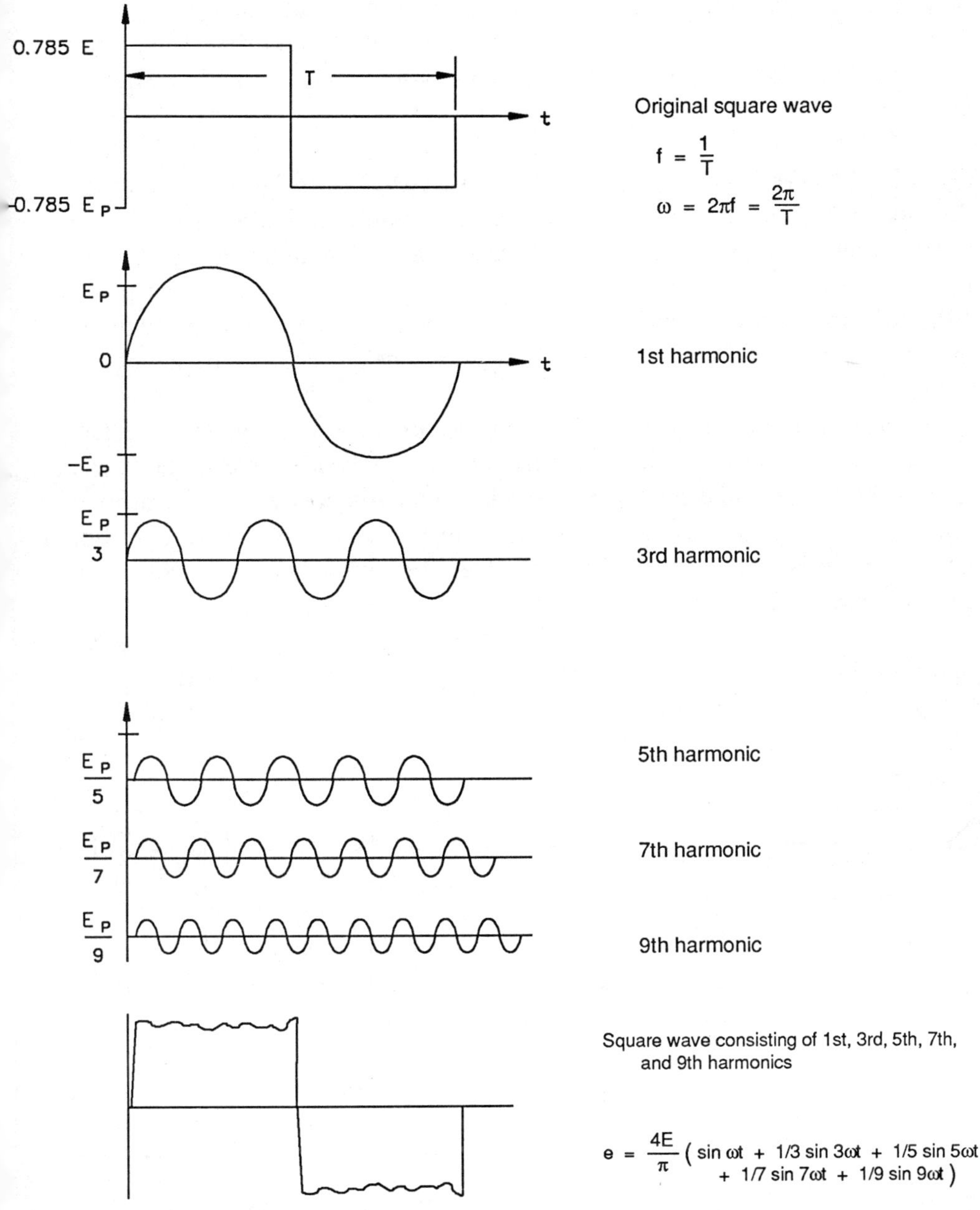

Figure 3-8 Fourier analysis of a square wave

mum of 10 V (p-p) by means of a continuous, variable control calibrated for connection to a 600-ohm load. Frequency is set by push-button range switches and is continuously variable within the selected range using a multi-turn precision potentiometer. The frequency can also be controlled by application of a direct voltage to the VCO input to permit frequency modulation of the output signal by means of an external voltage. In many cases, an internal logarithmic sweep generator enables the output frequency to be FM-modulated via an external signal source. TTL-compatible square-wave output is also standard on most instruments.

TESTING LOW-FREQUENCY AMPLIFIERS. While there are many applications for low-frequency signal generators, amplifier testing is by far the most common. An amplifier may be tested by applying an ac signal of varying magnitude and frequency to the input and analyzing the output for phase shift, gain frequency response, etc. This method of testing is long, tedious, and ineffective, however, in determining the transient characteristics of the amplifier.

A better method is to subject the amplifier under test to square-wave test signals. It can be shown mathematically (Fourier analysis) that a square wave consists of a sinusoidal signal of a fundamental frequency with an infinite amount of odd harmonics. The application of a square wave to an amplifier may therefore be considered as the equivalent of applying simultaneously an infinite number of sine waves.

Figure 3-9 depicts the test setup for performing a square-wave test on an amplifier. With the oscilloscope adjusted for proper display (refer to Chapter 5), the frequency response of an RC amplifier may be approximated by

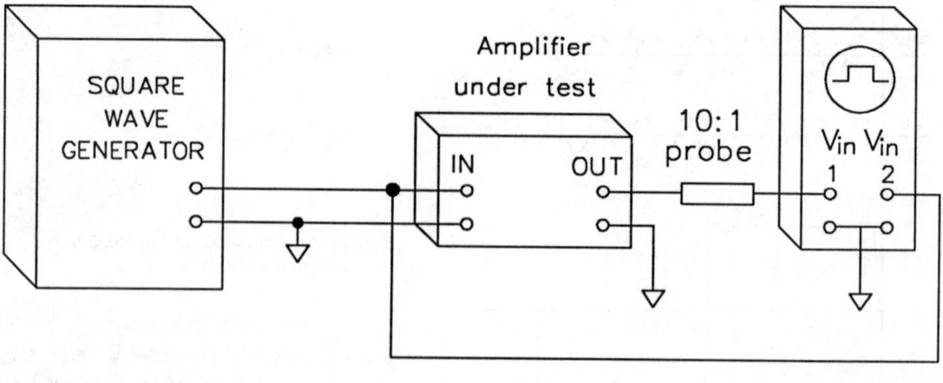

Figure 3-9 Test setup for amplifier square-wave testing

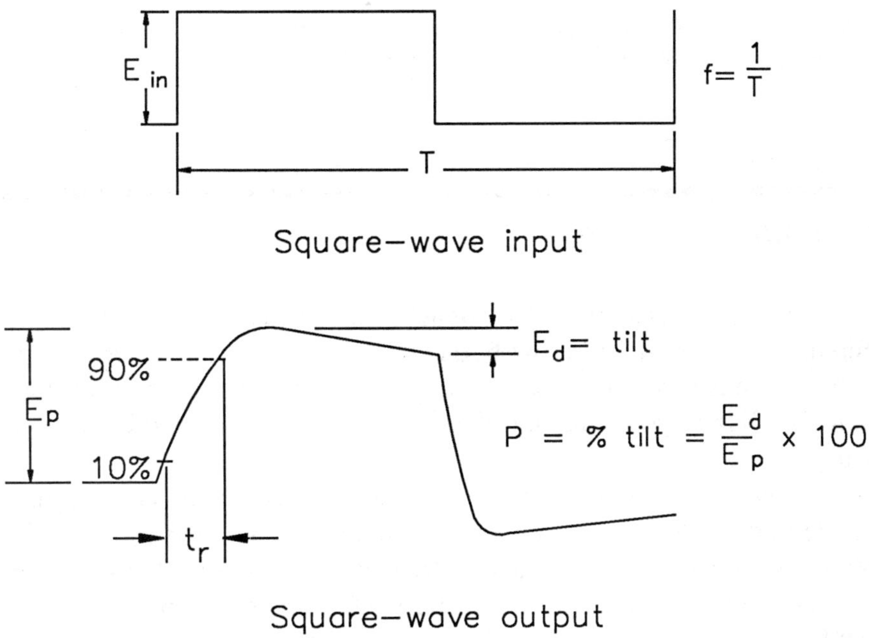

Figure 3-10 Square-wave form analysis

measuring the rise time and tilt of the output square wave. The rise time of the square wave is approximately related to the amplifier's high-frequency response by the formula:

$$f_h = \frac{0.35}{t_r}$$

where f_h is the upper 3-dB frequency roll-off in MHz and t_r is the rise time in microseconds. The tilt of the square wave, the amount of droop across the top of the waveform, determines the low-frequency response and is determined by:

$$f_l = \frac{P \times (\text{pulse repetition rate})}{100\pi}$$

where P equals the droop rate in percent, and P = (Ed/Ep) ×100.

Depending on how exacting the frequency response measurements are, the repetition rate of the square-wave output should contain at least up to the ninth harmonic for accurate results. If the amplifier under test has a specified frequency response of 20 Hz to 20 kHz, for example, the minimum repetition rate of the pulse train for accurate high-frequency response measurements is 2.2 kHz (20 kHz/ninth harmonic). Low frequency response, on the other hand, is most accurately determined with a pulse rate of 20 Hz.

RF Signal Generators

The circuit techniques employed for RF signal generation are substantially different from those used in low-frequency instruments. An RF signal generator has a sinusoidal output with a frequency range somewhere in the 100 kHz to 40 GHz region, although the range cannot be covered by any single instrument.

Basically, the instrument consists of an RF oscillator, an amplifier, a calibrated attenuator, and an output level indicator. The RF oscillator has a continuous frequency control and a frequency range switch, which determines the output frequency of the instrument, while the amplifier contains a variable gain control for setting the output level of the signal. A calibrated attenuator and output level indicator guarantee precision output levels to the device under test at the stated load conditions.

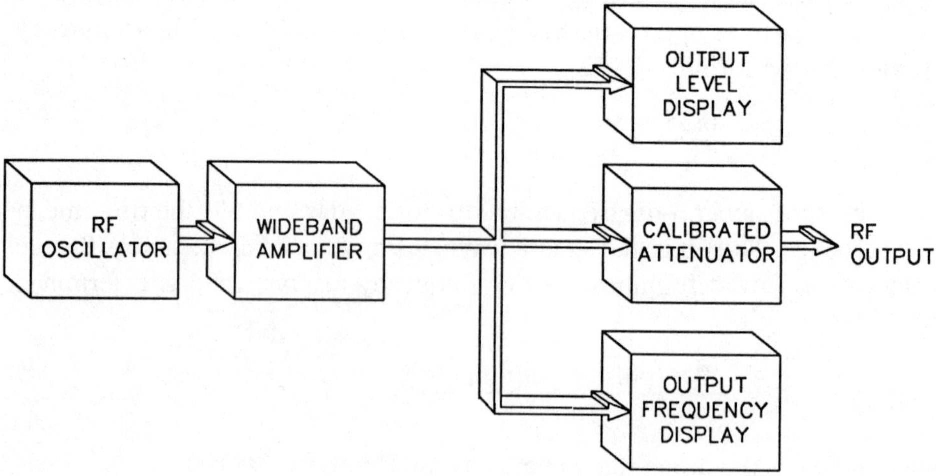

Figure 3-11 RF generator block diagram

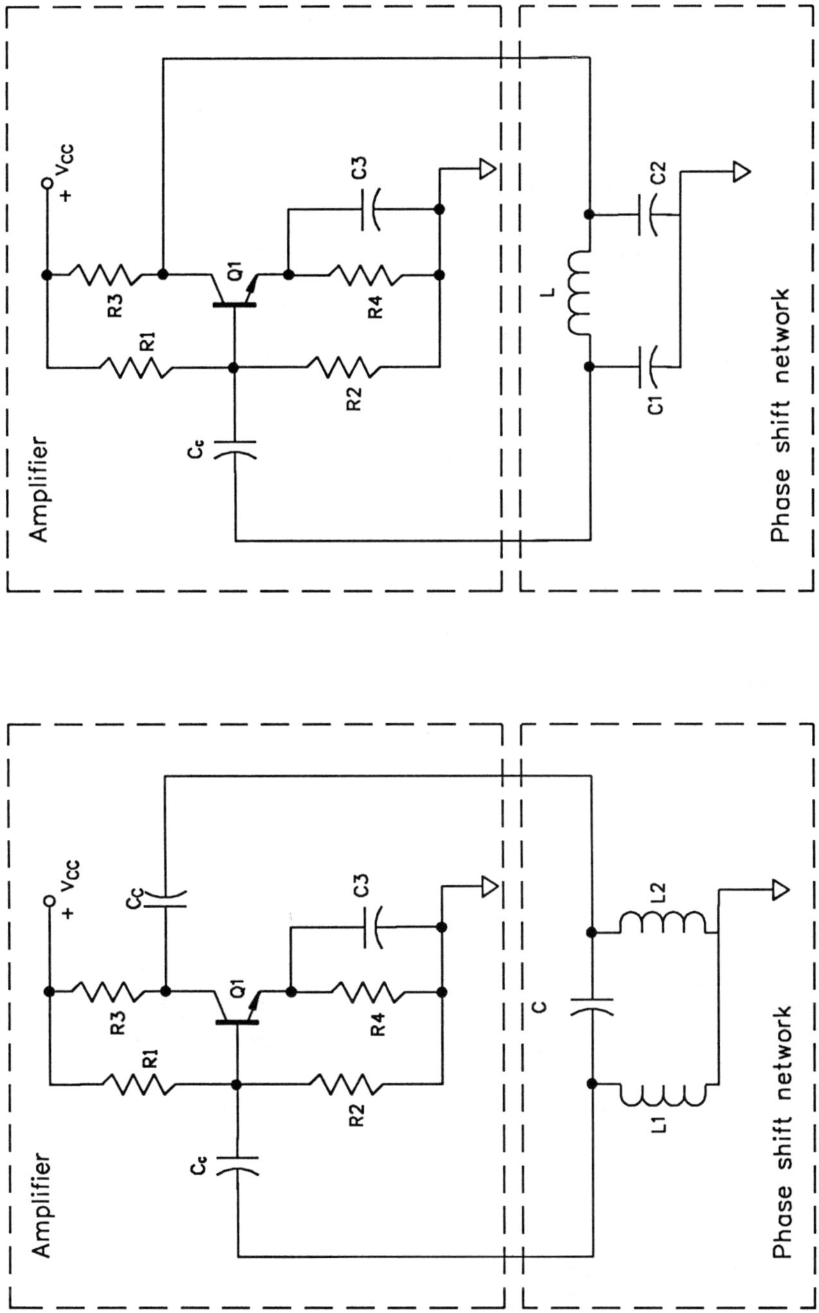

Figure 3-12 RF oscillator circuits

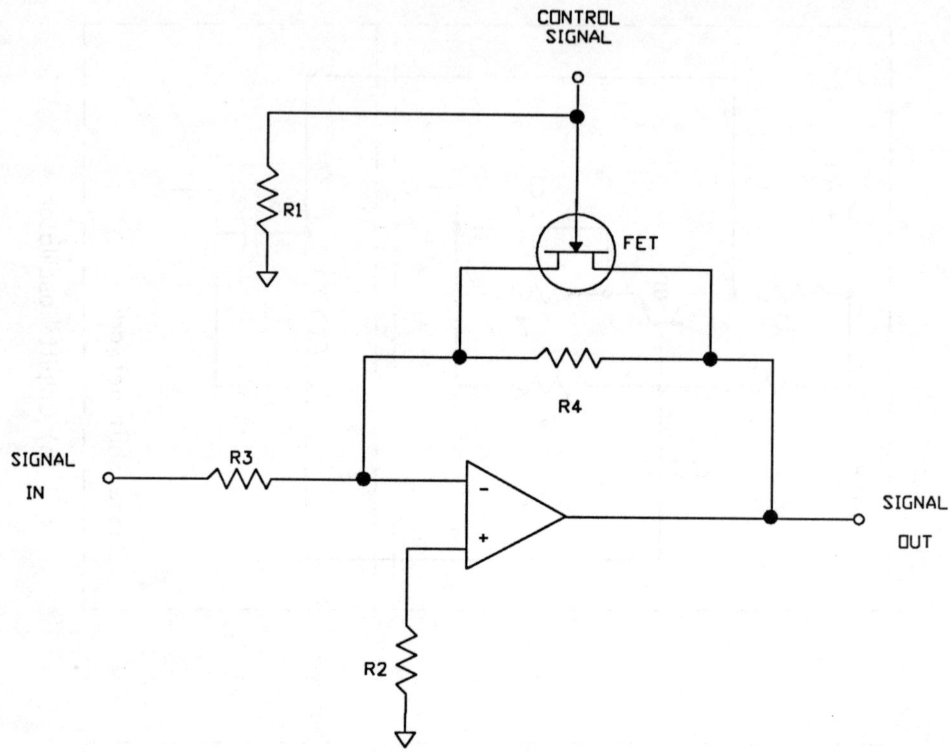

Figure 3-13 Amplitude modulation

The oscillator circuit used in simple, single-function RF signal generators is usually of the *Hartley* or *Colpitts* design. Both designs consist of a single-transistor amplifier and a phase-shifting LC network. The essential differences between the Hartley and Colpitts circuits are in the phase shift networks. The Hartley circuit has two inductors and a capacitor, whereas the Colpitts circuit uses two capacitors and one inductor. The Q1 amplifier transistor is configured to invert the input signal by 180° at the collector output.

The inverted output signal is then fed back to the base input through the phase shift network. The phase shift network causes the signal to shift another 180°, thus placing it back in phase and reinforcing the input. The frequency of oscillation is equal to:

$$f = \frac{1}{2\pi\sqrt{C_T L_T}}$$

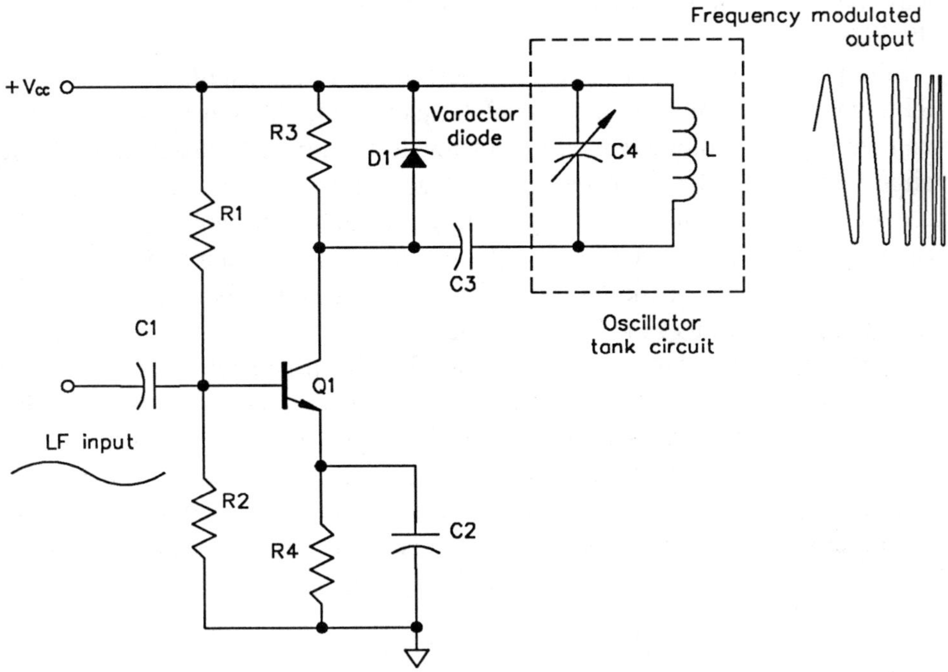

Figure 3-14 Frequency modulation

For the Hartley circuit shown in Figure 3-12, C_T = C, and L_T is the total inductance of L1 and L2, including mutual inductance. For the Colpitts circuit, L_T = L, and C_T is the total capacitance of C1 and C2 in series. The oscillating frequency of the circuit can be altered by changing the component values in the phase shift network. Frequency adjustment is most conveniently done when C_T is made the variable component.

Most RF signal generators include facilities for *amplitude modulation* (AM) and *frequency modulation* (FM) of the output signal. Amplitude modulation causes the output voltage of the oscillator to change in accordance with a control signal. Amplitude modulation is easily accomplished at the amplification stage. One method of amplitude modulation is shown in Figure 3-13. With no signal applied to the FET, the gain of the amplifier stage is:

$$A_V = \frac{R3}{R4}$$

As signal is applied to the gate of the FET, the resistance between its source and drain decreases, thus lowering the gain of the amplifier. In this way the amplitude of the RF output is increased or decreased in phase with the FET input signal.

Frequency modulation is usually performed at the oscillator stage in an RF signal generator. One method of frequency modulation uses a variable-voltage capacitor diode, called a varactor, in the tuned feedback loop of the circuit. Varying the reverse-bias voltage across the varactor alters its capacitance, which in turn changes the C_T value of the tank circuit. As the capacitance of the varactor changes in step with the applied signal, the resonant frequency of the tank circuit varies, and thus modulates the output frequency according to the low-frequency input signal.

The output impedance of a laboratory RF signal generator is typically 50 ohms. When the output of the generator is terminated into a pure resistive 50-ohm load, standing waves are prevented. Proper impedance termination is also required for proper output-level indication. The actual output voltage from the calibrated attenuator is the indicated level only when the instrument is correctly loaded. When load impedance is greater than 50 ohms, the output voltage (e_i) across the load is calculated using the formula:

$$e_i = \frac{e_O R_L}{R_S + R_L}$$

where e_o is the indicated output voltage, and R_S and R_L are the source and load impedances, respectively.

A typical laboratory RF generator has a frequency range of 10 kHz to 1000 MHz, and tuning is accomplished using either manual controls (range selection, course and fine tuning) or by direct entry via a keyboard. Most in-

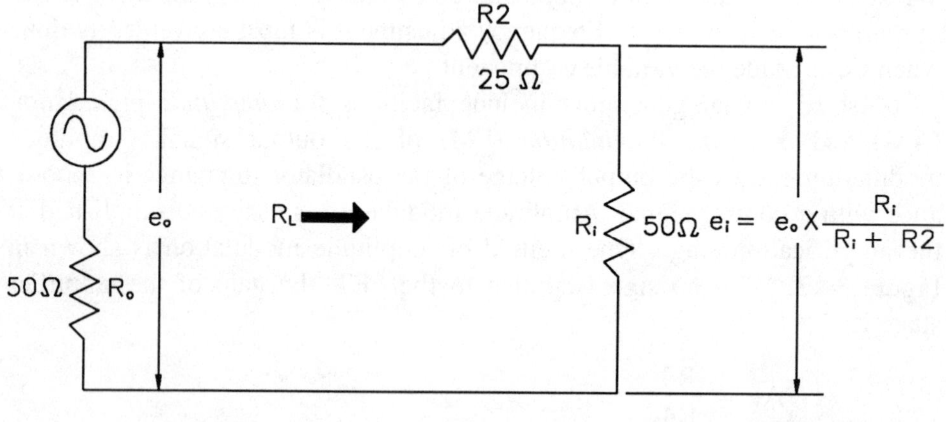

Figure 3-15 Calculating output voltage at an improperly terminated output

struments are also GPIB programmable (see Chapter 7). Output levels up to +19 dBm (4V e.m.f.) into a 50-ohm load are available, and the output is reverse-power-protected to prevent damage in the event of the accidental application of power up to 50 watts. The power-output indicator is generally of the digital type, as is the frequency-output indicator. Comprehensive AM, FM, and pulse modulation facilities are provided for testing all types of receivers and radar modules.

SWEEP GENERATORS. A common application for an RF signal generator is in the testing and alignment of IF and video amplifiers. Such applications require a special type of RF generator called a *sweep generator*. Basically, a sweep generator is an RF signal generator with a frequency-modulated output. The output begins at some preselected frequency and gradually increases in frequency in linear fashion to some cutoff frequency, at which point the increase halts abruptly and the sweep begins with the lowest frequency again.

A block diagram of a sweep frequency generator is illustrated in Figure 3-16. The instrument consists of six major sections: a ramp generator, a voltage-controlled oscillator, a variable-gain amplifier, an automatic-gain control, a calibrated attenuator, and an level output indicator.

The ramp generator applies a linear ramp voltage, called a sawtooth waveform, to the input of a voltage-controlled oscillator (VCO) through a differential amplifier. As the ramp voltage increases, the frequency output of the VCO increases proportionally. The discriminator circuit connected from the VCO output to the differential amplifier input forms a feedback loop that monitors and controls the VCO output frequency. The discriminator generates an output voltage that is proportional to the VCO output frequency. Because the discriminator's input is the swept frequency from the VCO, its output is a ramp voltage identical in shape and speed to the waveform generated by the ramp generator. If the VCO output frequency is lower than it should be at any instant, the instantaneous output voltage from the discriminator drops below the instantaneous level of the ramp voltage from the ramp generator. This results in an increased output voltage from the differential amplifier, which forces the VCO output frequency to rise. Similarly, if the VCO output frequency becomes higher than intended, the discriminator output voltage rises above the level of the sweep generator, causing the differential amplifier output voltage to decrease proportionally, thus lowering the VCO frequency. In this way, the output frequency of the VCO is stabilized.

The VCO output is then fed to a variable gain amplifier for level control. It is very important that the output voltage remains constant at whatever level it is set over the entire range of output frequencies. Because of inherent

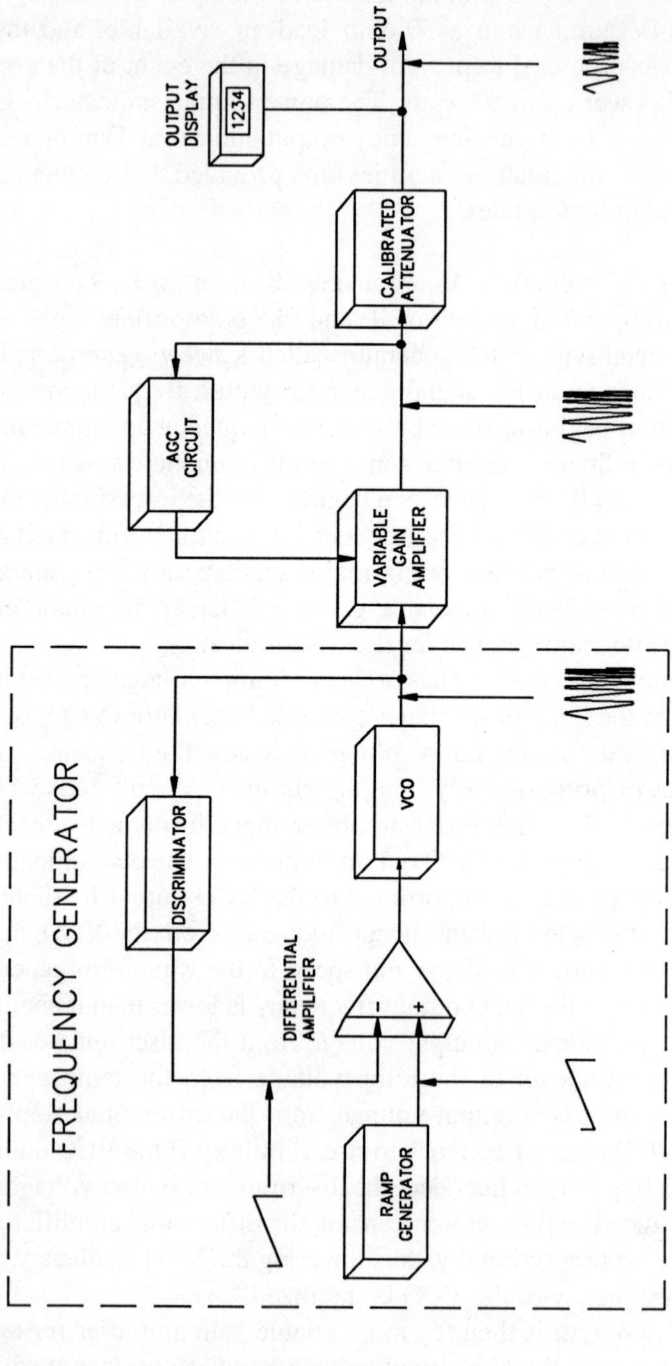

Figure 3-16 Sweep generator block diagram

properties of the VCO, however, the output signal tends to decrease as frequency increases. The output voltage level is stabilized by the action of the automatic gain control circuit (AGC) and the variable gain amplifier. The AGC circuit produces a voltage proportional to the power output of the variable gain amplifier. This voltage is compared to an internal reference voltage in the AGC circuit, and the difference between the two is applied to the variable gain amplifier. Where the power output is lower than required, the amplifier gain is increased. If the power output is too high, the gain is decreased.

A calibrated attenuator adjusts the signal level for output. The output level is monitored and displayed on a level output indicator, which may be an analog meter or digital readout device. In most sweep generators, both the ramp voltage and the frequency modulated output are available to the user.

RECEIVER ALIGNMENT. A typical application of an RF sweep generator is the alignment of radio receivers. With reference to the simplified block diagram of an FM receiver in Figure 3-17, the first stage is the tuned RF section, which in theory lets only one station at a time enter the mixer of the receiver. Actually, the RF section is coarsely tuned, with the fine tuning accomplished by the narrow bandwidth of the IF section. The local oscillator heterodynes the incoming RF to an IF signal into the mixer stage. The IF amplifiers are followed by an AGC detector, which controls the gain of one of the IF amplifiers. The signal then goes through a discriminator, or FM detector, where the audio is demodulated and fed to the audio amplifier and speaker.

An RF signal generator with frequency sweep is normally used to align the stages of the receiver so that only one station is heard at a time. The purpose of the signal generator is to simulate a signal received out of free space by the antenna of the receiver. A measuring instrument, usually an oscilloscope, is connected to the audio output of the receiver to measure the quality of the received signal and align the RF and IF sections.

An FM sweep generator is first connected to the receiver input through an antenna coupler, and its RF output is adjusted to a dead spot on the dial. The receiver is tuned to the same frequency as the generator. The generator is then modulated internally at a very low rate of about 40 to 50 Hz with a variable frequency oscillator. The maximum frequency sweep of the output is set much greater than the IF bandwidth of the receiver.

The ramp voltage of the sweep generator is fed to the external sweep input (Y-axis) of the oscilloscope. As the frequency output of the signal generator increases, a trace is scanned across the face of the oscilloscope. The

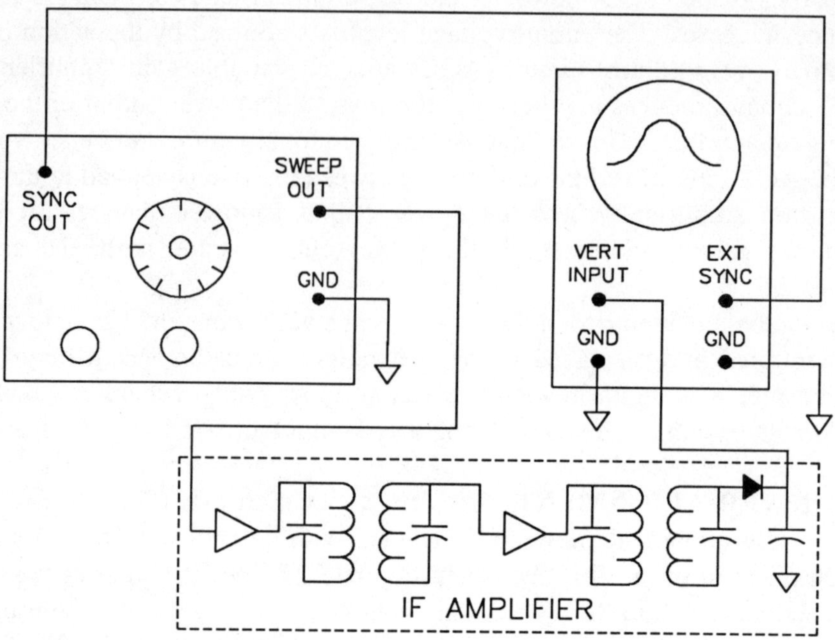

Figure 3-17 Receiver alignment test setup

frequency response of the receiver under test appears as a waveform on the oscilloscope screen. The alignment coils of the IF and discriminator section are then adjusted so that the waveform conforms to the specifications of the receiver.

Frequency Synthesizers

For unwavering frequency stability over a wide range of frequencies, nothing can match the *frequency synthesizer*. Synthesized signal sources derive a large number of accurately known frequencies from one fixed-frequency precision oscillator. Their increased tuning range and frequency stability makes them the preferred choice for most laboratory applications.

Frequency synthesis refers to the generation of a frequency that is related algebraically to one or more reference frequencies. Three basic methods are used to synthesize frequencies: heterodyne, multiplier, and divider.

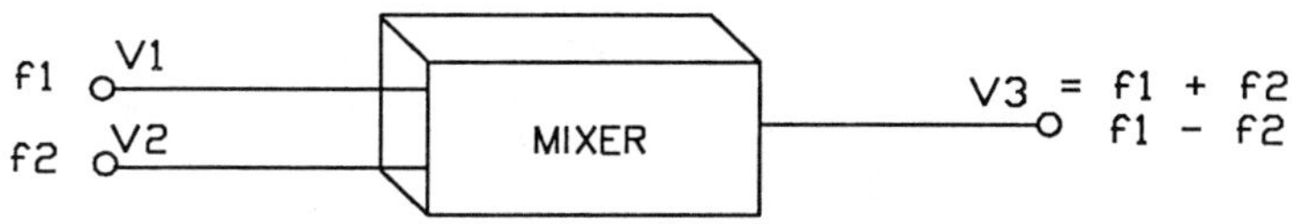

Figure 3-18 Product mixer

The two basic approaches to frequency synthesis are known as *direct* (or true) and *indirect*. Direct synthesis simply performs a series of arithmetic operations on the signal from the frequency standard to achieve the desired output frequency.

INDIRECT SYNTHESIS. The indirect synthesizer method is usually taken as one that has at least two oscillators and often employs a phase-locked oscillator. The indirect method uses tunable oscillators that are phase-locked to harmonics of signals derived from the standard.

The *heterodyne* method is an indirect synthesizer that uses two reference frequencies and an electronic mixer. The output of an ideal mixer is the product of the two inputs. If the two inputs are defined as:

$$V_1 = f_1 \sin \omega_1 t$$

$$V_2 = f_2 \sin \omega_2 t$$

then the output is:

$$V_3 = k f_1 f_2 \sin \omega_1 t \sin \omega_2 t$$

where k is the proportionality constant. The Fourier components of V_3 are obtained by:

$$\sin A \sin B = \frac{1}{2}[\cos(A - B) - \cos(A + B)]$$

The frequencies present at the output are the sum, $\omega_1 + \omega_2$, and difference, $\omega_1 - \omega_2$, of the two input frequencies. Their amplitudes are kf_1f_2. An appropriate band-pass or low-pass filter can then be used to select either the sum or difference frequency. The process is called *heterodyne*. By having one frequency fixed and the other variable, a wide range of frequency out-

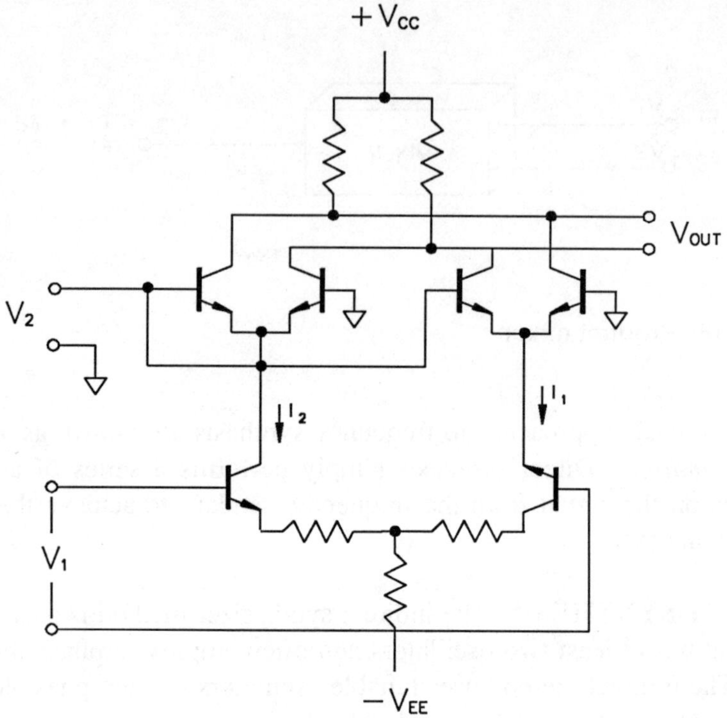

Figure 3-19 Double-balanced mixer

puts is available over a broader range than can be accomplished using a single oscillator.

One way to implement the mixer function is with the differential amplifier circuit shown in Figure 3-19. The two emitter-coupled pairs on top are variable-gain differential amplifiers with their outputs subtracting, so that when the two gains are equal, the net gain from V_2 to the output is zero. The gains of the two amplifiers are varied by changing their emitter currents:

$$V_{out} = \frac{qRc}{kT}(i_2 - i_1)V_2$$

The emitter currents i_1 and i_2 are varied differentially by V_1, the signal applied to the bottom pair of transistors. The net result is identical with the requirements for an ideal mixer.

One of the many methods of frequency multiplication involves passing the input frequency through a nonlinear device such as a transistor, rectifier,

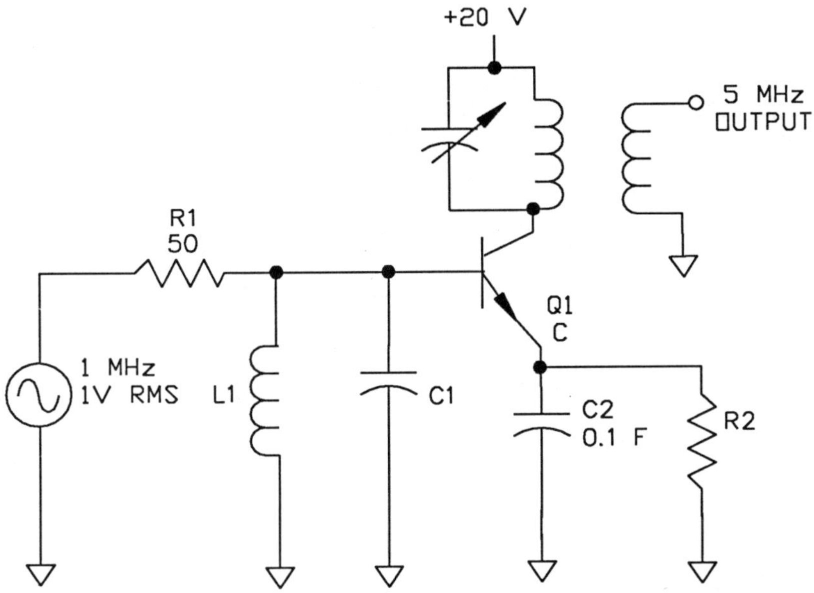

Figure 3-20 5-MHz class C frequency multiplier

or voltage-variable capacitor (varactor), and selecting the desired harmonic with a band-pass filter. A simple X5 multiplier is shown in Figure 3-20.

The input circuit is tuned to the fundamental reference frequency, and the collector circuit is tuned to the desired harmonic. The input voltage is sufficiently high that the transistor is cut off most of the time and only conducts on the peaks of the input waveform. This current, which contains many higher harmonics of the input frequency, is then filtered by the tuned collector circuit to give the fifth harmonic, resulting in a X5 multiplier. The frequency multiplier of Figure 3-20 is adequate for fixed-frequency operation when the harmonic number is not very great. When greater flexibility and higher harmonics are desired, however, the phase-locked oscillator is preferred.

PHASE-LOCKED LOOP. The *phase-locked loop*, or PLL, consists of three sections: a phase comparator, a low-pass filter, and a voltage-controlled oscillator (VCO). The arrangement is shown in Figure 3-21. When used in the capacity of a frequency multiplier, a *frequency divider*, indicated by the dashed lines, is included in the feedback loop coming from the VCO output.

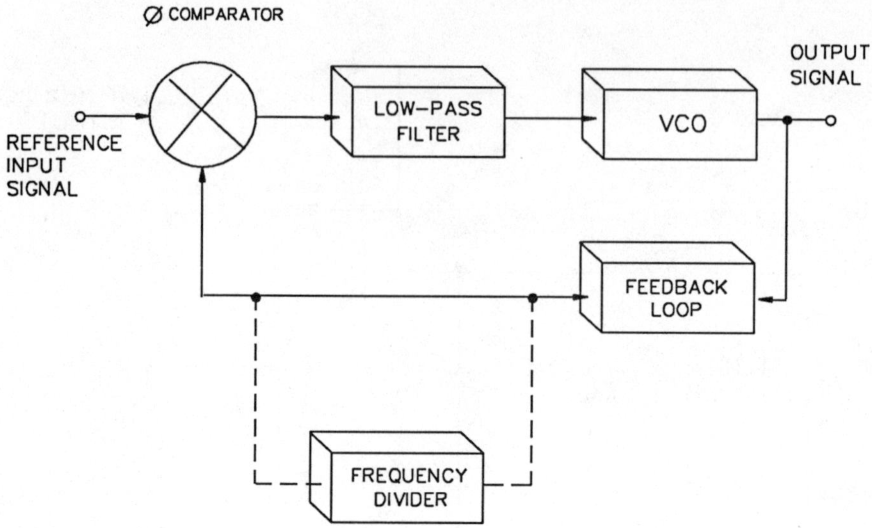

Figure 3-21 Phase-locked loop block diagram

In normal operation, the VCO oscillates at a center frequency established by an external RC network. The center frequency of the VCO is normally selected to be within the range of the desired output frequency. This frequency is referenced to the phase comparator.

The phase comparator accepts two inputs, one from the input reference frequency and a frequency sample from the voltage-controlled oscillator. The phase comparator detects the phase difference caused by a difference in frequency between the input signal and the VCO output. The difference frequency is translated into an output signal that is input to the low-pass filter.

The low-pass filter, sometimes called an integrator, may be a simple RC network or an elaborate multi-pole active filter. Either way, the low-pass filter converts the input signal into a dc voltage by averaging the output of the phase comparator to generate a dc voltage that is fed to the VCO.

Suppose, for now, that the loop feeding the VCO is closed with no input signal to the phase comparator. In this mode, the VCO will oscillate at its center frequency, drifting if it will at every opportunity. Now consider the application of a stable input frequency to the phase comparator. If the input frequency is within the PLL capture range or a harmonic thereof, a voltage is developed at the output of the low-pass filter, which will bias the VCO to lock onto the incoming frequency.

Next, assume that the reference input signal frequency increases. As shown in Figure 3-22, there will be an instantaneous phase change between

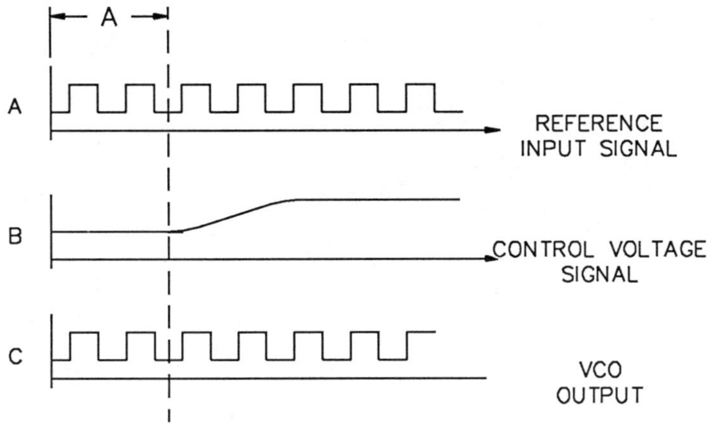

Figure 3-22 VCO control waveforms

the reference input signal and the VCO's output frequency, causing an increase in the output voltage of the low-pass filter. The filtered voltage is fed to the VCO's control input, which causes the VCO output frequency to increase until it catches up with the reference frequency. A decrease in input reference frequency causes the process to reverse itself.

An example of a phase-locked oscillator employed in frequency synthesis is shown in Figure 3-23. The VCO frequency varies between 19.80 MHz and 29.79 MHz in 10-kHz increments, and is controlled by BCD digital input to the divide-by-N frequency divider, and also to the digital-to-analog converter for pretuning the oscillator close the the frequency called for by the digital data. The VCO's frequency is divided by a number between 1980 and 2979, resulting in a 10-kHz signal at the divider output.

This signal is compared with the 10-kHz stable reference in the mixer, which generates a voltage proportional to the phase difference between these two inputs. The error voltage is fed back to the VCO to close the loop. The shaping amplifier is inserted to remove the nonlinearity caused by the variable-voltage capacitor (varactor) that tunes the oscillator.

DIRECT FREQUENCY SYNTHESIS. *Direct synthesis* is most readily accomplished by frequency division using digital techniques. Binary divider circuits and logic gates can be combined to make a frequency divider of great flexibility.

The objective of the direct synthesis generator is to obtain a very large number of switchable frequencies from a single stable source. Frequencies are

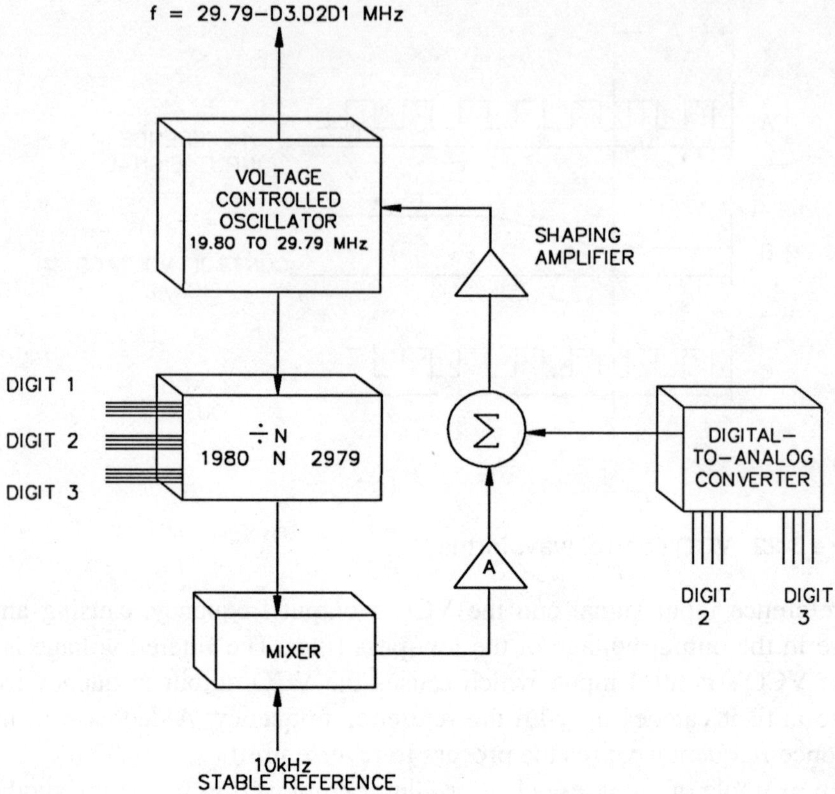

Figure 3-23 PLL frequency synthesizer

selected from a keyboard in decimal fashion, as on the keyboard of an adding machine. One synthesizer, for example, provides frequencies from 0.01 Hz to 50 MHz in digital increments as fine as 0.01 Hz, for a total of 5 billion discrete frequencies. A simplified block diagram of the overall instrument is shown in Figure 3-24.

The instrument is divided into two major sections, the driver block and the synthesizer unit. The driver block contains a frequency standard, a spectrum generator, and appropriate selection networks to provide a series of fixed frequencies between 3 and 39 MHz to the synthesizer unit. The synthesizer unit contains harmonic generators and suitable mixers, dividers, filters, and amplifiers to derive the desired output frequency as a function of the fixed frequencies.

The fine-resolution portion of the instrument is particularly interesting and serves to illustrate the method of synthesis used. As shown in the right-hand portion of the block diagram, there are seven identical mixer-divider

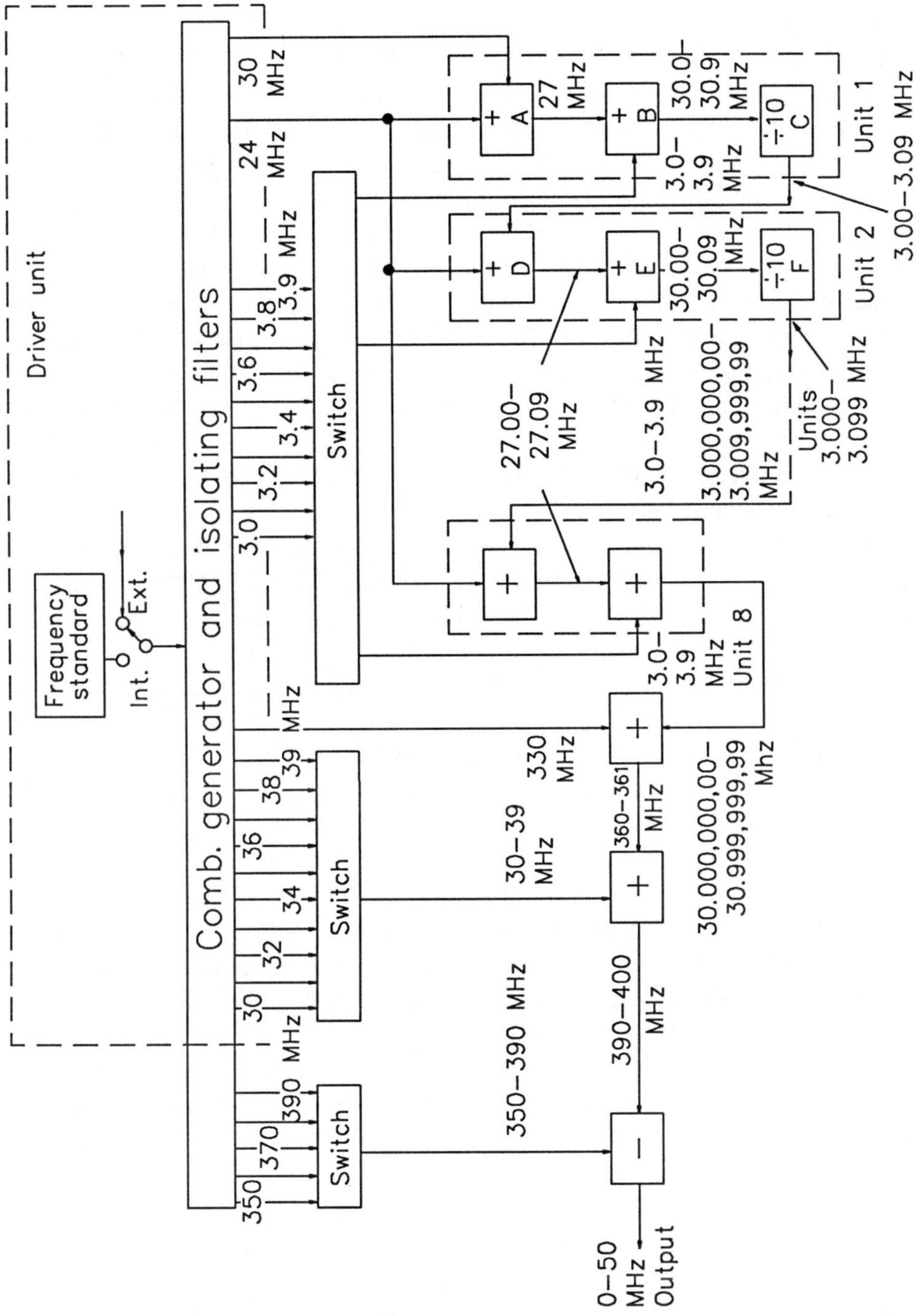

Figure 3-24 Direct synthesizer block diagram

units, each of which corresponds to a place, or decimal position, in the final output frequency number. In each of these units, and in the eighth unit as well, a frequency of 24 MHz is used as a carrier input.

In the right-hand unit, which produces what ultimately becomes the highest resolution digit, the 24-MHz carrier is added to a 3.0-MHz frequency in frequency mixer A to produce 27.0 MHz. In B, the 27.0-MHz signal is added to a frequency of 3.0 to 3.9 MHz, depending on the setting of the front panel controls. This produces an output at B that is a frequency of 30.0 to 30.9 MHz, which is divided in C to produce 3.00 to 3.09 MHz. This frequency is applied to the second unit, where it adds with the 24-MHz carrier as before, and the process repeats. If the process is followed through, it will be seen that the frequencies noted in the block diagram are obtained at the outputs of the various adders and dividers. In essence, each mixer-divider unit, through a frequency-division process, moves a given digit one place to the right for the final frequency of between 30.00000000 MHz and 30.99999999 MHz, depending on the output frequency selected.

Next, the signal is added to a frequency of 330 MHz, and the resultant is again added to an appropriate frequency between 30 and 39 MHz, to yield a frequency of between 390 and 400 MHz. One of the five frequencies from 350 to 390 MHz is then subtracted from this to yield the desired 0.01-Hz to 50-MHz output frequency range of the instrument. A frequency change of 100 kHz at the input to unit 1 produces a change of 0.01 Hz at the output of the whole system, because of the repeated division by 10.

One of the distinctions between direct and indirect synthesis is the use of large multiplying numbers in indirect synthesizers. When the frequency is multiplied, so is the phase noise. To reduce the multiplied noise, a filter is needed either between the multiplier and the newly formed output or at the output alone. In direct synthesizers, multiplication factors as large as 3000 cannot be filtered properly. But when a phase-locked loop is used as the multiplier, as in the indirect method, the filtering is accomplished automatically by the inherently narrow-loop bandwidth.

An example of the indirect synthesis technique is illustrated in the block diagram of Figure 3-25. In this example, the synthesizer instrument covers an output range from 20.000001 MHz to 30.000000 MHz in 1-Hz increments. This synthesizer contains five phase-locked loops — three of which are divide-by-N types and two of which are summing loops (mixers) — with an output frequency equal to the sum of the two input frequencies.

A typical commercial RF synthesizer generator has an output frequency extending from 0.15 MHz to 50 MHz in eight ranges. The frequency error is less than 1% and the output voltage is typically 50 mV into 75Ω.

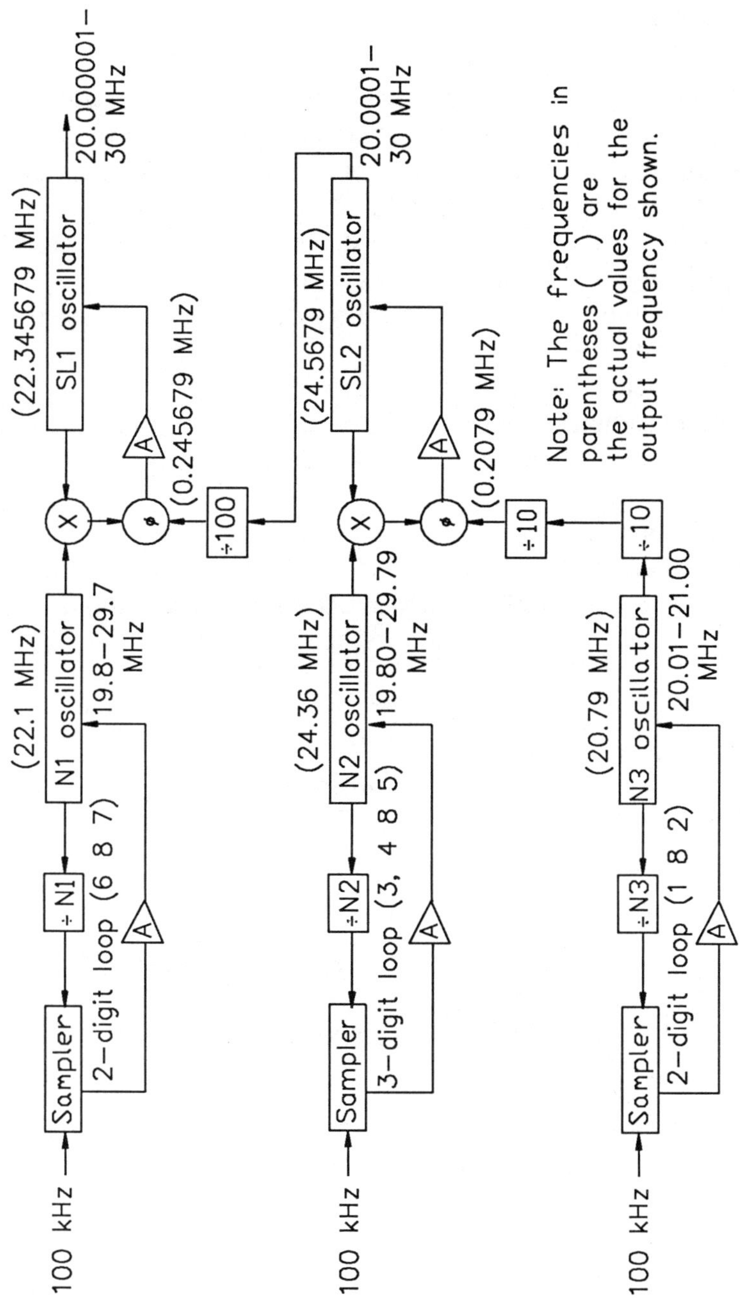

Figure 3-25 Indirect synthesizer block diagram

Pulse Generators

Since the advent of digital electronics in the 1940's, the *pulse generator* has become an increasingly important laboratory instrument. Originally designed for radar applications, the pulse generator finds many applications because of the great flexibility of its output. Some of these applications are:

- Transient analysis of linear systems, such as overload recover of amplifiers and damping of servo systems.
- Radar system testing, in which the pulse generator is used to simulate the video of a radar return.
- Semiconductor component testing.
- Communications system testing, including worst-case analysis of communications receivers with bandwidths from below 1 MHz to radar systems.
- Digital computer testing, design, and analysis.

In the selection of a pulse generator for laboratory or field service, the quality of the output is of primary importance. High-quality test pulses are

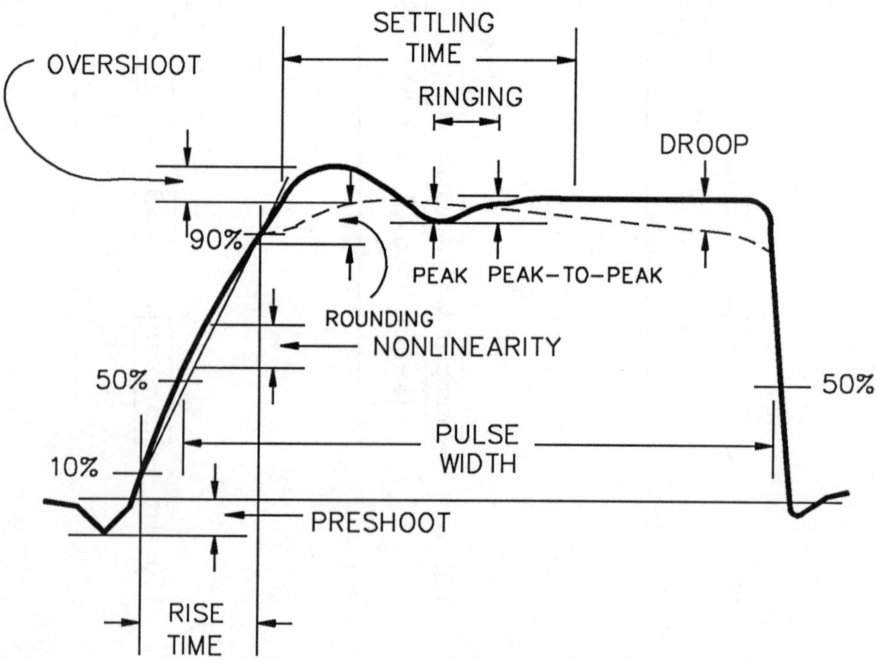

Figure 3-26 Pulse characteristics

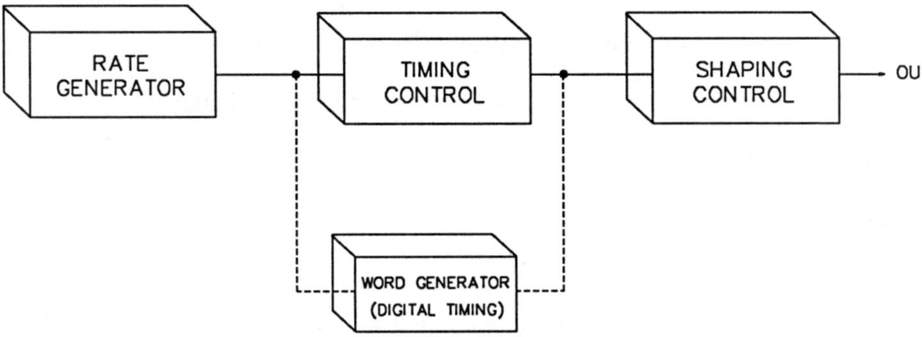

Figure 3-27 Pulse generator block diagram

necessary to ensure that any degradation in the quality of the signal may be attributed to the circuit under test alone and not the instrument itself.

The pertinent characteristics of a pulse generator output signal involve six parameters: base line, amplitude, transition time, width, repetition rate, and duty cycle. Other factors affecting pulse generator performance are overshoot, ringing, and sag.

Of the listed parameters, the transition times are most important. The leading edge transition time is generally referred to as the *rise time*. Rise time is the time it takes for the pulse signal, either voltage or current (according to the test being performed), to increase from the zero reference voltage level to its required amplitude. Rise time is measured from 10% to 90% of pulse amplitude level. The trailing edge of the transition time, called the *fall time*, is the time it takes for the signal level to decrease from 90% to 10% of pulse amplitude. Rise and fall times of the pulse generator should be significantly faster than the circuits or systems to be tested, by a factor of at least three.

Pulse width, repetition rate, and duty cycle are generally interrelated, and refer to the pulse train, in most cases. The pulse width is the time in seconds or fractions of a second between the transition times. The duty cycle is the width of this time compared to the time the waveform is at the base-line reference value. It is the ratio of pulse width to pulse period expressed as a percentage, where the pulse period is determined by the repetition rate of the waveform. If the period is 1 second and the pulse width is 0.1 second, then the duty cycle is

$$\text{Duty cycle} = \frac{0.1}{1} = 10\%$$

The range of the pulse repetition rate is limited by the sum of the duty cycle and pulse width.

Glossary of Pulse Characteristics

Amplitude: Voltage level from base line to which pulse switches.

Base line: Reference dc level at which pulse starts.

Droop: Decrease in pulse amplitude with time.

Duty cycle: Ratio of pulse width to pulse period, expressed as a percent. If the period is 1 second and the pulse width is 0.1 second, then the duty cycle is (0.1/1.0) x 100% = 10%.

Jitter: Uncertainty in turning off pulse.

Rise time: Measured from 10% to 90% of pulse amplitude.

Nonlinearity: Vertical variation in transition from straight line through 10-90% points, expressed as a percentage of pulse amplitude.

Offset: Amount of dc that reference level is shifted from ground.

Overshoot: Amplitude perturbation immediately following transition time, expressed as a percentage of pulse amplitude.

Polarity: Direction of pulse amplitude with respect to the base line.

Preshoot: Amplitude perturbation immediately preceding transition, expressed as a percentage of pulse amplitude.

Repetition rate: How often the pulses occur. Measured in hertz or seconds.

Ringing: Amplitude perturbation following overshoot, usually a damped sinusoid, expressed as a percentage of pulse amplitude.

Settling time: Time for overshoot to be within a specified percent of pulse amplitude.

Fall time: Measured from 90% to 10% of pulse amplitude.

Transition time: Time for pulse to switch from one level to another.

Width: The amount of time between the 50% point on the leading edge to the 50% point on the trailing edge.

Table 3-1

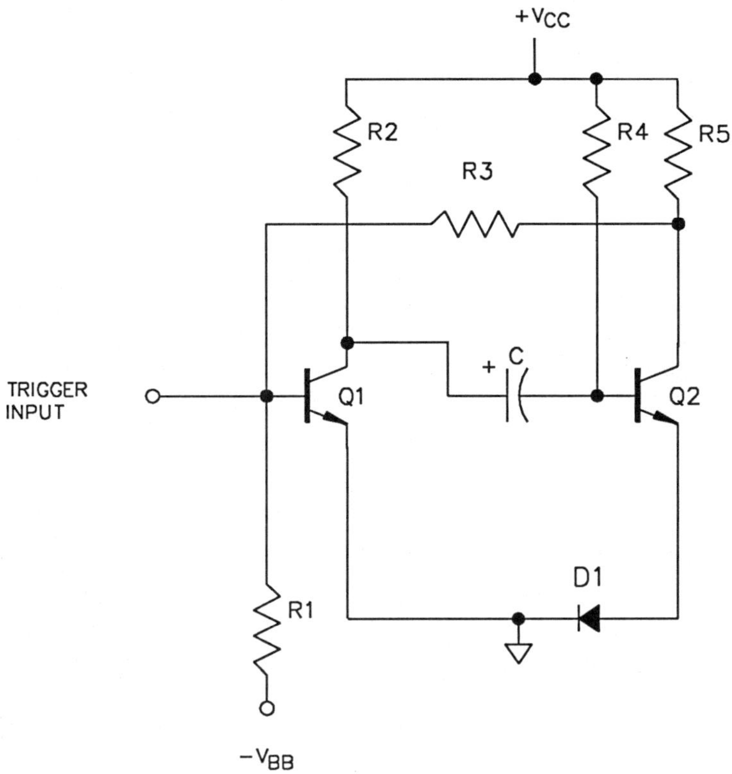

Figure 3-28 Monostable multivibrator

An instrument that is capable of generating a pulse as described above is considerably more complex than the sine-wave generators. The pulse generator is made up of three basic blocks: rate, timing, and output shaping. The block diagram in Figure 3-27 gives the relative position of each major function and control of a pulse generator.

The heart of the pulse generator is the timing control circuit. The timing control circuit actually generates the pulse that will ultimately be shaped and output by the instrument. The timing control circuit is generally a monostable multivibrator circuit with external controls. The monostable, or one-shot, multivibrator has but one stable state. It is a hybrid between the bistable flip-flop and the astable square-wave multivibrators. Consider the circuit in Figure 3-28 for the following discussion.

At rest, transistor Q2 is held on by the base current through R4, while Q1 is held off by R3, R1, and $-V_{BB}$. If either transistor is triggered out of its stable state through application of a signal to the base of Q1, regeneration

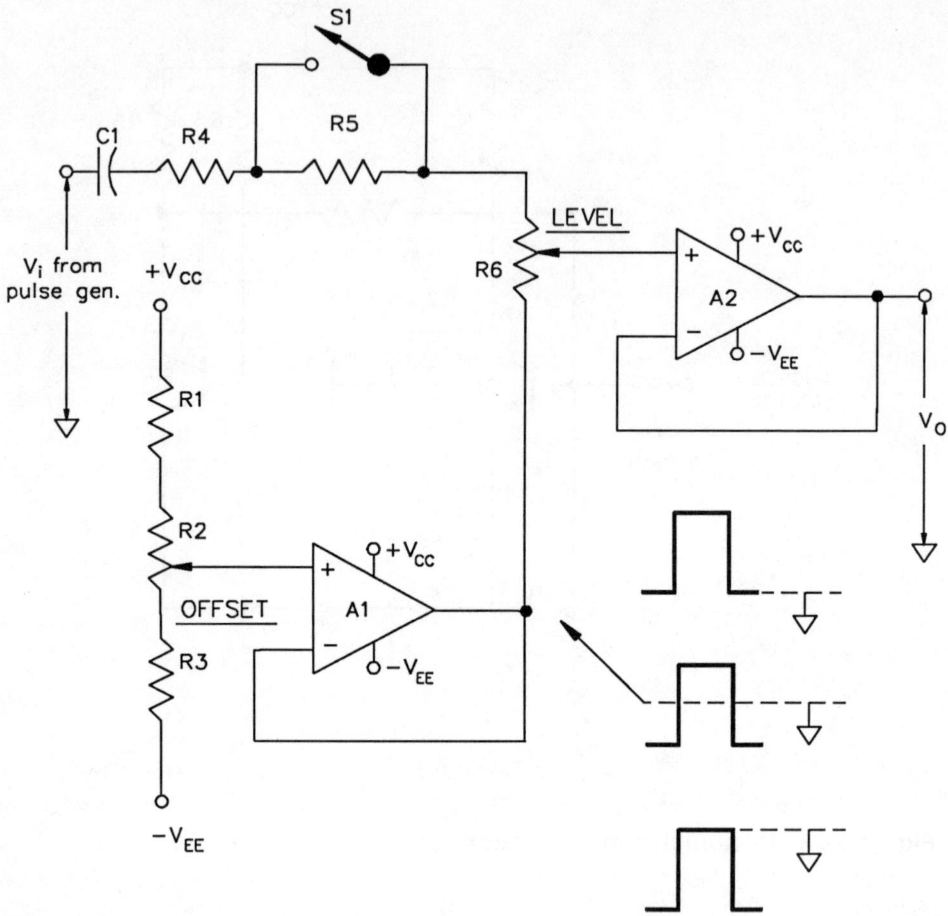

Figure 3-29 Shaping circuit

can occur, and the astable state (Q1 on, Q2 off) will exist for a time determined by R4C. In this mode, Q2 is held off and Q1 is held on. After C has charged to $V_{BE(ON)}$, Q2 begins to conduct. The resulting negative excursion at the collector of Q2 is coupled to the base of Q1 through R3, thus forcing transistor Q1 off and completing the cycle until another trigger signal comes along. Diode D1 prevents negative pulses from triggering the circuit.

Each time the monostable multivibrator is triggered, it generates a single pulse whose width is determined by:

$$\text{Pulse width} \approx 0.7 \times R4 \times C$$

The trigger pulse is derived from the rate generator, which is generally a modified astable generator. In the normal pulse mode, the rate generator is free-running at a rate specified by the user for the test conditions at hand. Each time the multivibrator outputs the leading edge of a square wave, the timing control circuit generates an output pulse.

The shaping control circuit inputs this raw pulse and transforms it into an instrument output pulse by giving it its characteristic pulse parameters of polarity, amplitude, and transition times. In some instruments, the shaping circuit is composed of operational amplifiers, as illustrated in Figure 3-29. In Figure 3-29, resistors R4, R5, and R6, together with operational amplifier A2, represent an output attenuator. The output level may be adjusted through the setting of S1 or R6, or any combination thereof. DC offset of the pulse is accomplished by an input voltage from A1 to A2. When the moving wiper of R2 is adjusted, the output pulse level moves from above ground to below ground.

For critical pulse applications, however, the operational amplifier does not provide the fast transition times required. In these instruments, the shaping circuit may be another monostable multivibrator, with the transition times determined by RC networks attached to transistors Q1 and Q2, or a TTL-, ECL-, or HCMOS-type gate driver with guaranteed parameters.

TESTING WITH PULSE GENERATORS. Pulse and square-wave generators most often are used with an oscilloscope as the readout device, which provides both qualitative and quantitative evaluation of the performance of the system or device under test. Pulse generators with fast rise times are widely used in the development of digital circuitry. Teamed with a suitably fast oscilloscope, these generators allow evaluation of gate switching times.

With many circuits under test having a rise time equal to that specified by the manufacturers for the test equipment involved, the rise time displayed by the oscilloscope may be different than the actual rise time. The error is a function of oscilloscope bandwidth, which is determined by the formula:

$$t_{r\ scope} = \frac{0.35}{BW}$$

If the bandwidth of the oscilloscope under question is 100 MHz, for example, then the rise time is

$$t_{r\ scope} = \frac{0.35}{1 \times 10^8} = 3.5\ \text{ns}$$

The pulse generator rise time is measured by the oscilloscope, and the error is calculated using the following equation:

$$t_{r\,gen} = \sqrt{t_{r\,measured}^2 - t_{r\,scope}^2}$$

Knowing these two factors, $t_{r\,gen}$ and $t_{r\,scope}$, the true rise time of the circuit under test can be derived from:

$$t_r = \sqrt{t_{r\,measured}^2 - t_{r\,scope}^2 - t_{r\,gen}^2}$$

To illustrate, consider a test setup using a pulse generator with 2-ns rise time and an oscilloscope with a bandwidth of 100 MHz. If the measured rise time of the device under test is 10 ns, then the actual rise time is:

$$t_r = \sqrt{t_{r\,measured}^2 - t_{r\,scope}^2 - t_{r\,gen}^2}$$

$$= \sqrt{10^2 - 3.5^2 - 2^2}$$

$$= \sqrt{100 - 12.25 - 4}$$

$$= 9.15 \times 10^{-9}$$

or 9.15 nanoseconds.

4. OSCILLOSCOPES

The most versatile of all laboratory instruments is the cathode ray oscilloscope. The oscilloscope provides a two-dimensional visual display of the electrical parameters under measurement. This general-purpose display presents far more information than is available from other test and measure instruments.

With an oscilloscope, one can visually inspect a signal to see how much of the signal is direct current and how much is alternating current, while simultaneously analyzing its noise content. Many other quantities — such as frequency, phase angle, pulse width, rise time, and delay time — can also be displayed on an oscilloscope screen.

The oscilloscope is successful at so many tasks primarily because it is able to display a plot of changing voltage in a vertical axis against a time standard in the horizontal axis. The result is a graphical representation of the signal in standard Cartesian-coordinate format, from which time and quantity measurements may be derived.

Cathode Ray Tube

The device that makes all this possible is the *cathode ray tube,* or CRT. Developed in the early 1930's as a means for displaying radar signals, the CRT is an imaging device that converts electrical impulses into visible light.

The CRT is basically a large evacuated glass bottle with an electron source at one end and a fluorescent viewing screen at the other. An *electron gun* sup-

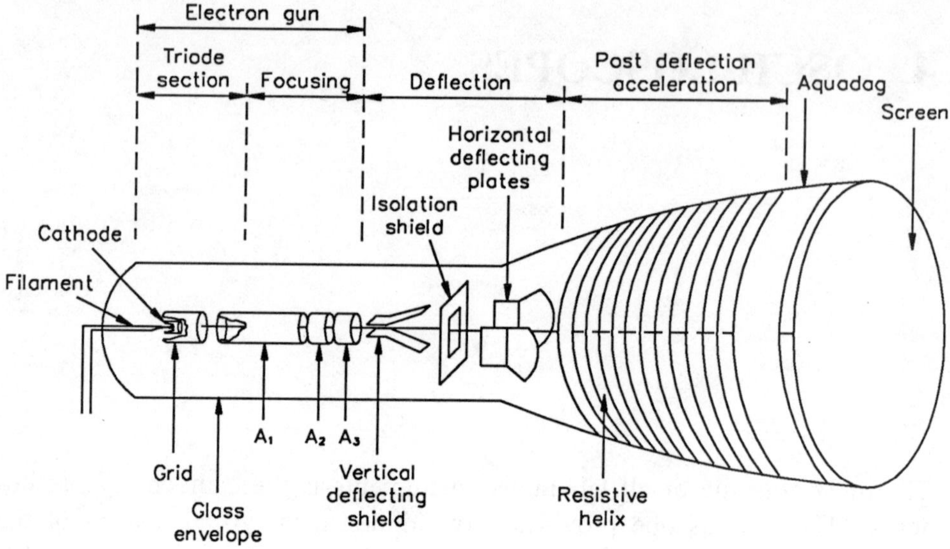

Figure 4-1 Cathode ray tube

plies the electron source, focuses the electrons into a sharp beam, and aims the beam at the *fluorescent screen*, where the electrons impinge on the fluorescent material at a high speed, causing it to glow with a dot of light that is visible thorough the clear glass from the outside of the tube. Waveform patterns are created on the fluorescent screen by deflecting the electron beam horizontally and vertically, using externally applied voltages and a *deflection system.*

ELECTRON GUN. The electron gun assembly consists of a *heater* (or filament), a *cathode*, a *control grid*, and an *anode*. The purpose of the cathode is to generate the electrons required for CRT operation. The cathode is a small, pencil-like nickel cylinder that surrounds the heater element. The cathode is aligned in the neck of the CRT with the small, closed, circular end pointing toward the screen. Over the closed end of the cathode is deposited a layer of electron-rich cesium oxide. When the cesium oxide is heated to a high temperature by the heater coil, it liberates electrons. These electrons literally boil from the surface of the oxide and swarm like a hive of bees around the end of the cathode.

Positioned between the cathode and screen within the neck of the CRT are the grid and anode, respectively. The anode is biased with a positively charged accelerating voltage that attracts and accelerates the electrons toward the fluorescent screen.

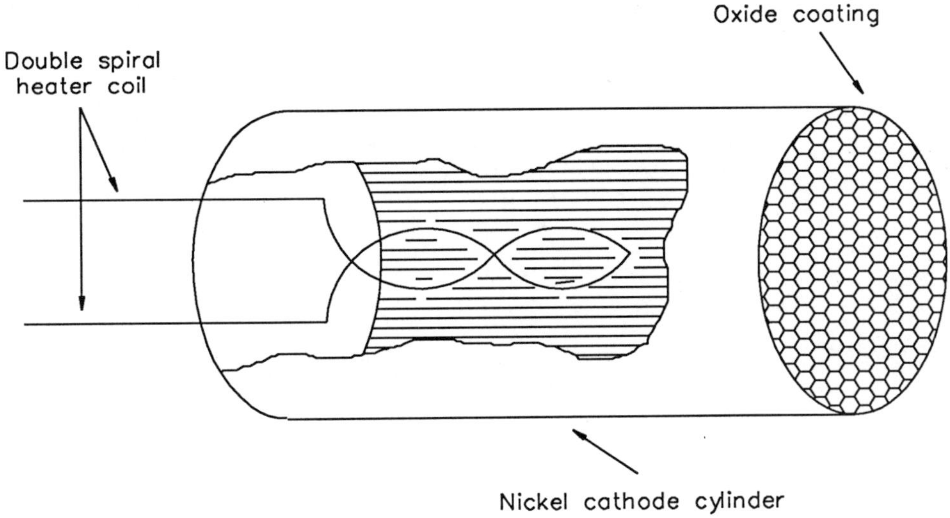

Figure 4-2 Heater/cathode assembly

The grid, which is actually a nickel cup with a small hole machined in its top, fits over and around the cathode assembly, almost completely enclosing it. The electrons must pass through this hole on their way to the accelerating anode and onto the screen.

A small negative-differential voltage, adjustable between 0 and 50 volts, is applied across the grid and cathode. The negative grid potential controls the electron flow from the cathode through mutual repulsion, in which like charges repel, and thus controls the number of electrons directed toward the screen. A zero-differential voltage allows a large number of electrons to accelerate toward the screen area, causing the screen to glow brightly; a larger-differential voltage allows fewer electrons to pass, producing a dim glow. Consequently, the bias voltage is used as a screen brightness control.

ELECTRON LENS. The anode is actually composed of three elements commonly referred to as the *electron lens*. Their function is to focus the electrons to a fine point on the screen of the tube. The first anode (A_1) is cylindrically shaped, open at one end and closed at the other, with a hole in the center of the closed end facing the cathode. A_1 provides the accelerating field to draw the electrons from the cathode.

A_1 is also the *focusing* anode. At first, it might seem that the focusing anode, because it is positive with respect to the cathode, would merely attract electrons in a random manner. Because of the construction and alignment of

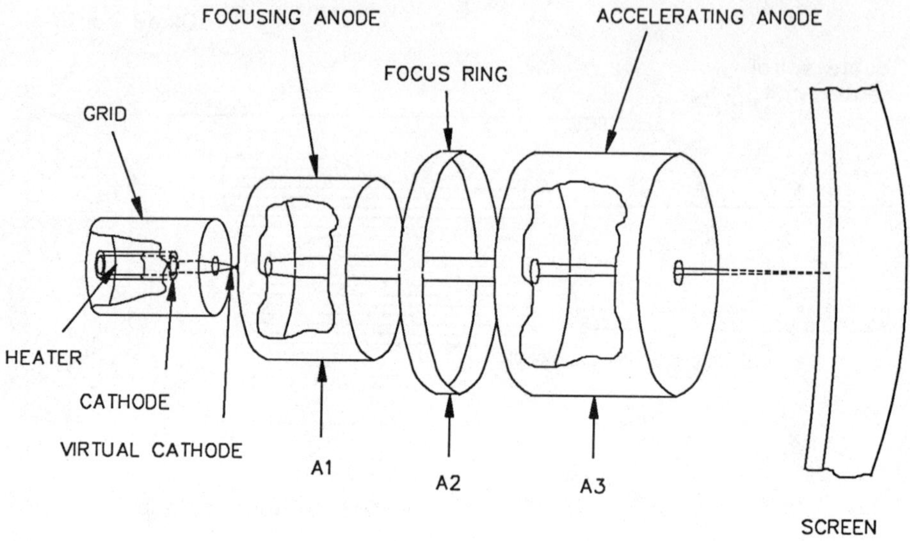

Figure 4-3 Anode assembly

the cathode and focusing-anode elements, however, this is not the case. Since the anode is cylindrical, it exerts an equal pull all around. The net effect is to force the beam as close to the center axis as possible, creating a crossover point midway between the anode and cathode called the *virtual cathode* (as distinguished from the physical cathode). The virtual cathode represents a point source of electron emission. The hole in A_1 further serves to limit the dispersion angle of the emerging electron beam.

In practice, the cathode is biased with a negative potential of approximately –2kV. The focusing anode is then returned to ground, thus establishing a strong positive potential on the anode element. Also referenced to ground is the third anode element, A_3.

Sandwiched between A_1 and A_3 is the *focus ring*, A_2. A negative voltage approximately equal to that of the cathode is applied to the focus ring. The result of the potential difference between the three anodes is that equipotential lines are set up, as shown in Figure 4-3. These lines represent force fields along which the voltage potential is constant. Line 1, for example, may have a potential of –700 volts along its entire length, while the potential along Line 2 is –500 volts.

Because of the nature of the virtual cathode, the electrons enter A_1 as a divergent beam. On crossing the equipotential lines, the electrons experience a force that changes their direction of travel toward right angles with respect to the equipotential lines. The shape of the lines within A_1 produces a convergent force on the divergent electron beam, and that within A_3 produces a divergent force on the beam. The net effect is identical to that exerted on a beam of light passing through a refracting glass lens: Those electrons passing through the center of the focusing system are affected less than those nearer the outer edges. Adjusting the bias voltage of $A2$ changes the contour of the convergent and divergent forces, which, in turn, determines the shape of the lens and the focal point of the electron beam.

The negative potential on $A2$ tends to slow down the electrons as they pass through the focus ring, but they are accelerated again by $A3$ to a constant velocity. The focused beam now passes through a system of deflection plates and strikes the CRT screen, causing a small dot to appear on the face of the tube.

FLUORESCENT SCREEN. The screen of a CRT is formed by coating the inside face of the tube with a fluorescent material commonly referred to as a *phosphor*. When an electron beam strikes the fluorescent compound, its electrons are stimulated and raised to a higher energy level. As the electrons drop back down into their former (lower) energy level, the pumped energy is released in the form of light, much in the same way laser light is created from pumped electrons. The brightness of the glow is proportional to the number of electrons striking the phosphor and their velocity. Maximum brightness is obtained with a thin, dense beam traveling at very high speeds. The glow may persist from a few milliseconds to several seconds, depending on the type of screen phosphor used.

Although the term phosphor implies that the screen material contains the element phosphorus (the phosphorus mineral-bearing rock in which the fluorescent effect was first observed), most phosphors used in CRTs today are composed of zinc silicate or zinc sulfide and contain no phosphorus whatsoever. Screen color and persistence are a characteristic of phosphor design.

Because these phosphors are insulators, the screen would soon develop a negative potential as the primary electrons accumulated, were it not for secondary emission. The secondary electrons — electrons that bounce off or are knocked loose of the screen — are collected by a graphite coating of *aquadag* deposited around the inside neck of the tube, thus preventing a negative potential from accumulating on the screen.

In some newer CRT designs, a fine screen of thin-film aluminum is deposited on the back surface of the phosphor in place of the aquadag. The metal mesh permits high-speed primary electrons to pass through the grid, while collecting the slower-speed secondary electrons on their rebound. The aluminum film also improves the brightness of the display by reflecting the emitted light forward.

DEFLECTION SYSTEM. The electron gun and its associated focusing system is designed to produce a sharp spot on the screen. Before this spot can display information, though, it is necessary to move it to any required point on the screen. This is the function of the *deflection system.*

Basically, the deflection system consists of four metal plates, aligned in parallel pairs perpendicular to each other, located just beyond the electron gun in the neck of the CRT. When viewed straight on, the plates resemble a box that has its sides aligned along the horizontal and vertical axis of the face plate. Appropriately, the plate pairs are identified as horizontal and vertical deflection plates.

The deflection plates use electrostatic control to manipulate the electron beam. The electrostatic deflection method is based on the law of electric attraction, where unlike charges attract each other and like charges repel. If the horizontal and vertical deflection plates have no charge to them (i.e., they are grounded), the electron beam passes between each pair unaltered and strikes the screen in the center of the face plate. This produces a bright glow in the center of the screen. When a charge is applied to any one of the plates, however, it affects the direction of the electron beam, causing the beam to strike different points on the screen.

Each plate pair is treated as a differential system. When one plate of a plate pair has a positive charge applied to it, its counterpart has an equal negative potential applied to it. Equal but opposite charges negate the effects of beam acceleration. Consider the situation in which a positive or negative charge is applied to one plate and its opposing plate is grounded. In this case, the potential at the center of the charge space between the plates is greater or less than zero volts, depending on polarity, causing horizontal acceleration or deceleration of the electrons and thus altering the velocity of the beam. Application of equal positive and negative deflecting voltages of +E/2 and –E/2 to the deflecting plates, on the other hand, establishes a zero voltage space zone between the plates, leaving the electron beam's speed unaffected, while still influencing the direction of the beam with a potential equal to E. To prevent interaction between the horizontal and vertical plate pairs, an isolation shield returned to ground is established between the deflection pairs.

The electron beam's sensitivity to deflecting voltages is dependent on the physical location of the deflection plates and the velocity of the electrons. Physical location of the deflection plates is inherent to the design of the tube and is unalterable. Beam velocity, on the other hand, can be controlled by externally applied voltages.

The deflection system is most sensitive when the electrons are traveling slowly. Slow electron velocity is achieved by reducing the positive voltage to the accelerating anode. Unfortunately, beam velocity directly affects the

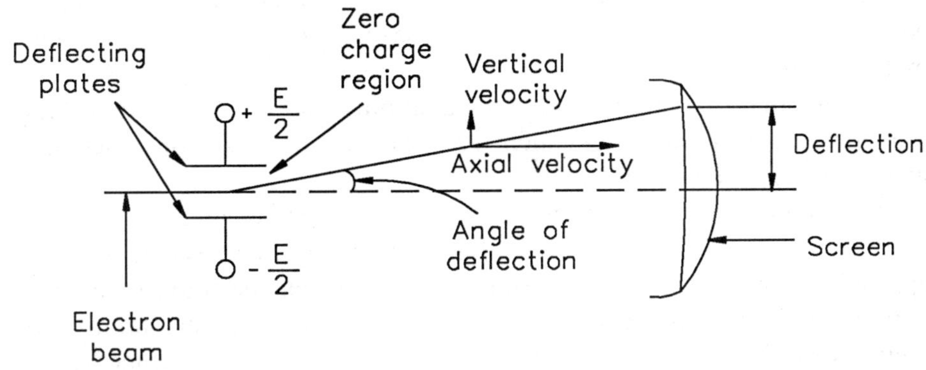

Figure 4-4 Deflection system

brightness of the screen image: The slower the beam, the dimmer the glow. It is for this reason that post-deflection acceleration is provided. Added energy is imparted to the electrons after they pass through the deflection plate by application of a very high voltage potential to the aquadag collector ring. This potential is on the order of about +12kV. Electrons leaving the deflection plate area experience a continuous accelerating force as the attracting voltage increases with traveling distance, with the attracting force being highest at the screen. This design allows maximum control of the electron beam in the vicinity of the deflection plates through slower electron velocity, while achieving the required speed for maximum screen brightness.

The electron beam's sensitivity to deflecting voltages can be expressed in two ways. The voltage required to move the beam one centimeter at the screen (V/cm) is referred to as the *deflection factor*. The deflection produced by a 1-volt change across the deflection plates (cm/V) is called the *deflection sensitivity*.

Vertical Deflection Amplifier

Within the realm of everyday testing, however, the test voltages involved are very small compared to the voltages required to produce a noticeable change in the position of the electron beam on the screen. Consequently, amplification is required. Each deflection amplifier must meet two criteria. The first is to increase the input signal to a level at which it is within the normal operating voltage range of the deflection plates. Second, it must have differential output to the deflection plates to prevent beam acceleration and deceleration by the deflection plates.

A block diagram of a typical vertical amplifier section is shown in Figure 4-5. Notice that the amplifier contains several elements, including attenuators, gain switches, and buffer circuits.

The attenuator circuit provides control of input coupling and vertical deflection factor. Typically, a vertical deflection amplifier has a gain factor of approximately 2000. That is, for every 1-volt change at the amplifier input, the output responds with a corresponding 2000-volt change. This amplification factor is much too large for many measurements, and a high-impedance attenuator proceeds the input stage. The attenuator resistors are selected to produce a wide range of deflection sensitivity. At the most sensitive switch setting (least attenuation), a 2-millivolt input typically produces a 1-centimeter deflection on the screen. At the greatest attenuation position, a 10-volt input is required to produce the same deflection.

The signal applied to the input attenuator can be ac-coupled or dc-coupled, depending on the setting S1. When S1 is set for dc coupling, the signal is applied directly to the input of the high-impedance attenuator stage. When ac coupled, the input signal passes through dc-blocking capacitor C1. The blocking capacitor prevents the dc component of the input signal from being applied to the attenuator circuit.

Following the attenuator is a buffer amplifier. The buffer amplifier, beginning with U10 in Figure 4-6, presents a high-impedance load to the signal from the attenuator and a low-impedance output to the vertical preamplifier. The buffer amplifier incorporates a dual signal path to combine stable dc performance with responsive high-speed performance.

In the slow path, the input signal is applied to both the gate of source-follower Q13 and the inverting input of U10, through the divide-by-two network composed of R3 and R5. Transistor Q13 and emitter-follower Q18 isolate the input signal from the loading of the vertical preamplifier. A portion of the output signal is applied to the noninverting input of U10 through divider network R46, R47, and R48. Operational amplifier U10 compares the two input voltages and changes the conduction level of current-source Q15 to correct for any errors at the source of Q13. Capacitor C10 limits the bandwidth of U10 so that the slow path responds only to frequencies below 100 kHz. In the fast path, input signals are coupled through R6, C6, Q13, and Q18 to the circuit output, thus providing high-speed performance. By adjusting R47, the gain in both paths is matched.

The buffer output drives a vertical preamplifier, which converts the single-ended signal into a differential signal for application to the vertical deflection

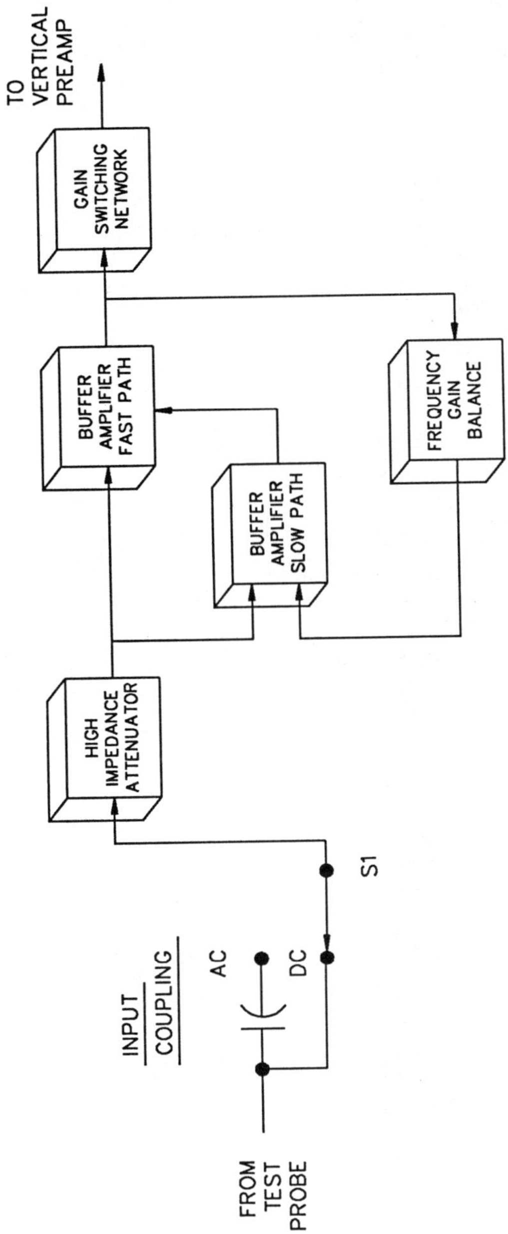

Figure 4-5 Vertical amplifier block diagram

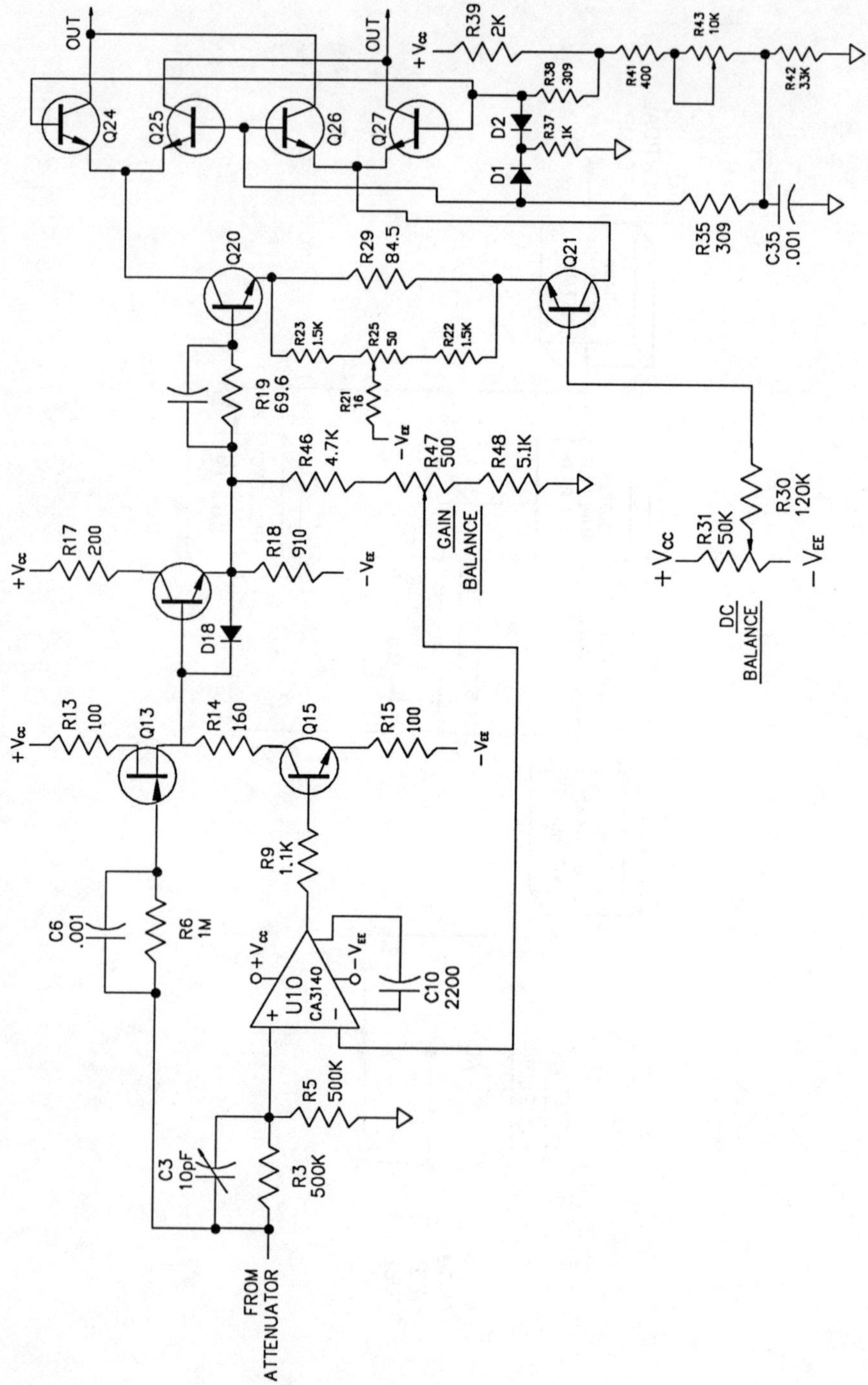

Figure 4-6 Buffer and differential driver amplifiers

plates. The signal is applied to the base of transistor Q20. Transistor Q21 is biased by the divider network, composed of R30 and R31, to a level that produces a null at the differential outputs with no signal applied to the Q20 input. Emitter current for the two input transistors is supplied by R21, R22, R23, and R25, with R29 serving as the gain-setting resistor between the two emitters.

The collector current through the two input transistors serves as emitter current for the two differential output transistor pairs, Q24-Q25 and Q26-Q27. Base bias current for the two pairs are derived from the divider network composed of R39, R41, R42, and R43. The current ratio in the output transistor pairs is determined by diodes D1 and D2. As R43 is rotated counterclockwise, the conduction level of the transistors connected to R35 increases. Because the transistor pair outputs are cross-wired, this increased conduction will subtract from the signal produced by Q24 and Q27, causing the overall gain of the stage to decrease. Potentiometer R25 adjusts the balance of the differential amplifier so there is minimal shift in output voltage as the gain changes.

The vertical output amplifier, Figure 4-7, provides final amplification of the input signals for application to the vertical deflection plates of the CRT. Signals from the preamplifier are applied to a differential amplifier composed of Q230 and Q231, with low- and high-frequency compensation provided by the RC network connected between the emitters. Overall gain of the stage is set by R233. The output stage of the vertical output amplifier utilizes two totem-pole transistor pairs, Q254-Q256 and Q255-Q257, that convert the collector currents of Q230 and Q231 to proportional output voltages. Resistors R256, R258, R257, and R259 serve as feedback elements and also as divider networks, so that each transistor in a pair drops half the final output voltage. These output signals are applied to the vertical deflection plates to produce up-and-down deflection of the electron beam.

FREQUENCY RESPONSE. The frequency response of the vertical amplifier often determines the general class of the oscilloscope. Analog oscilloscopes are classified into two general categories: low frequency and high frequency. The low-frequency oscilloscope has a frequency response of dc to 10 MHz, and is used most often in general service applications, such as television repair. High-frequency oscilloscopes have bandwidths up to 350 MHz, and are primarily found in the laboratory.

The frequency response of an amplifier is determined by both the gain of the amplifier and the phase angle of the signal. The frequency at which $A_V(\omega)$ is down $1/\sqrt{2}$ (3dB, $\pi/4$ phase angle) from its low-frequency value, $A_V(0)$, is the bandwidth.

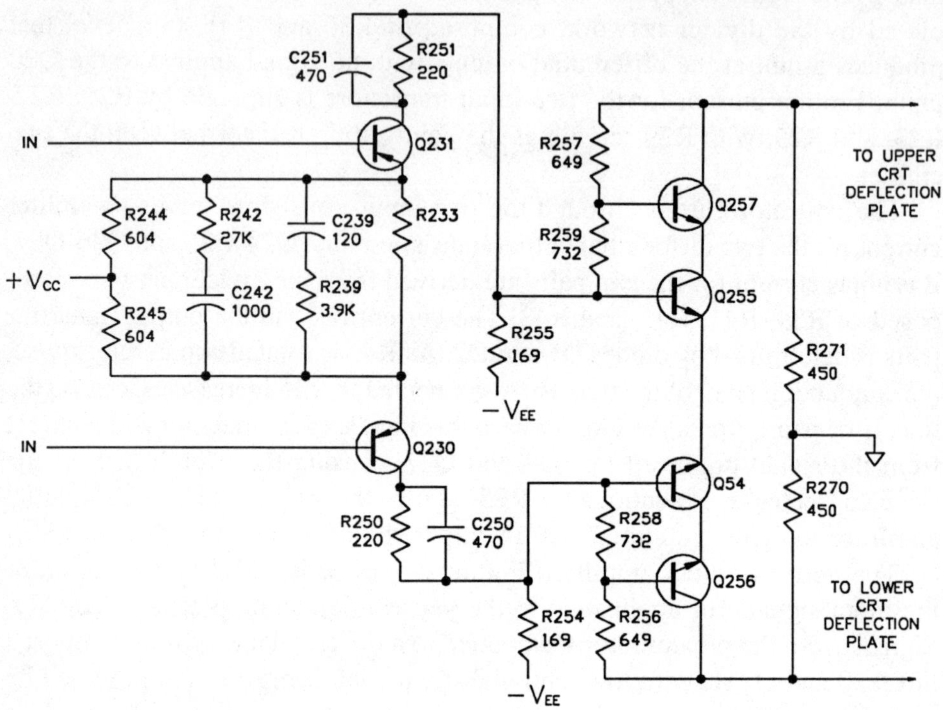

Figure 4-7 Vertical output amplifier

With high-frequency oscilloscopes intended for pulse measurements, the rise time (tr) of the amplifier is often specified. The relationship between bandwidth and rise time is:

$$\text{bandwidth (BW)} = \frac{0.35}{\text{tr}}$$

where bandwidth (BW) is in megahertz and tr is in microseconds. A high-frequency oscilloscope with a rise time of 1 ns has a frequency response of 0.35/1 ns = 350 MHz.

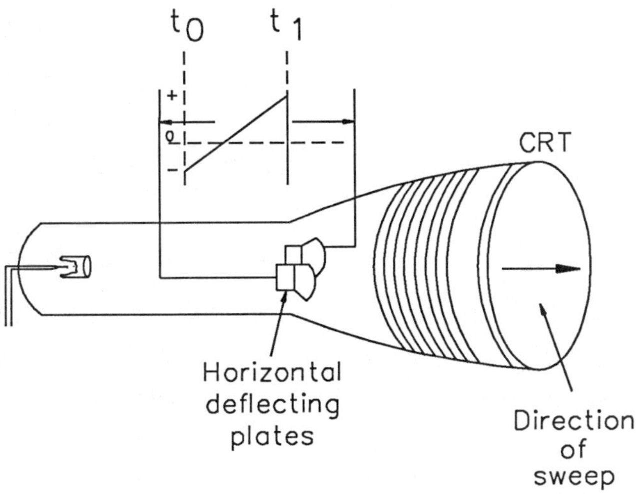

Figure 4-8 Horizontal sweep

Horizontal Sweep

The primary function of an oscilloscope is to display a plot of changing voltage in a vertical axis against a time standard in the horizontal axis. A time reference standard is created by sweeping the spot across the screen from left to right at a constant rate. The result is a linear line across the face of the screen that is used as a yardstick.

A linear sweep is generated when a sawtooth waveform is applied to the horizontal plates. The sawtooth waveform is unique in that it has an increasing linear voltage, called the ramp, followed by a sharp return to zero. When the sawtooth signal is differentially applied across the horizontal deflection plates, the electron beam is initially positioned to the extreme left of the CRT face, as represented by t_0 in Figure 4-8. As the ramp voltage increases, the spot is deflected toward the right at a linear rate. At the peak of the sawtooth ramp, t_1, the ramp voltage suddenly falls to zero volts, causing the beam to retrace back to its initial starting position. This part of the cycle is referred to as the *retrace period* and is normally blocked from the display by cutting off the flow of electrons from the electron gun with a strong negative bias to the CRT control grid.

Because the beam movement is proportional to the voltage at the deflection plates, the length of the sweep is determined by the applied voltage,

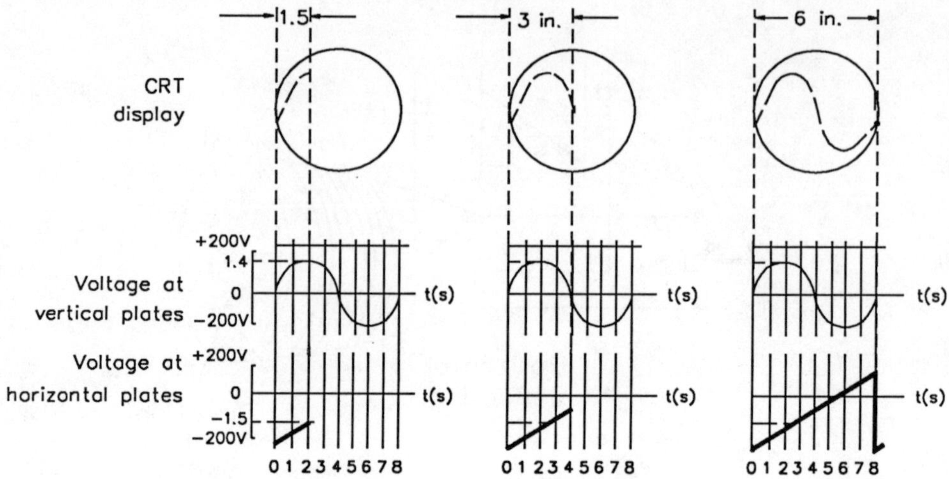

Figure 4-9 Waveform display

which is set by the gain of the horizontal amplifier. The amount of time it takes for the sawtooth ramp to complete a sweep is a function of sawtooth frequency. The resultant combination is a reference line of fixed length and time called the *timebase*. The timebase can now be marked in linear time divisions, from which time measurements are made.

Waveform Display

When an alternating voltage is applied to the vertical deflection plates, the spot on the face of the tube moves up and down continuously in response to the voltage. When this action is combined with a linear sweep, as described above, the result is a graphic representation of the waveform with respect to time.

Consider Figure 4-9, in which a sine wave is applied to the vertical deflecting plates and a sawtooth signal is applied to the horizontal plates. Assuming both waveforms begin at the same time, then at t(s) = 0 the vertical deflecting voltage is zero and the horizontal deflecting voltage is –200 volts. This places the spot to the far left of the screen on the center line. When t(s) = 2, the horizontal deflection voltage has increased to –150 volts, moving the beam two increments closer to the center of the screen. At the same time, the vertical input voltage has increased from zero to 140 volts, forcing the beam upward. The resultant screen position is represented by point 2. Point 3 represents t(s) = 2, and so on.

When t(s) = 4, both the vertical and horizontal signals are at zero volts, placing the dot in the center of the screen. Further changes in the vertical input, t(s) = 5, force the spot below the center line of the screen, while the horizontal position continues its advance across the screen. At t(s) = 8, the horizontal deflection voltage drops sharply back to -200 volts, returning the electron beam to the left side of the screen. The result is a plot of the vertical waveform with respect to time. When this waveform is rapidly repeated on the screen, phosphor persistence creates the illusion of a solid trace.

Horizontal Sweep Generator

The sawtooth voltage for the timebase is produced by a *sweep generator circuit.* The sweep generator circuit produces a linear voltage ramp that is amplified by the horizontal amplifier to provide horizontal deflection of the electron beam. The sweep generator also produces signals that are used to generate correct timing of the CRT blanking and intensity levels used for viewing the display.

The Miller sweep generator circuit shown in Figure 4-10 produces this ramp by maintaining a constant current through timing capacitors to obtain a linearly increasing voltage. Field-effect transistors Q704A and Q704B are matched devices with Q704B sourcing current for Q704A. Since the gate and source of Q704B are connected, the source current of Q704A is of such a magnitude that there is no voltage drop across its gate-source junction.

When the sweep is not running, Q701 is biased on to hold the timing capacitors in a discharged state. The low impedance of Q701 in the feedback path holds the Miller sweep output near ground potential. The voltage across Q701, in addition to the base-emitter voltage of Q706, prevents saturation of the output device.

The sweep ramp is initiated when Q701 is biased off by the trigger circuit, allowing the timing capacitors to charge at a rate determined by timing resistors R701 and R702, in accord with the setting of S701. One end of timing resistor R701 is connected to the input of the Miller integrator. Because of feedback from the circuit output through the timing capacitors, the integrator input voltage remains fixed and establishes a constant voltage across the timing resistors. This constant voltage, which produces a constant

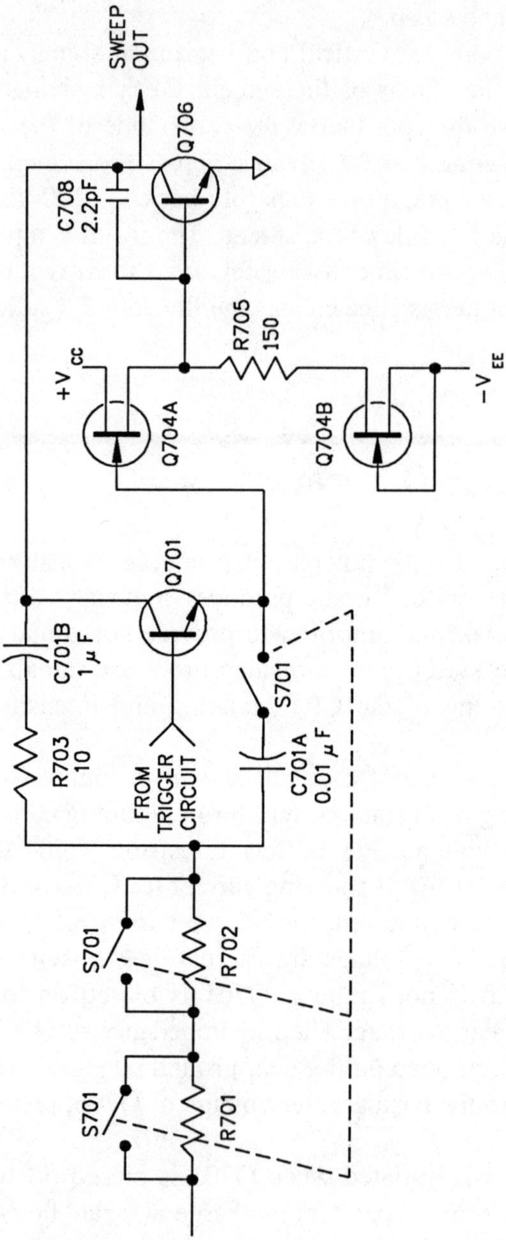

Figure 4-10 Horizontal sweep generator

current through the timing capacitors, results in a linearly increasing voltage at the output of the Miller sweep circuit.

When the output reaches approximately 12 volts, the sweep logic circuitry initiates a *holdoff period* in which Q701 is turned on and the sweep generator is reset. This holdoff period is necessary so that the timing capacitors can be fully discharged before another sweep starts. Capacitors C702 and C703 are always in the charging circuit and are used for high speed sweeps. Capacitor C701A, in series with C701B, is used for medium sweep speeds, and C701B is used for slow sweep speeds.

SWEEP TRIGGERING. To display a stationary pattern on the CRT, the sweep frequency must be a fixed multiple of the input signal; otherwise the pattern drifts uncontrollably. Sweep stability is achieved by synchronizing the start of the vertical input with the horizontal sweep generator, using a *trigger generator circuit.* The sweep generator logic, located within the sweep generator, initiates sweep generation as a function of incoming trigger signals and the trigger mode selected.

Basically, there are three types of sweep triggers: normal, auto, and delayed. Each has different display applications, discussed in the next chapter. The following technical discussion of sweep modes refer to the logic circuitry in shown Figure 4-12.

In the *normal mode*, the input to OR gate U532D is held high, causing the gate output to be high also. The output of U532D will then be low, and U506A will not be held reset. Simultaneously, the input to NOR gate U532A is held high, causing its output to be low, which places a low on the D input of U506A.

At the beginning of the holdoff period, the reset input of U506A is set high by U532B. This sets pins 3 and 4 of U630 low, forcing pin 6 of U630

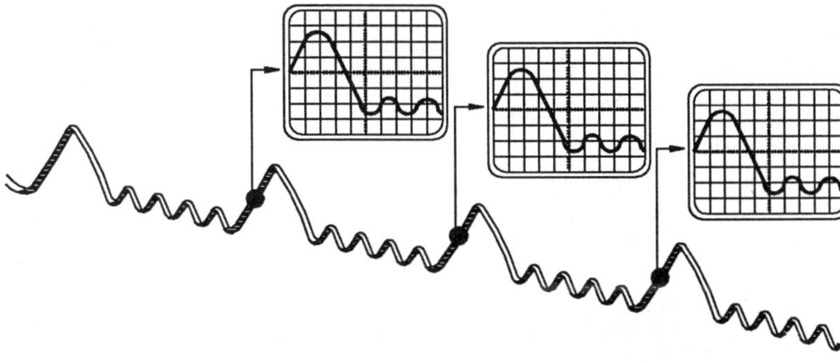

Figure 4-11 Trigger waveform

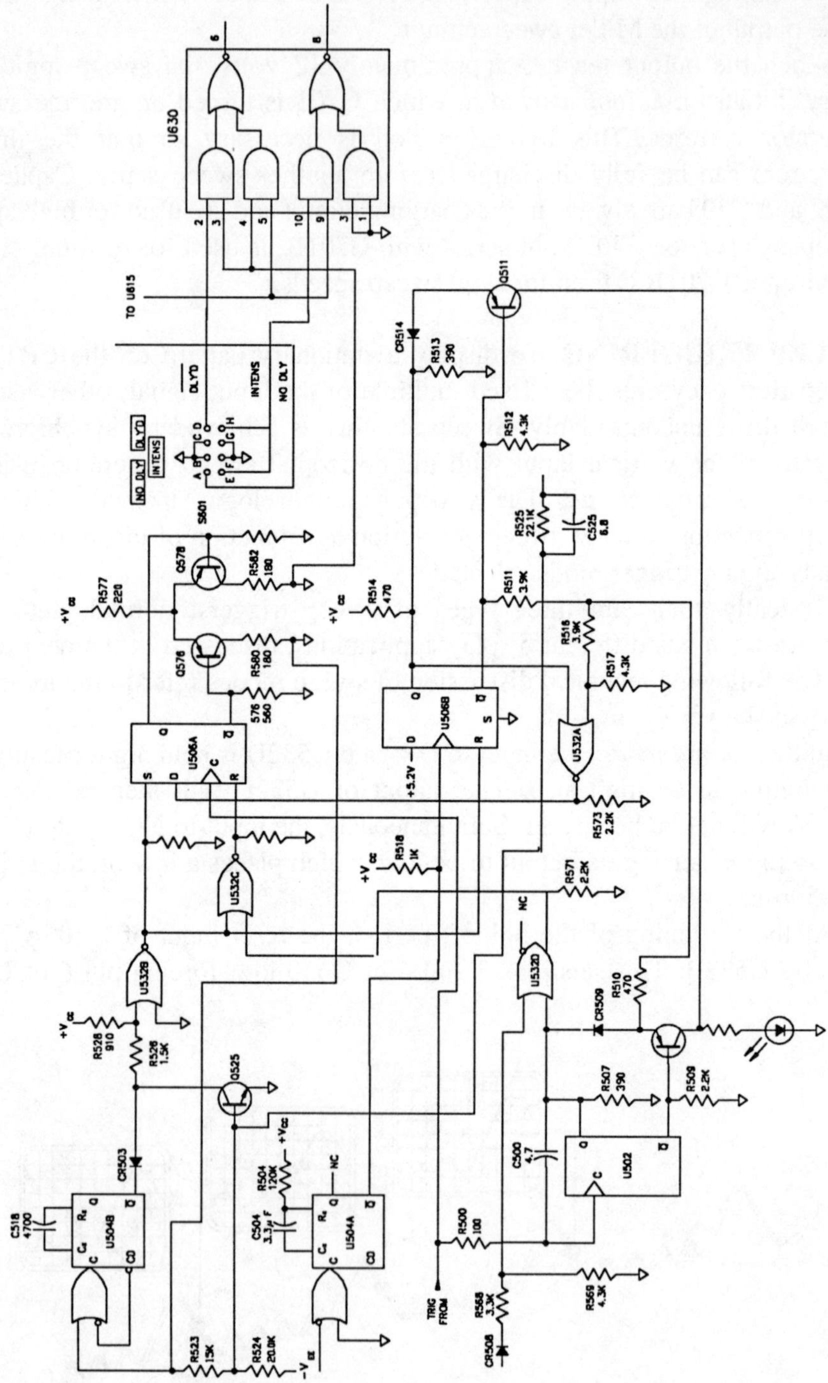

Figure 4-12 Block diagram/schematic trigger logic circuitry

high and holding the Miller sweep in a reset condition. At the end of the holdoff period, the set input of U506A goes low, thus enabling the next trigger signal to clock the low on the D input through the output. The low Q output from U506A forces pin 6 of U630 low, allowing the Miller sweep to generate the sweep ramp as described above. When the ramp voltage is about 12 volts, Q525 is biased on. The output of U532B then changes from low to high, setting U506A, thus triggering holdoff one-shot U504B and forcing pin 6 of U630 high. This starts holdoff and resets the Miller sweep, which turns off Q525.

With U504B triggered, the $\overline{Q}$ output changes from high to low and will stay low for a time duration determined by the holdoff period circuitry. During the time the U504B $\overline{Q}$ output is low, the set input of U506A is held high so that no trigger pulses can initiate a new sweep. Also during this period, capacitor C518 is charged at a rate determined by an external constant current source. When the voltage on Rx of U504B reaches the threshold voltage of Cx, the $\overline{Q}$ output goes high to end the holdoff period and release U506A from the reset conditions. The circuit is then enabled to generate another sweep once a trigger signal is again applied to the clock input of U506A.

In the *auto mode*, the sweep generator automatically generates a baseline sweep in the absence of a trigger signal. When the auto mode is selected, the input of U532D is held low (instead of high, as in the normal mode) by R569, allowing the output of the gate to follow the signal provided by the Q output of U502. If trigger signals are being received by U502, the output of U532D will be high, forcing the output of U532C low. Under these conditions, flip-flop U506A responds to trigger signals in the same manner as it would in the normal mode.

If no trigger signals are present at the clock input of U502, the output of U532D goes low. The output of U532C is then the inverse of the output signal of U532B so that U506A will be reset when the holdoff period ends, causing a sweep to be generated. With no trigger pulses applied to the circuitry, U506A will continuously set and reset at the end of each holdoff period to generate a baseline sweep.

DELAYED SWEEP. *Delayed sweep* provides a set amount of delay time between the start of the trigger event and the actual start of the sweep. This allows viewing of portions of a waveform that might otherwise scan off the face of the screen during normal presentation.

The delay circuit — composed of Q606, Q611, U615, U630, and associated components shown in Figure 4-13 — generates the timing and gate signals required to produce an intensified sweep display and to provide the

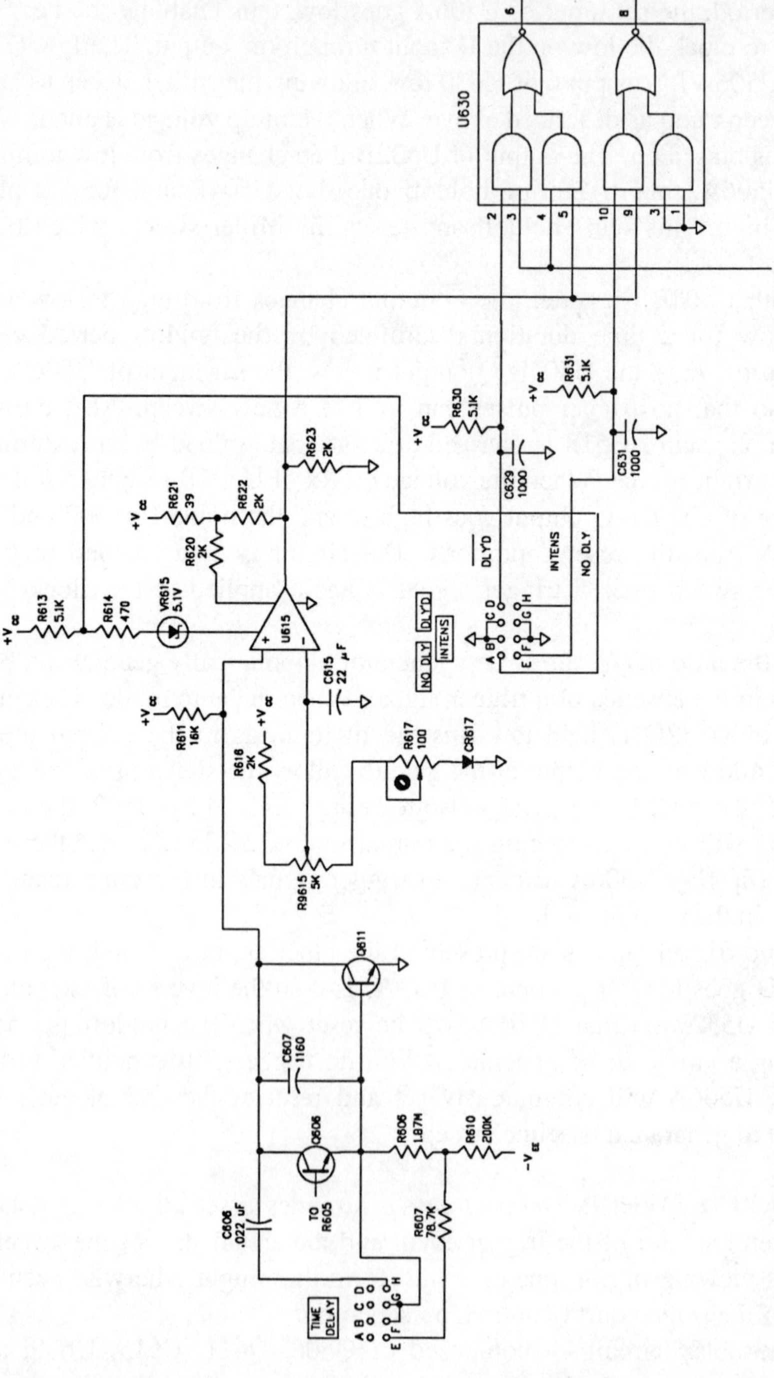

Figure 4-13 Delayed sweep

variable sweep delay. Switch S601 controls the display mode, which is defined as either no-delay, intensified-sweep, or delayed-sweep.

With S601 set to the no-delay position, pin 2 of U630 is held high. In this mode, the sweep start is controlled by the signal on pin 3 of U630, in the manner described above. When a trigger occurs in either the normal or auto mode, pin 3 of U630 goes high, causing pin 6 of U630 to go low and start the Miller sweep generator. In the no-delay mode, the delayed comparator U615 is disabled so that it does not fire during the sweep.

With S601 set to the intensified sweep position, pin 2 of U630 is held high, and the sweep starts in the same manner as in the no-delay mode. In this mode, however, the trigger that starts the sweep also starts the delay circuit by forcing the base of Q606 low.

The delay time is determined by comparing the variable voltage from the multiplier control R9615 with the delay voltage ramp. The delay voltage ramp is generated from a Miller integrator composed of Q611, R611, and timing components R606, R607, C606, and C607. The timing components are switched by S603 to select the different delays.

When the signal at the collector of Q576 (Figure 4-12) is high, Q606 is turned on, discharging C606 and C607 while holding the ramp output low. When the trigger occurs, Q606 is turned off and the timing capacitors charge at a rate determined by the timing resistors. When the ramp voltage exceeds the comparison voltage, the output of the delayed comparator U615 goes high, forcing pin 9 of U630 high. Since pin 10 of U630 is held high in the intensified mode, pin 8 of U630 is forced low at the end of the delay. The low voltage on pin 8 increases the Z-axis drive and display intensity. The nonintensified portion of the trace indicates the amount of total delay time. This sweep can be used to set the exact amount of sweep time required by the delayed-sweep mode.

In the delayed-sweep mode, the start of the sweep is delayed by the amount of time established by the delay circuit. In this mode, pins 2 and 10 of U630 are held low, thus disabling the nondelayed sweep and the intensified display. When the trigger occurs, pin 4 of U630 goes high enabling the delay and comparator to determine the sweep start. At the end of the delay period, the output of U615 goes high, forcing U630 pin 6 low. The low signal at pin 6 initiates the sweep.

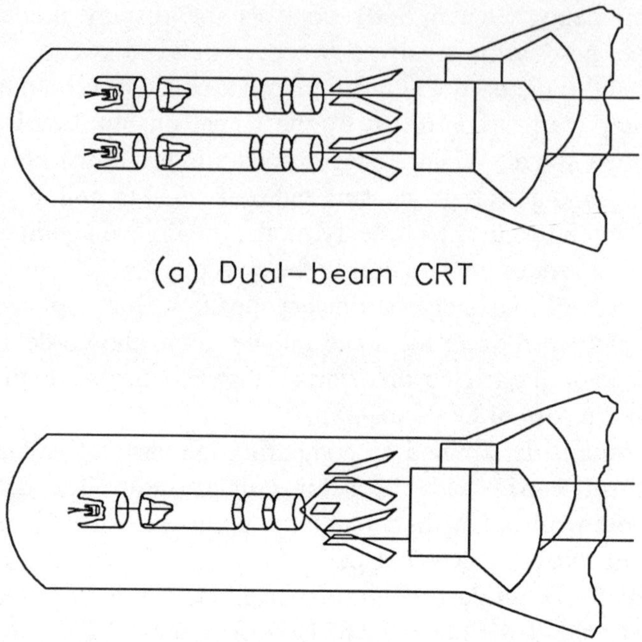

(a) Dual–beam CRT

(b) Split–beam CRT

Figure 4-14 Dual-trace CRTs

Dual Channel Oscilloscopes

Most modern oscilloscopes can display two waveforms simultaneously, which enables the operator to perform time, amplitude, and phase comparisons between two waveforms. These devices are commonly referred to as *dual-channel* or *dual-trace* oscilloscopes.

Dual-trace displays can be achieved in one of two ways. The obvious is to modify the CRT so that it projects two separate electron beams on the screen. One method is to incorporate two separate, but complete, electron guns in a single tube. In this case, the instrument is called a *dual-beam* oscilloscope because there are two separate electron beams, one for each waveform trace. In another type of dual-trace CRT, a single electron beam is split into two separate beams before passing through the deflector plates. This is referred to as a *split-beam* CRT. In both designs, only one set of horizontal deflection plates is used. Consequently, both beams move horizontally across the face of the screen in unison, using a single horizontal sweep generator. Each pair

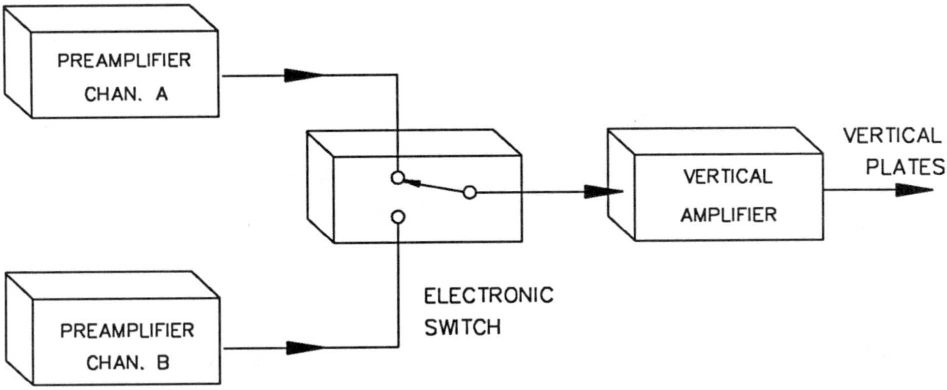

Figure 4-15 Block diagram of electronic-switching dual-trace oscilloscope

of vertical deflection plates, however, is driven by a separate vertical amplifier, causing each trace to display a different signal.

ELECTRONIC SWITCHING. The more common method of creating dual-trace images on an oscilloscope screen uses a single-beam CRT, with one set of horizontal deflection plates, one set of vertical deflection plates, and an *electronic switch.* The electronic switch is positioned between two vertical preamplifiers, identified as channel A and channel B, and a vertical driver amplifier, as illustrated in Figure 4-15.

The electronic switch is an ultra-fast relay, usually in the form of a flip-flop, that toggles between the outputs of the vertical preamplifiers, alternately feeding the two signals to the vertical driver amplifier that drives the vertical deflection plates, thus creating the effect of having two traces on the screen simultaneously. The switch may be operated on one of two modes: alternate and chop.

ALTERNATE MODE. In the *alternate mode*, the screen displays the input signal of one vertical channel for a full sweep, then that of the other channel for the next sweep. On the first timebase sweep, for example, the input signal to channel A may be traced on the screen. During the retrace period, the electronic switch flips position and feeds the output of channel B to the vertical deflection plates, thereby displaying channel B on the screen. The next sweep period again displays channel A, followed by channel B input once again. Thus the two inputs are alternately and repeatedly traced on the screen.

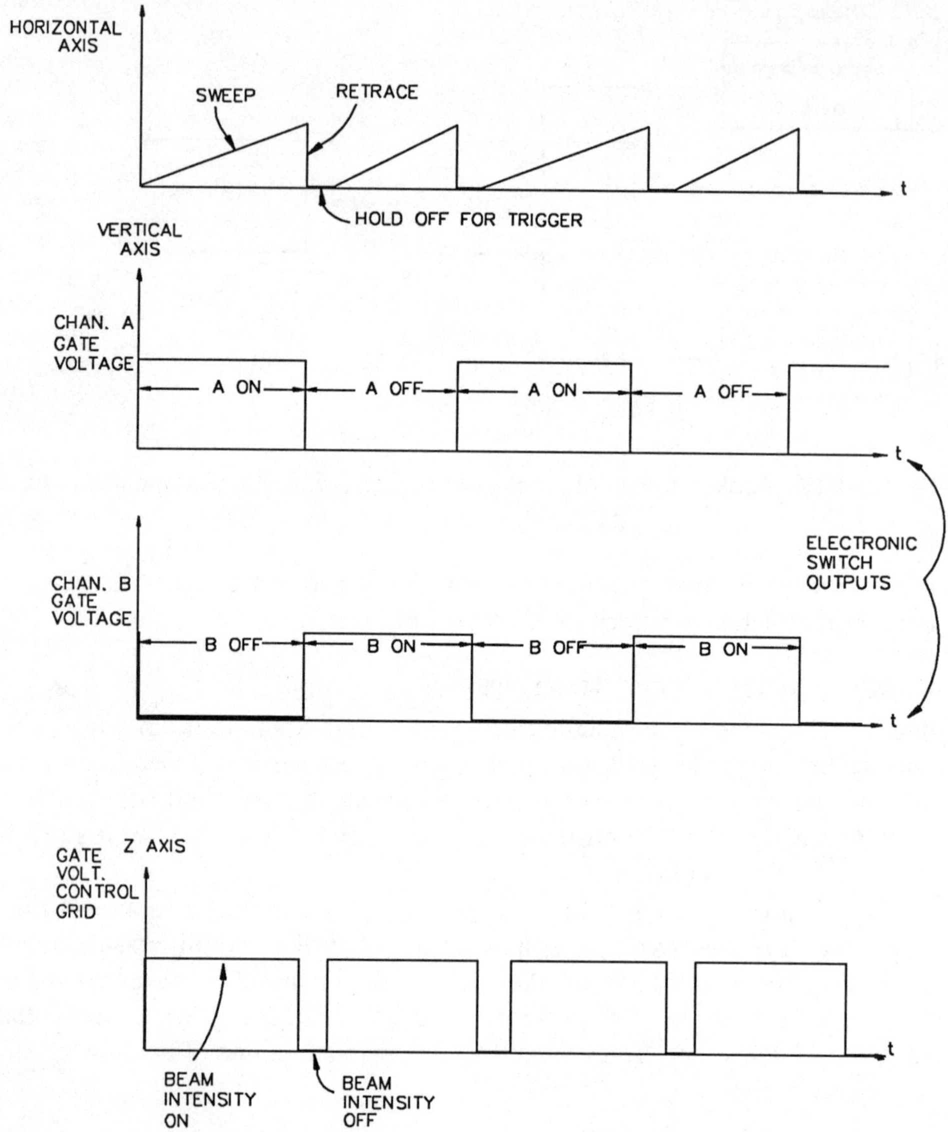

Figure 4-16 Alternate display timing

The repetition rate of the electronic switch is usually so high that the two waveforms appear to be displayed simultaneously. To keep the images separate, each preamplifier output contains a dc offset voltage that forces the trace above (channel A) or below (channel B) the center line of the screen, making identification easy. This mode of operation is commonly used for viewing high-frequency signals, where sweep speeds are much faster than

CRT phosphor decay time. When low-frequency signals are to be viewed, signals that require a slow sweep, the alternate mode generally produces objectionable flicker that makes waveform comparison difficult.

CHOPPED MODE. In the *chopped mode*, the electronic switch alternates between channel A and channel B at a rate considerably faster than the sweep frequency, taking small samples of each waveform as it does. Consult Figure 4-17 for the following discussion. During the t_0 portion of the horizontal sweep, the electronic switch connects the channel A preamplifier output to the vertical driver amplifier, which begins tracing the channel A waveform on the face of the tube. Within a microsecond or less, the electronic switch toggles to the t_1 condition, which connects channel B output to the vertical driver amplifier. This causes the vertical driver amplifier to shift the electron beam, according to the dc offset of channel B, and display a sampling of the B channel on the screen. At t_3, another sample of channel A is displayed on the screen, followed by a t_4 sampling and display of channel B.

The result is a broken-line tracing of the original channel A and channel B waveforms displayed on the screen within a single sweep period. Although 50 percent of each trace is missing, the gaps are so small that our eyes are tricked into believing two solid traces exist.

The disadvantage of the chopped display is that short-lived transients may be cut out of the picture entirely during the switching period, possibly leading to a false representation of the waveform. Furthermore, as the input frequency increases, fewer samples per cycle are taken, and the pattern grows coarser and harder to distinguish. The trick is to run the switching oscillator much faster than the input frequency; with this method, the samples are smaller and more representative of the waveform. There is a limit, however, to how fast the electronic switch can scan the inputs and still return a true value. Among the problems encountered are slew rate and settling time, both of which impose an upper frequency limit on the sampling rate. Consequently, chopped-mode displays are most desirable when medium- and low-frequency waveforms are displayed.

The display mode that is best for any particular application varies according to the complexity of the waveforms involved and the accuracy of the measurement. As a rule of thumb, use the chopped mode for sweep rates of 1 ms per division or more, and the alternate mode for anything faster than 1 ms. By and large, though, the mode selection is made automatically by the instrument, according to the setting of the sweep generator timebase switch.

The alternate and chopped scanning techniques may be expanded to provide three, four, or even eight vertical input channels on a single tube. When working in the alternate mode, each input is scanned, one at a time,

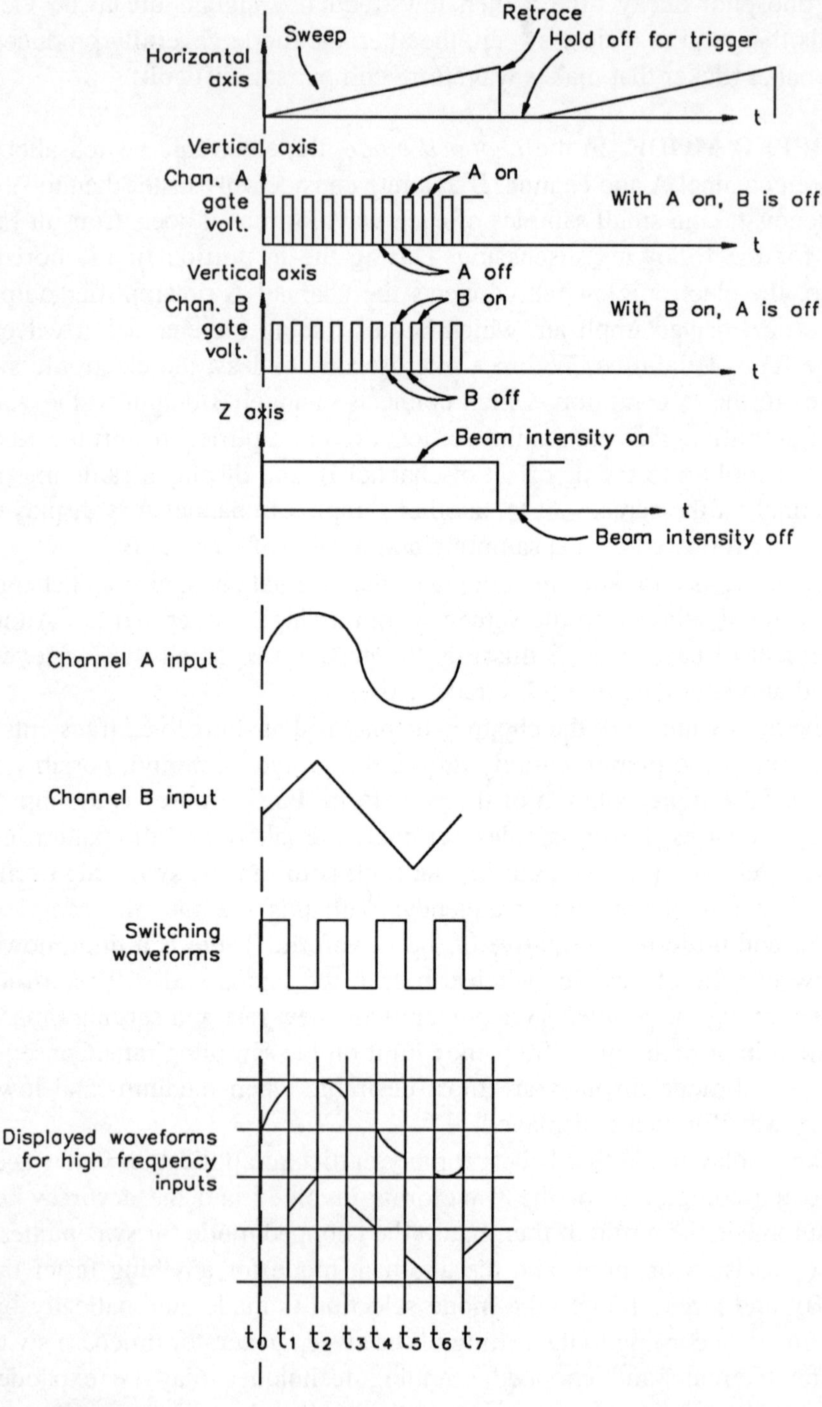

Figure 4-17 Chopped display timing

and displayed on the screen in a round-robin manner. In the chopped mode, samples of each input are taken in order, creating the multiple images. In some designs, dual-beam CRTs are combined with one or both of the above techniques to increase the number of vertical input channels. The number of vertical traces a CRT can support is limited by the repetition rate of the waveform, its complexity, and the size of the screen.

SWEEP TRIGGERING. Dual-trace oscilloscopes have unique triggering requirements because there is more than one waveform from which to obtain a trigger signal.

Generally, the trigger is taken from the channel A input. As with a single-trace oscilloscope, the trigger signal is processed by the trigger generator circuit to start the sweep at a predetermined point. Alternatively, the trigger may be taken from the channel B input. Selection of the trigger input is with a front panel control. When in this mode, the input signal is sampled by the trigger generator at the preamplifier output before it enters the electronic switch.

In some cases, the trigger may be taken from channel A and channel B alternately. In this mode, the trigger signal is obtained after the input is processed by the electronic switch. Consequently, the sweep is triggered by whichever input the electronic switch happens to be sampling at the time. Alternate channel triggering is useful in comparing two waveforms that may not be synchronized to each other.

Consider the case where the vertical input signals to be compared are of a slightly different frequency, as shown in Figure 4-18. Note that the frequency

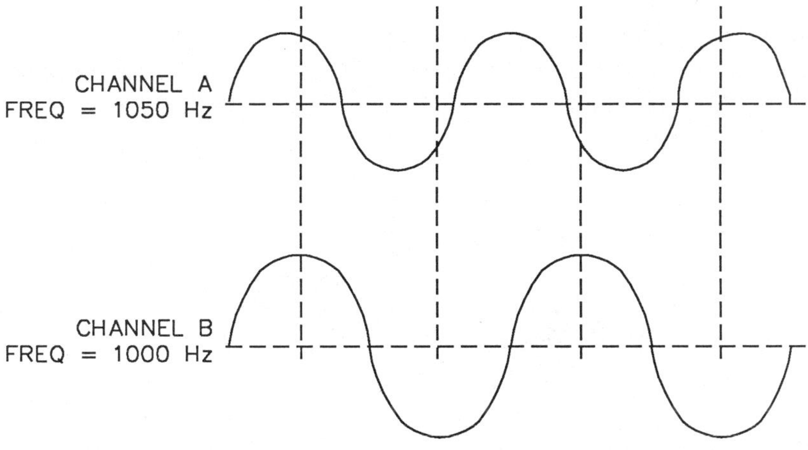

Figure 4-18 Two sine waves, at 1 kHz and 1050 Hz

of the two signals differs by 50 Hz. If the sweep is triggered from the channel A input, the waveform display of channel B would drift across the screen because of the frequency difference, making waveform comparisons impossible. And vice versa: Triggering from channel B would cause the channel A display to drift uncontrollably. If we alternately trigger on each individual waveform after it is selected by the electronic switch for display, both images appear stable. Be aware that the alternate triggering method cannot be used with the chopped input mode. It is applicable only to the alternate display mode.

Oscilloscope Probes

Virtually all general-purpose and laboratory oscilloscopes use a *probe* to make a direct, electrical connection to the device under test. Probes connect the measurement test points in the device under test to the inputs of the oscilloscope. Although a simple piece of wire could be used to make the input connection, unshielded conductors suffer from many shortcomings that would not let the user fully realize the capabilities of the instrument.

A typical oscilloscope probe consists of a length of coaxial cable with a plastic hand probe at one end and a BNC connector at the other. The coaxial cable protects the low-level input signals from outside interference while providing an electrical match for the oscilloscope's input. This includes 60-Hz power, radio, and television signals that would otherwise also make the journey along the probe wire to the oscilloscope input.

A coaxial cable consists of a single conductor surrounded by a braided wire shield. When the shield is grounded, any noise the cable encounters along the way is intercepted and returned to ground before it has a chance to reach and affect the center conductor. There are several types of coaxial cable

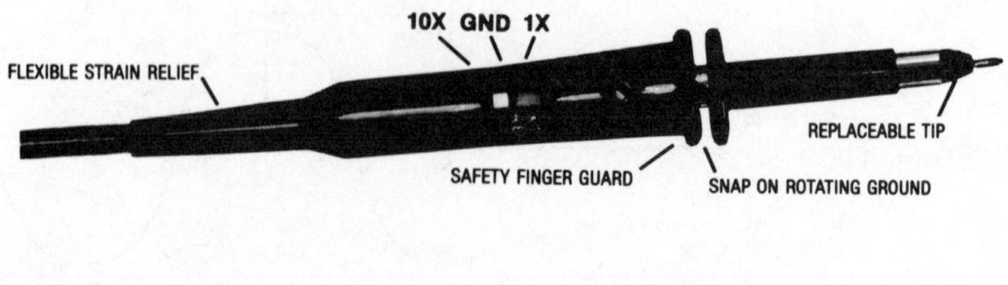

Figure 4-19 Test probe (courtesy Probe Master, Inc.)

used for scope probes, among the most popular are RG58/U, RG59/U, and thin-wire RG188A/U. Unfortunately, the use of coaxial cable introduces problems of its own. The most disturbing effects are *circuit loading* and frequency-sensitive *signal attenuation.*

Circuit loading modifies the environment of the signals in the device under test to an extent determined by total added impedance. To avoid loading by the coaxial cable and the oscilloscope input, the total lump impedance should always be much larger than the signal source. The loading effect is generally stated in terms of impedance at some specified frequency, and is made up of a combination of resistance, capacitance, and inductance. Obviously, source impedance has the largest impact on probe loading. A device under test with near zero output impedance, for example, is not affected in the same way a 1-megohm, 50-pF signal line would be.

Consider the case of a pulse generator with a source impedance of 50 ohms at 20 pF. If the initial rise time of the output pulse is ideally assumed to be zero (tr = 0), the output rise time of the generator would be limited by the associated output resistance and capacitance of the generator. This integration network produces an output rise time equal to 2.2 RC, or

$$t_{r1} = 2.2R1C1 = 2.2 \text{ ns}$$

as illustrated in Figure 4-20.

If this output is now monitored by an instrument with an impedance of 1 M in shunt with 30 pF, the instrument's input capacitance and series resistance would add to and alter the generator's parameters. Because the instrument's 1-M input resistance is much greater than the generator's 50-ohm output resistance, it can be ignored. The resultant circuit parameters and rise-time calculation now reads:

$$\begin{aligned} t_{r2} &= 2.2R_1\,(C_1 + C_i) \\ &= 2.2\,(50)\,(50) \times 10^{-3} \\ &= 5.5 \text{ ns} \end{aligned}$$

as illustrated in Figure 4-20. This is nearly double the rise time of the original output pulse, and could affect more than just an oscilloscope reading if the rise time is now beyond the limits of the device-under-test system design.

The input impedance of a general-purpose oscilloscope *at the front panel* is typically 1 M in shunt with 30 pF. The coaxial cable can add as much as another 100 pF to the total input capacitance for a total load impedance of 1 M in parallel with 130 pF.

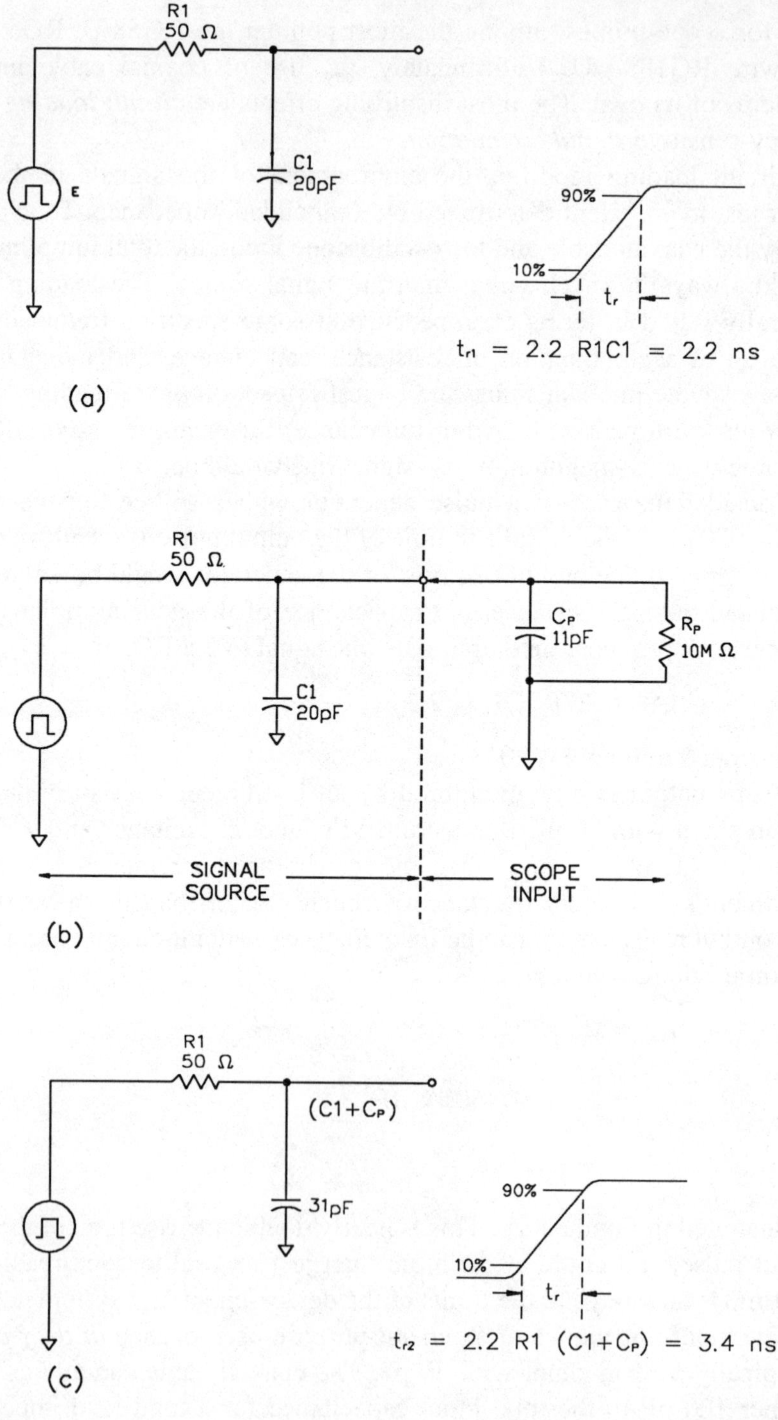

Figure 4-20 Loading effects on pulse generator rise-time characteristics

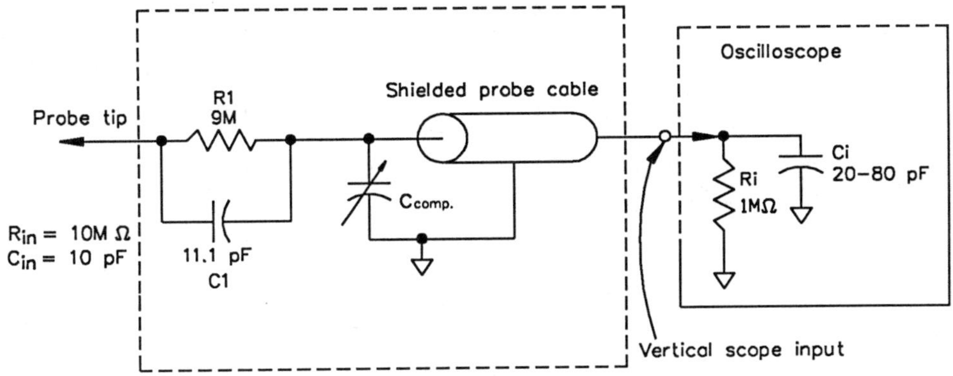

Figure 4-21 10:1 compensated probe

The effects of both resistance and capacitance loading can be reduced by inserting a resistor in series with the center conductor of the coaxial cable. The series resistor presents a higher input resistance and lower shunt capacitance at all frequencies by isolating the probe and front panel impedance from the device under test. The resistor is placed where it is most effective: at the very tip of the probe, just before it makes contact with the test point. The load now appears largely resistive.

The combination is called a compensated probe or *attenuator probe*. Attenuator probes attenuate the input signal according to the isolation resistor to input resistance ratio. Typical probes available have 10:1, 50:1, and 100:1 attenuation ratios. A circuit of a typical 10:1 compensated probe is shown in Figure 4-21. A 9-M resistor (R1) is placed in series with the input terminal, and a fixed capacitor of 11.1 pF (C1) is connected in parallel with input resistor R1. A variable capacitor, C_{comp}, is placed across the open end of the coaxial cable after the input network. This capacitor, which provides frequency compensation, will be discussed later.

The total impedance of the test probe is the sum of input resistance R1 and vertical input impedance R_i, shunted by a capacitive network consisting of C1, C_{comp}, cable capacitance C_{CC}, and input capacitance C_i, or

$$Z = R1 + R_i; (C1)(C_{comp} + C_{CC} + C_i)/C1 + (C_{comp} + C_{CC} + C_i)$$

$$= 9M + 1M; (11.1)(20 + 100 + 30)/(11.1) + (20 + 100 + 30)$$

$$= 10\text{ M}; 10.3\text{ pF}$$

for an equivalent impedance of 10 megohms shunted by 10.3 pF — quite an improvement over the 1-megohm, 130-pF load with which we began.

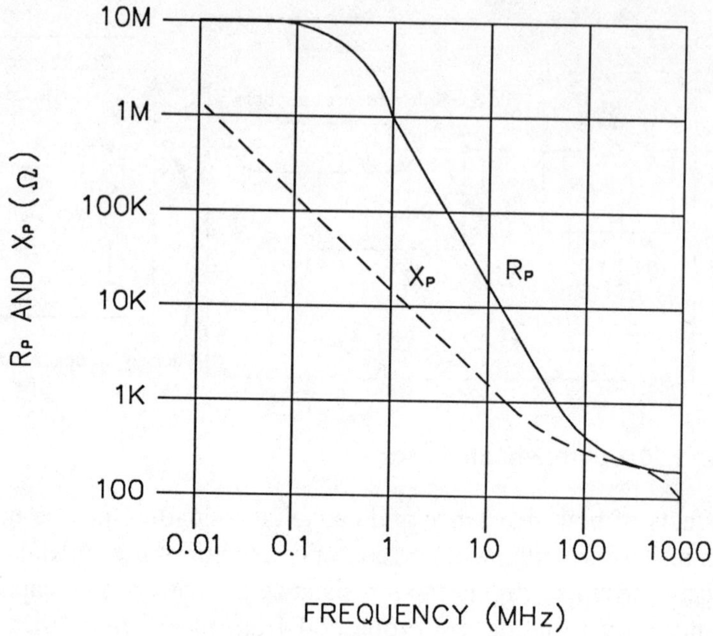

Figure 4-22 Probe load impedance versus frequency

The penalty is the reduction in signal amplitude. At low and medium frequencies, the capacitive reactance impedances are too large to be effective and the oscilloscope input voltage is

$$V_i = V_S \frac{R_i}{R_S + R1 + R_i}$$

$$= V_S \frac{1M}{50 + 9M + 1M}$$

$$= V_S / 10$$

where V_i is the voltage seen at the oscilloscope input and V_S is the voltage at the source of the device under test. For higher source impedances (R_S), the attenuation factor will be less; with a 1-M source impedance, the attenuation factor is nine (9).

As the signal frequency increases, the capacitive reactance grows smaller, thus placing an additional load on the device under test. When probing continuous wave (CW) signals, the probe's capacitive reactance at the operating frequency must be taken into account.

The total impedance, as seen at the probe tip, is designated R_P and is a function of frequency. In addition to the capacitive and resistive elements, designed-in inductance elements serve to offset the pure capacitive loading to some degree. Figure 4-22 shows a typical plot of X_P and R_P versus frequency. These curves can be derived from the formula:

$$X_P = \frac{1}{2\pi fC}$$

For example, a standard 10-megohm probe with a tip capacitance of 11 pF will have a capacitive reactance at 50 MHz of:

$$X_P = \frac{1}{(6.28)(50)(11) \times 10^{-6}}$$

$$= 289.52$$

or about 290 ohms. Depending on the source impedance, this loading could have a major effect on the signal amplitude and even on the operation of the device under test.

Another way of estimating the effect of probe tip capacitance on the rise time of a source is to take the ratio of probe tip capacitance to the known or estimated source capacitance.

$$\% \text{ change} = \frac{C_{\text{probe tip}}}{C_{\text{source}}}$$

Table 4-1 is a representation of probe tip capacitance as it relates to attenuation factor.

FREQUENCY ATTENUATION. Another effect of using coaxial cable for the input medium is *frequency attenuation.* Part of the attenuation can be attributed to dielectric absorption, but cable capacitance is largely responsible for the signal loss. The cable capacitance acts like a high-frequency shunt,

Attenuation Factor	Tip Capacitance
X1	54 pF
X10	11 pF
X100	2 pF

Table 4-1 Attenuation and Tip Capacitance

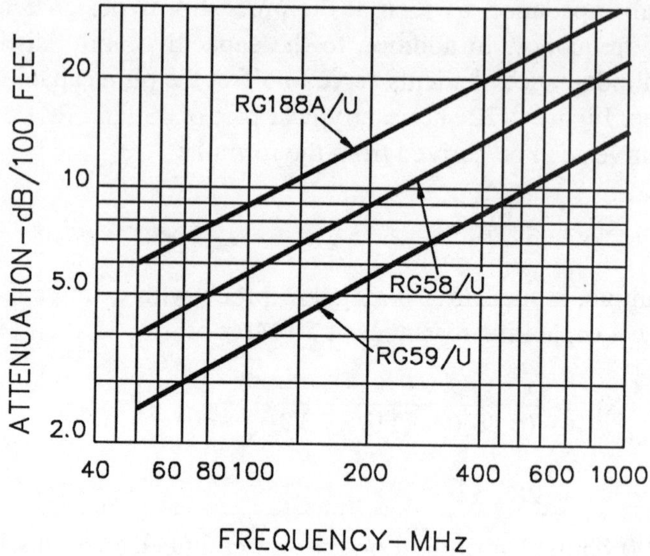

Figure 4-23 Signal attenuation through coaxial cable

progressively shunting more signal to ground as the input frequency increases and the capacitive reactance decreases.

To overcome the loss, a high-frequency boost is required. Ideally the boost would increase with the frequency in proportion to the loss sustained by the coaxial cable.

Fortunately, a series capacitor offers the perfect solution. As illustrated in Figure 4-21, the compensating capacitor C1 is placed in parallel with isolation resistor R1. Because the compensating capacitor is typically a very low value, 11.1 pF in this case, it has little effect on the lower frequencies, which must make their way through the isolation resistor. The higher frequencies, however, find the capacitive reactance path easier going, and bypass the isolation resistor. Since capacitive reactance is an inverse function of frequency, higher frequencies encounter less resistance than lower frequencies. By the time the high-frequency and low-frequency components find their way to the vertical input, however, the boosted high frequencies have been attenuated by the coaxial cable to the level of the low frequencies, and the waveform appears in its original form. Trimmer capacitor C_{comp} is used as an adjustment to offset the compensating effect of C1 for perfect signal reconstruction.

The attenuation of the signal that results from the capacitance acting alone can be calculated from:

$$V_i = V_S \frac{X_{C2}}{X_{C1} + X_{C2}}$$

$$V_1 = V_S \frac{1/\omega\, C2}{1/\omega\, C1 + 1/\omega\, C2}$$

$$= V_S \frac{C1}{C2 + C1}$$

where C2 is the lump sum of C_{comp}, C_{CC}, and C_i.

OSCILLOSCOPE PROBE BANDWIDTH. Like oscilloscopes, probes have bandwidth limitations. Bandwidth is the frequency cutoff point at which the input signal to the vertical amplifier is down by 3dB. Oscilloscope vertical amplifiers are designed for a Gaussian roll-off at the high end. The frequency response of a probe, on the other hand, is closely related to rise time response, and is expressed as:

$$BW\,(MHz) = \frac{350}{t_r\,(ns)}$$

It is important to note that the amplitude measurement is –3dB (30%) down in amplitude at the specified bandwidth limit. A 50-MHz oscilloscope, for example, is really only true to 35 MHz (70% of bandwidth), after which signal amplitude must be degraded accordingly.

Bandwidth is a function of rise time (t_r), which in turn is a function of input impedance. Basically, rise time is the amount of time it takes to charge the RC components of the test probe to the nominal voltage value. The resistive component is largely the probe input resistance, R1. Input capacitance is a combination of distributed coaxial cable capacitance C_{CC} and vertical amplifier input impedance C_i. As the input resistance increases, the time constant of the RC network also increases, thus reducing bandwidth.

Consequently, the value of the probe input resistor, R1, is a compromise between isolation characteristics, signal attenuation, and bandwidth. Typically a value of 9 M is selected for the series resistor (in a 10:1 probe) because it reduces input loading by an order of one magnitude while maintaining a respectable bandwidth.

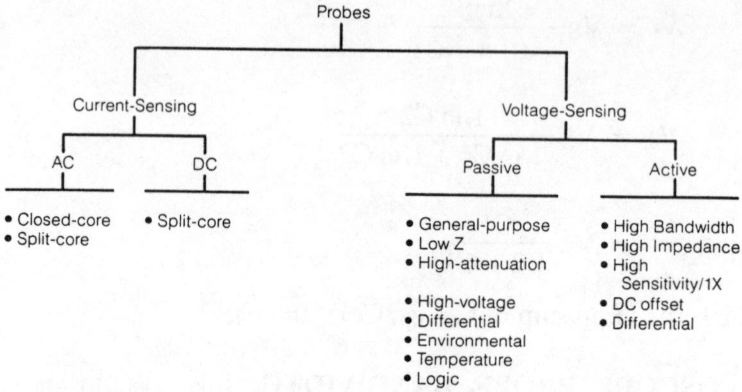

Figure 4-24 Basic probe types

Types of Oscilloscope Probes

Oscilloscope probes are generally divided into two types: voltage-sensing and current-sensing. Voltage-sensing probes, which are the more commonly used, are further divided into active and passive types. These two main types differ in their internal circuitry and mode of operation.

PASSIVE VOLTAGE-SENSING PROBES. Passive voltage-sensing probes are built using only passive circuit components, such as resistors, capacitors, and inductors. Passive probes contain no active components (transistors, etc.) and require no external power for operation. The 10X probe detailed in the discussion above is an example of a passive voltage-sensing probe.

Probe Type	Attenuation Factor
1X	no attenuation
10X	10-to-1
50X	50-to-1
100X	100-to-1
1000X	1000-to-1

Table 4-2 Probe attenuation factor

Passive voltage-sensing probes are further classified by their attenuation of signals. The attenuation is a factor of input voltage at the probe tip to the voltage seen at the vertical amplifier's input when working with the input impedance of the oscilloscope.

Attenuation in itself is neither good nor bad. High-attenuation probes are essential for measuring high-voltage signals while protecting the oscilloscope input circuits. Too much attenuation, on the other hand, may reduce a low-voltage signal to the point of being unmeasurable.

Passive probes have the advantage of ruggedness at a relatively low cost. They are also simpler to use than active probes and less prone to electrical damage caused by input overloads. Their main limitation is their lower input impedance. As the input frequency increases, the loading effects of a passive probe also increase. Isolating the effects of loading using a series input resistor creates additional problems by attenuating the input signal.

LOW Z PROBES. Low Z passive probes are specially designed for measuring very-high-frequency signals. The benefits of a Low Z probe (also called low capacitance probe) include constant and predictable loading through a wide range of frequencies, and a standard 50-ohm impedance matching output.

A Low Z passive probe offers a very low tip capacitance at the expense of relatively high resistive loading. A typical 10X Low Z probe has an input capacitance of about 1 pF and a resistive load of 500 ohms. The circuit diagram and electrical equivalent of a 10X Low Z probe is shown in Figure 4-26. In this configuration, the probe appears as a 10X resistive, high-frequency voltage divider. According to transmission line theory, the 50-ohm oscilloscope input represents a pure resistive load, with no C or L component. No frequency compensation is necessary because the network is not a capacitive divider.

Low Z probes are typically high-bandwidth devices with response up to 3.5 GHz and rise times to 100 ps. Unfortunately, the very low input impedance may affect the amplitude of the pulse by simple resistive divider action between the source and the load, making the Low Z probe better suited for rise time measurements than voltage measurements. Because of its resistive loading effects, Low Z probes perform best on devices under test with a source impedance of 50 ohms or lower.

The 50-ohm output impedance of the Low Z probe also provides the convenience of standard termination with transmission-line characteristics. This design cancels the effects of cable capacitance, thus allowing the use of longer cables with no affect on bandwidth. Note that these probes operate into 50-ohm inputs only, and are typically teamed with fast (500 MHz to 1

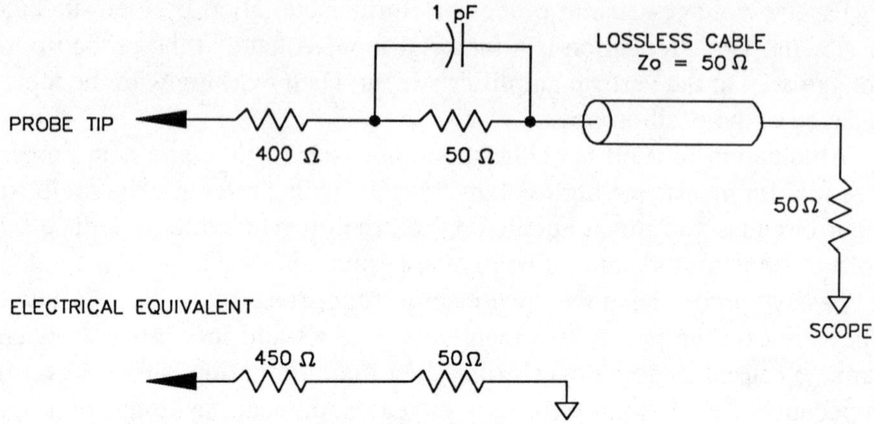

Figure 4-25 Schematic of Low Z probe

GHz) real-time oscilloscopes or with oscilloscopes employing sampling techniques (see Chapter 6).

DIFFERENTIAL PROBES. Signals that are not referenced to ground, such as the voltage drop across a collector resistor, require differential measurement techniques and differential vertical input amplifiers. Differential amplifiers measure only the voltage difference between two points and reject the unwanted common mode signal (a signal or dc voltage common to both test points).

To preserve the high common mode rejection ratio (CMRR) of the differential inputs, it is necessary to use differential sampling techniques. This extends to the type of oscilloscope probes used. Two methods for differential sampling exist.

The first requires the use of two identical test probes, one for each differential input. The two probes must be matched in length for equal transit-time transfer of the signal, otherwise false readings will be obtained. The loading factor (cable impedance) and attenuation of the cables must also be matched to within 1% to ensure maximum CMRR performance. Probes of this type are normally sold in sets as a matched pair.

The alternative is a single differential probe with two inputs. Probes of this type have two test points embedded in a single probe head. Because this design permits exact matching of probe parameters, it provides the best CMRR performance of any differential probe combination.

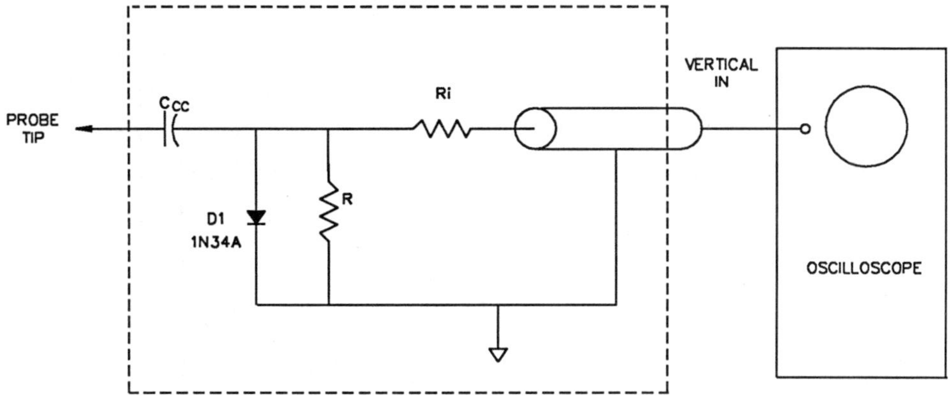

Figure 4-26 Demodulator probe

SPECIALTY PROBES. A special category of passive voltage-sensing probes include high-voltage (40 kV) probes and environmental probes, which operate over a very wide temperature range. Logic probes sense logic states in digital circuits. Temperature probes are used to measure temperature at the probe's tip. The demodulator probe is used for converting an RF signal into a dc voltage, with its output voltage directly proportional to the peak RF signal. This type of probe is primarily used for signal tracing and alignment of receivers.

ACTIVE VOLTAGE-SENSING PROBES. Active voltage-sensing probes are built with active circuit components, such as transistors or operational amplifiers. The active components form a buffer amplifier circuit that effectively isolates the device under test from the oscilloscope.

Modern active probes use field-effect transistors (FET) or FET input operational amplifiers for the probe's input stage. In fact, some active voltage-sensing probes are marketed as "FET probes." The amplifier circuit functions as a voltage follower with a gain of one. Because the amplifier provides unity gain, the probe experiences no signal attenuation.

Typical input impedance of an active probe is 10 M with 1- to 2-pF tip capacitance; the attenuation factor is 1X. Active probes are primarily used for high-frequency measurements of extremely low-level pulse signals. The active probe is also capable of driving a 50-ohm transmission line, making it a good alternative to the Low Z probe in high-frequency measurements.

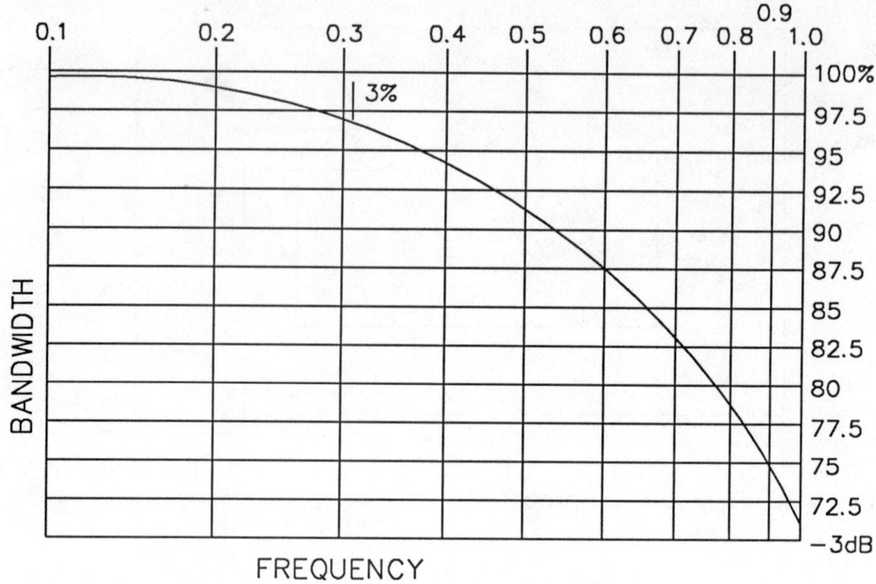

Figure 4-27 Active probe frequency response curve

Among the disadvantages of the active voltage-sensing probe are higher cost, a limited dynamic signal capability (compared to the range offered by a passive probe), and the need for a power supply. Power to drive the active probe can be derived from the oscilloscope, or the probe may contain its own power source.

Because it is an amplifier, the bandwidth specifications of the active voltage-sensing probe are different than those of a passive probe. In this respect, active probes are like oscilloscopes or other amplifiers that are ranked by bandwidth. When working with an active probe and oscilloscope, both with set bandwidths, the combination may not always be what it seems. A 100-MHz active probe coupled to a 100-MHz oscilloscope, for example, does not result in a 100-MHz bandwidth. Total bandwidth is the square root of the sum of the squares, or:

$$t_{r\ \mathrm{system}} = \sqrt{t_r^2\ \mathrm{probe} + t_r^2\ \mathrm{scope}}$$

where

$$t_r = \frac{350}{\mathrm{BW\ (MHz)}}$$

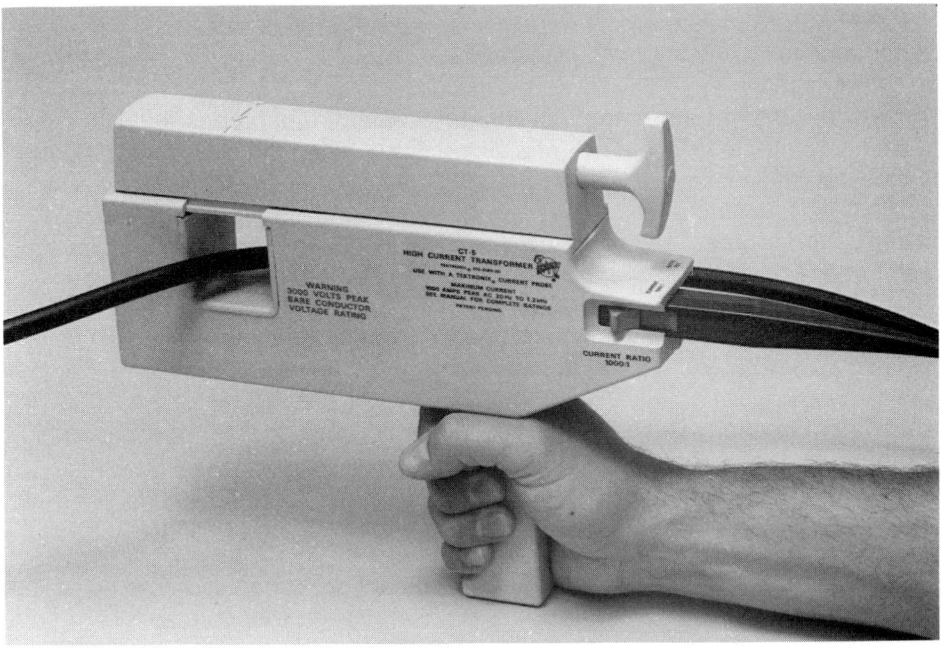

Figure 4-28 Split-core current probe

For convenience, the bandwidth is expressed in rise time with t_r in nanoseconds (ns). A 1-ns pulse at the probe tip of a 100-MHz active probe, for instance, would appear as 3.5 ns on a 100-MHz oscilloscope.

CURRENT-SENSING PROBES. Current-sensing probes use transformers (or a combination of transformers and Hall Effect devices) to convert flux fields to voltage signals. Flux fields are the electromagnetic fields that surround a wire when a current flows through the wire. The greater the current, the more intense is the flux field.

Current probes may be classified as either ac- or dc-responding. Ac current probes are basically transformers that sense the change in the flux field and convert it to a voltage. Most current-sensing probes take measurements without breaking connections to the device under test, by allowing part of the

GLOSSARY OF PROBE TERMS

Aberrations: A measurement system's response to a pulse input. Specified as a percentage deviation from a flat response.

Amp-second product: A specification used with current probes. It defines the maximum area under the curve of a single pulse, when the peak amplitude exceeds the continuous rating of the probe.

Attenuation factor: The ratio of a probe input signal amplitude to its output signal. For example: 10X is the attenuation factor of a 10-to-1 attenuation probe.

Bandwidth: A measurement of the system's highest useful operating frequency defined as 3 dB (0.707 volt) of the maximum signal amplitude.

Capacitive loading: The effect of probe tip capacitance on the circuit under test.

Circuit loading: The effect that a probe produces on the normal operation of a circuit under test; generally brought about by loss of oscillation, etc.

Duty cycle: The ratio of pulse width to signal period, expressed as a percentage.

Input attenuation: The decrease in signal amplitude resulting from the probe/scope attenuation ratio.

Input capacitance: The total capacitance at the probe tip.

Input resistance: The total probe and oscilloscope seen by the device under test.

Insertion impedance: The effect on a circuit of connecting a current probe within a circuit; includes both C and L influences.

Table 4-3 Glossary of probe terms

split transformer core to slide back and then clipping the probe over the wire carrying the current being measured. Some special-purpose, high-frequency current probes, however, require a wire to be threaded from the device under test through the closed core of the probe. Ac probes have both an upper- and lower-frequency roll-off (upper- and lower-frequency bandwidth) that must be taken into account when making measurements.

Maximum input current rating: A specification for CW current probes, and the maximum current the probe will accept safely. Maximum input current is derated with frequency.

Maximum peak pulse current: The maximum peak pulse current that a probe can safely accept, under any conditions .

Maximum voltage rating: The maximum voltage the prove can accept without power dissipation damage or breakdown in the cable connector.Maximum input voltage is derated with frequency.

Period: The time required for one cycle of a signal, if the signal repeats itself.

Phase shift: Results from adding inductive or capacitive elements to the signal path, resulting in retarding or advancing a time point on a waveform.

Probe compensation: The adjustment of probe capacitance so that the probe attenuates all input signals equally across a range of frequencies.

Propagation delay: The time, in picoseconds or nanoseconds, taken by a wavefront to travel a given distance in a conductor.

Sensitivity: A function of system gain stated in volts per division or amps per division.

Temperature rating: The temperature range over which the probe will operate within the published electrical specifications.

Time constant: The product of a circuit's resistance and capacitance. It usually determines the 3 dB roll-off point for ac-coupled devices.

Voltage/frequency derating: The maximum voltage a probe can safely handle at a given frequency.

Table 4-3 (continued) Glossary of probe terms

Dc current probes employ active circuit elements, such as Hall Effect devices, to generate the desired output. A Hall Effect device produces an output in the presence of magnetic fields, as opposed to the transformer's method of sensing only flux changes, making it an essential component of dc current probes. Dc current probes combine transformers and Hall Effect devices to convert the flux fields into voltages. The Hall Effect device provides the dc response in the dc current probe, while the transformer takes over at a predetermined crossover point to provide extended high-frequency response. Dc current probes, therefore, have a frequency response from dc up to some upper 3 dB roll-off point. dc current probes are typically of the split-core variety.

5. OSCILLOSCOPE MEASUREMENTS

The oscilloscope is basically a voltage device. The two most basic measurements made with an oscilloscope are amplitude and time; from these two quantities all other measurements are derived. Because the oscilloscope is a voltage-measuring device, voltage is displayed as vertical amplitude on the CRT screen, while time is displayed as horizontal displacement.

Before measurements can be made, however, one must be acquainted with the various oscilloscope controls. One must know how each control reacts and relates to the display, and which controls interact. The first section of this chapter explains basic control operation, followed by an in-depth look at measuring techniques.

Oscilloscope Controls

As outlined in the last chapter, the oscilloscope is divided into four basic sections: CRT display, vertical amplifiers, horizontal sweep generator, and trigger generator. Each of the four sections has a corresponding set of front panel controls. These controls are manipulated by the operator to produce the desired image on the screen, much in the same way that a color TV is adjusted for best picture.

The oscilloscope shown in Figure 5-2 made by Tektronix is a dual-channel, 60-MHz portable oscilloscope designed as an easy-to-use, lightweight, general-purpose instrument. The Tektronix 2213A is representative of the

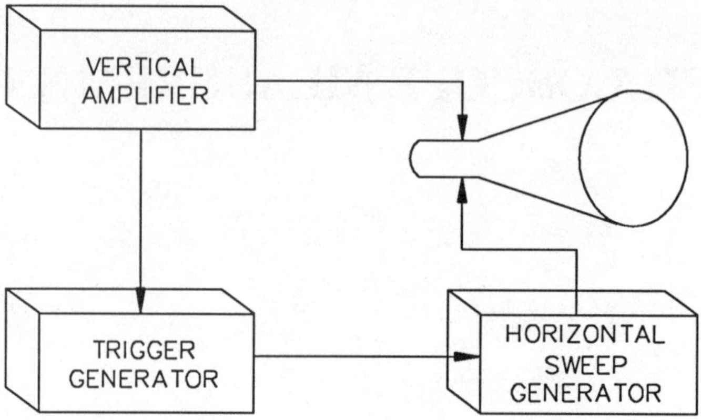

Figure 5-1 Oscilloscope block diagram

type of oscilloscope one is likely to encounter for general-purpose lab and field measurements.

The heart of the system is the CRT, which is located to the left of the instrument. Scribed on the face of the CRT is a *graticule*. The graticule is a grid of lines used for measuring the amplitude (in vertical divisions) and the time period (in horizontal divisions) of any displayed waveform. Typically,

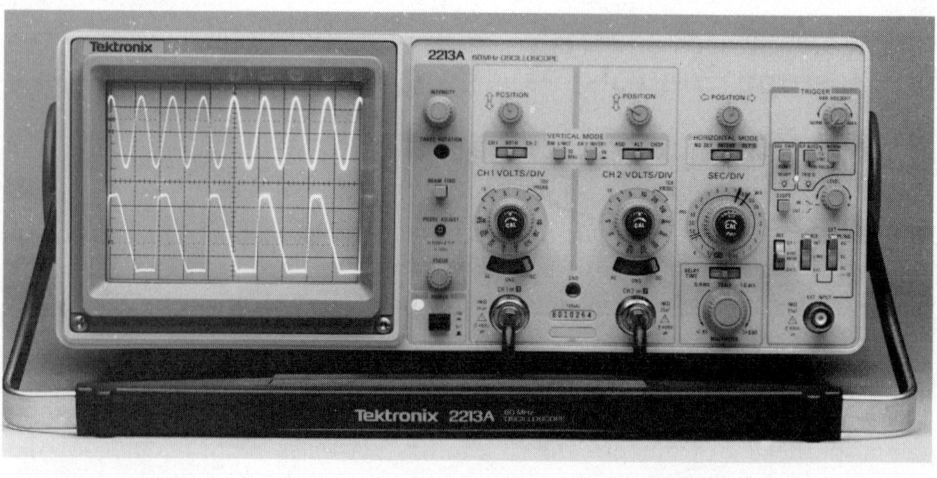

Figure 5-2 Tektronix 2213A

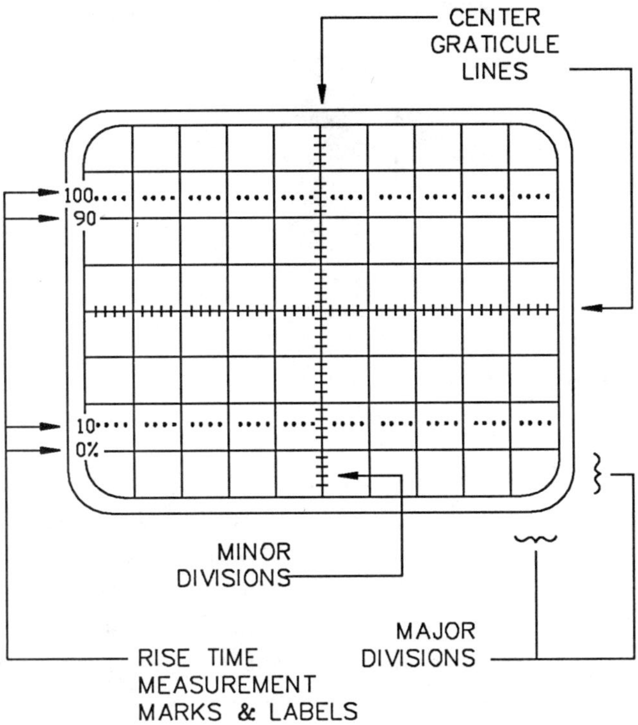

Figure 5-3 Graticule

the graticule is etched or silk-screened on the inside of the CRT faceplate. Putting the graticule inside — on the same plane as the trace drawn by the electron beam and not on the outside of the glass — eliminates measurement inaccuracies caused by *parallax* errors. Parallax errors occur when the trace and the graticule are on different planes, as is the case when the graticule is etched into a hard piece of plastic placed over the CRT face, and the observer is shifted slightly from the direct line of sight.

Though different-sized CRTs may be used, graticules are usually laid out in an 8 × 10 pattern. Each of the eight vertical and ten horizontal lines block off *major divisions* of the screen. The labels on the front panel controls always refer to *major divisions.* The tick marks on the center graticule lines represent minor divisions, or subdivisions. Sometimes the graticule may support other reference points and nomenclature, as is the case with the percentile marks on the left of the Tektronix CRT faceplate.

DISPLAY SYSTEM CONTROLS. Common controls for the display system include INTENSITY and FOCUS. Less often, you may find a BEAM FINDER and TRACE ROTATION control.

Figure 5-4 CRT system controls

The INTENSITY control adjusts the brightness of the trace for use in varying ambient light conditions and with different types of waveforms. When viewing a square wave, for instance, it may be beneficial to change the trace brightness to look at different parts of the waveform, because the slower horizontal components will appear brighter than the faster vertical components. The sweep rate also affects screen brightness. As different sweep speeds are selected, the beam's on-and-off times change and the electron beam has more or less time to excite the phosphor. On most scopes, intensity is a manual adjustment. With some instruments, however, an automatic brightness circuit compensates for changes in sweep speed.

The electron beam is focused on the CRT faceplate by an electron lens within the tube. The FOCUS control controls the voltage going to the electron lens's focusing anodes, which changes the focal length of the electron beam (see Chapter 4 for further discussion), and the sharpness of the dot on the screen. In many cases, the focus of the screen changes with respect to screen brightness.

The BEAM FINDER is a convenience control that allows you to locate the electron beam anytime it is off screen. When the BEAM FINDER button is pressed, the voltage to the horizontal and vertical deflection plates is reduced so that the beam always appears within the graticule. By seeing in which quadrant the beam appears, one can tell in which direction the image is off the screen and make the appropriate adjustments using the horizontal and vertical POSITION controls. The BEAM FINDER also overrides the INTENSITY control when engaged.

The TRACE ROTATION control is a potentiometer that electrically aligns the horizontal deflection of the trace parallel with the horizontal lines of the fixed graticule. The TRACE ROTATION control compensates for variations in the earth's magnetic field, which causes the horizontal scan line

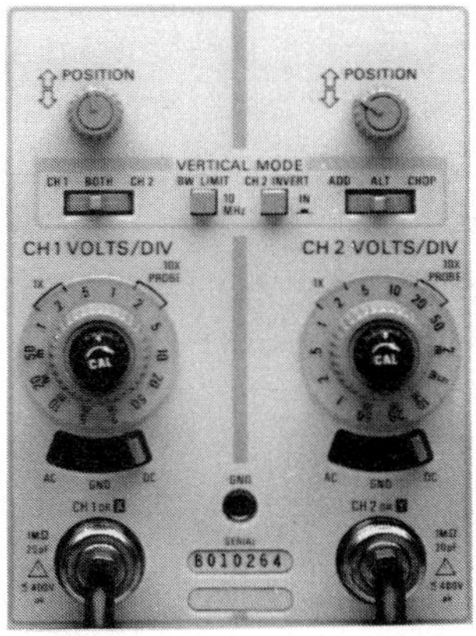

Figure 5-5 Vertical system controls

to deviate from its true path. To avoid accidental adjustment of this control while the scope is in use, this control is normally recessed behind the front panel and is accessible only by screwdriver.

VERTICAL SYSTEM CONTROLS. Panel controls commonly associated with the vertical input are vertical POSITION, input sensitivity, input coupling, and VERTICAL MODE. The 2213A supports two vertical channels, labeled CH1 and CH2. Some manufacturers label the two vertical inputs as CHANNEL A and CHANNEL B, which would correspond to the Tektronix CH1 and CH2 inputs, respectively. Each channel has its own individual set of panel controls, plus a small selection of controls shared by each channel.

The two vertical POSITION controls move their respective traces (CH1 or CH2) above and below the center reference line, according to their setting.

The input sensitivity of the vertical amplifier is determined by the large rotary switch below the POSITION controls. They are labeled CH1 VOLTS/DIV and CH2 VOLTS/DIV, representing channels 1 and 2, respectively. The VOLTS/DIV switch controls the attenuation factor of the vertical input amplifier and extends the input voltage range of the vertical amplifier from millivolts to several volts.

Using the VOLTS/DIV switch to change sensitivity also changes the *scale factor*, the value of each major division on the screen. Each setting of the control knob is marked with the scale factor for that channel. For example, with a setting of 5V (i.e., 5 divisions per volt), each of the eight vertical major divisions represents 5 volts, and the entire screen can show 40 volts from top to bottom. With a VOLTS/DIV setting of 2 mV, each major division represents 2 millivolts and the screen can display 16 millivolts from top to bottom.

The VOLTS/DIV calibration may be changed by rotating the vernier knob at the center of the VOLTS/DIV knob. The vernier control is labeled CAL and has a small arrow on it. The vernier control provides continuous volts per division adjustment, so that the display amplitude may be increased as desired. A variable sensitivity control is useful for making quick amplitude comparisons on a series of signals. You could, for example, take a known signal of almost any amplitude and use the CAL control to make the waveform fit exactly between major division graticule lines. Maintaining the same calibration, additional signals may be compared to see whether they have the same amplitude, and if not it is relatively easy to see whether the new input is less or greater than the original.

The input coupling switch located directly beneath the VOLTS/DIV knob determines the type of input coupling to the vertical amplifier. Dc coupling inputs the signal directly to the vertical amplifier. In the ac position, a capacitor is inserted in line with the input, thus blocking the dc component from affecting the vertical amplifier's input. The middle position of this switch is marked GND for ground. Choosing the GND position disconnects the vertical amplifier from the input connector and references the vertical input to ground, but does not short the circuit under test to ground. Switching from ac or dc to GND and back again is a handy way to measure signal voltage levels with respect to chassis ground. The vertical input connectors are directly below the input coupling switches.

The VERTICAL MODE switches, located directly below the POSITION knobs, select the vertical mode of operation for either single-trace or dual-trace operation. The left-most switch chooses the channel to be displayed. In the CH1 position, only channel 1 is displayed; in the CH2 position, only channel 2 is displayed. In the BOTH position both traces can be seen simultaneously. The right-most VERTICAL MODE switch selects the type of dual-trace display for the BOTH setting. CHOP is the chopped mode, ALT is alternate, and ADD is the sum mode, as outlined in Chapter 4. To make difference measurements, the CH2 INVERT button is engaged to invert the polarity of channel 2 before it is mixed with channel 1. When the CH2 INVERT switch is out, both channels have the same polarity.

HORIZONTAL SYSTEM CONTROLS. The horizontal system controls are located to the right of the vertical amplifier section. They include a horizontal POSITION, HORIZONTAL MODE, SEC/DIV, TIME delay and MULTIPLIER, and magnifier control.

Like the vertical POSITION control, the horizontal POSITION control moves the waveform on the screen. Instead of up or down, however, the horizontal POSITION adjustment moves the trace from side to side.

The horizontal SEC/DIV switch selects the rate at which the electron beam sweeps across the screen. Changing the setting of SEC/DIV permits you to look at longer or shorter intervals of the vertical input signal. Like the vertical system VOLTS/DIV switch, the SEC/DIV markings refer to the graticule scale factors. If the SEC/DIV setting is 1 ms, then each major division on the horizontal graticule line represents 1 ms of time, with a total screen time of 10 ms. Typical sweep rates range from 0.5 seconds to .05 μs (50 ns) per division.

The SEC/DIV control has two variable speed controllers associated with it. The vernier control knob located in the center of the SEC/DIV control provides continuous time/division adjustment of the sweep rate. Labeled CAL, this control slows the sweep speed by at least 2.5:1, making the slowest sweep 0.5 sec x 2.5, or 1.25 seconds. The second sweep setting involves the 10X magnifier. When the CAL knob is pulled out, the gain of the horizontal amplifier is increased by a factor of 10, thus providing a sweep rate that is ten times faster than the SEC/DIV setting indicates. The upper sweep limit for the Tektronix 2213A with the 10X magnifier engaged is a very fast 5 ns per division in the magnified sweep mode. Although the screen displays only a small portion (1/10) of the magnified waveform at any one time, different segments of the 10X sweep may be viewed by adjusting the horizontal POSITION control.

A feature found on most laboratory-quality instruments is delayed sweep. Delayed sweep is a technique that adds a precise amount of time delay between the trigger point and the beginning of a scope sweep. Delayed sweep is used as a method for displaying portions of a waveform not obtainable by other methods. When the signal to be measured is so far from the only available trigger point that it will not show on the screen, the sweep can be delayed by up to 20 ms in anticipation of the expected event. The amount of delay is

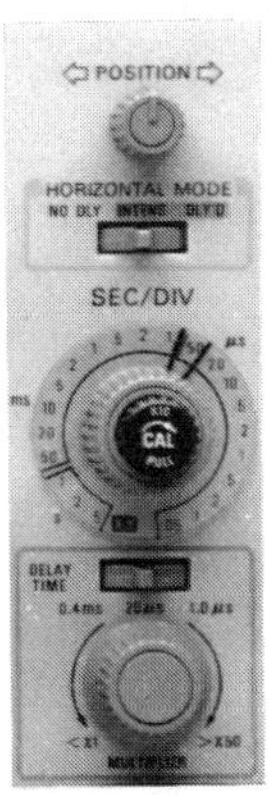

Figure 5-6 Horizontal system controls

determined by both the DELAY TIME switch, with its 1-μs, 20-μs, or 0.4-ms selection, and the DELAY TIME MULTIPLIER control. The multiplier increases the value of the DELAY TIME switch by as much as 50 times, giving a maximum delay time of 20 milliseconds.

The actual delay time can be read from the screen using the intensified sweep mode. Using a longer timebase than desired for the waveform, the sweep is triggered normally and the waveform displayed on the CRT screen. With the HORIZONTAL MODE switch in the INTENS position, a segment of the waveform is highlighted using the DELAY TIME function. The nonintensified portion of the trace indicates the amount of total delay time. The HORIZONTAL MODE switch is then switched to the delayed mode (DLY'D) and the sweep speed increased to magnify the highlighted portion of the waveform.

Sometimes an oscilloscope may contain two independent horizontal sweep generators called dual-time-base generators. In dual-time-base instruments, one sweep is triggered in the normal fashion, and the start of the second sweep is delayed with a delayed-sweep circuit. To distinguish these two sweeps, the *delaying sweep* is called the A sweep and the *delayed sweep* is called the B sweep. The length of time between the start of the A sweep and the start of the B sweep is called the *delay time*.

The dual-time-base delayed-sweep mode is engaged by the HORIZONTAL MODE ALT switch (not shown in Figure 5-4). When in the ALT mode, the oscilloscope displays the A sweep intensified by the B sweep and the B sweep delayed. As the B SEC/DIV switch is set faster, the intensified zone on the A trace gets smaller, and the B sweep expands by the new speed setting. Moving the B DELAY TIME POSITION dial and changing the point at which the B sweep starts causes the intensified zone to move across the A trace as the B waveform changes. This sounds more complicated in words than it is in practice, and a few minutes of hands-on experience quickly clears the air.

TRIGGERING SYSTEM CONTROLS. Because of its complex nature, the trigger system offers a wide variety of trigger controls. The controls are located to the far right of the front panel. They include the variable trigger holdoff (VAR HOLDOFF), MODE, SLOPE, LEVEL, and SOURCE.

The trigger sources are grouped into two categories that depend on whether the trigger signal is provided internally or externally. The source has no influence on how the trigger circuit operates, but internal triggering usually means the sweep triggers from the same source that is being displayed. Two switches on the front panel, labeled SOURCE and INT, determine the trigger source. Do not confuse the INT position of the SOURCE lever with the INT

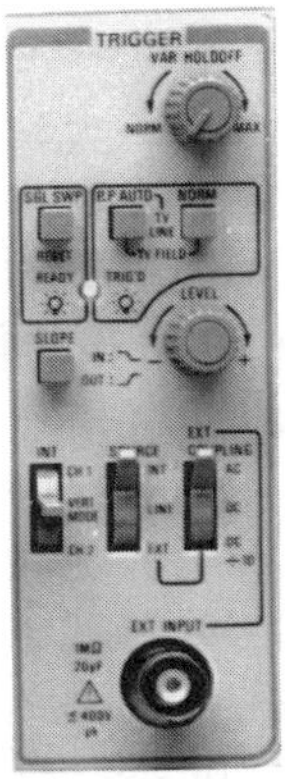

Figure 5-7 Trigger system controls

selector switch located to its left. The internal triggering sources are enabled only when the SOURCE lever is moved to the INT position.

The internal channel to be triggered on is now selected with the INT switch. The CH1 position of the INT lever internally routes the output of the channel 1 preamplifier to the triggering circuit. The sweep may now be triggered by the signal input to CH1. Likewise, the CH2 position uses the channel 2 waveform for its trigger.

The VERTICAL MODE setting of the INT switch obtains its signal from the VERTICAL MODE switches located in the horizontal sweep system of the front panel (see Figure 5-6). This mode of operation is used to select between the chopped or alternate modes of dual-trace display. If the horizontal VERTICAL MODE switch is set at CH1, then only the signal on channel 1 can trigger the sweep; in the CH2 position, channel 2 initiates the sweep. In the alternate (ALT) mode, the trigger generator looks for triggers alternately on the two channels. If the vertical mode is ADD or inverted ADD (difference mode), then the algebraic sum of CH1 + CH2 is the triggering signal. To prevent erratic triggering of the sweep generator when in the CHOP mode, the trigger generator takes its trigger signal from the algebraic sum (ADD) of the vertical inputs.

EXTERNAL TRIGGERS. Because triggering on the vertical input signal or signals is not always desirable, an external triggering option is also available. To use an external trigger, set the SOURCE switch to the EXT position and connect the triggering signal to the front panel connector marked EXT INPUT. Occasions when external triggering is used often occurs in digital design and repair where several events are synchronized with a master clock.

External input trigger signals may be ac or dc coupled, depending on the setting of the COUPLING switch. In the DC position, all elements of the triggering signal (both ac and dc) are input to the trigger generator. The DC÷10 position reduces the external trigger signal by a factor of ten. To isolate the dc component from the trigger source, a capacitor is inserted in the input line when the COUPLING lever is in the AC position.

The LINE trigger input is unique in that it receives its trigger source from the 60-Hz ac line. This mode is appropriate when the waveform conforms to ac line timing. Examples include devices like power supplies and power controllers.

HOLDOFF PERIOD. The LEVEL and SLOPE controls define the trigger point. The SLOPE control determines whether the trigger point is found on the rising (positive-going) or the falling (negative-going) edge of the input signal. The LEVEL control specifies the voltage at which the trigger point occurs. The LEVEL control voltage is continuously variable from negative, through zero, to positive, allowing a wide range of triggering values, both positive and negative.

Not every trigger event can be accepted as a trigger. The trigger generator does not recognize a trigger during the sweep or the retrace, and for a short time thereafter called the *holdoff period*. The holdoff period provides additional time beyond the retrace to ensure a stable display.

Sometimes, however, the normal holdoff period is not long enough to sustain a stable display. This possibility exists when the trigger signal is a complex waveform with many possible trigger points on it. Although the waveform is repetitive, the trigger generator may mistake portions of one waveform for the beginning of the next. Digital pulse trains are a good example of such a situation. Because one pulse is very much like another, there

Trigger Source	Trigger Switch Position		Vertical Switches
	SOURCE	VERTICAL MODE INT	MODE
Channel 1 only	INT	CH1	don't care
Channel 2 only	INT	CH2	don't care
Channel 1 only	INT	VERT MODE	CH1
Channel 2 only	INT	VERT MODE	CH2
Alternate CH1/CH2	INT	VERT MODE	BOTH/ALT
CH1 + CH2	INT	VERT MODE	ADD
CH1 – CH2	INT	VERT MODE	ADD/CH2 INVERT
External	EXT	don't care	don't care
Line	LINE	don't care	don't care

Table 5-1 Possible trigger combinations

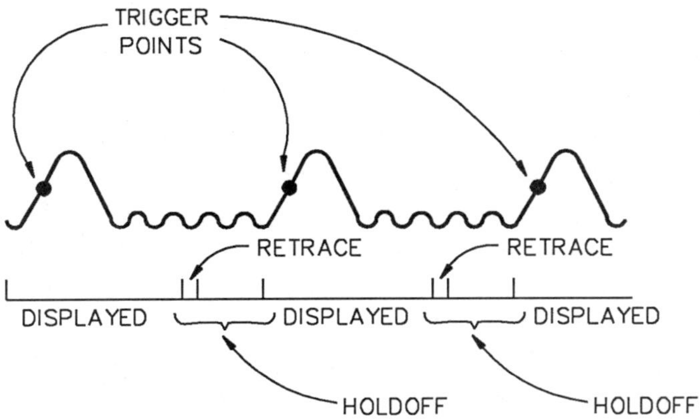

Figure 5-8 Trigger holdoff time

is a good likelihood the trigger generator could mistakenly take its cue from the wrong point.

The variable holdoff control, labeled VAR HOLDOFF on the front panel, allows the trigger generator to ignore false triggering points by extending the holdoff period. To operate, the VAR HOLDOFF knob is simply rotated until a stable image appears on the screen.

TRIGGER OPERATING MODES. All commercial oscilloscopes can operate in one of three triggering modes: automatic, normal, and television mode. Automatic triggering is the most frequently used.

In the automatic mode, labeled P-P AUTO on the front panel, a trigger signal is generated whenever a signal matching the requirements of SLOPE, LEVEL, and VAR HOLDOFF is encountered. The automatic mode is a signal-seeking mode that deals with percentages rather than absolutes. When the LEVEL control is set halfway between zero (middle position) and +, for example, the auto mode triggers on a positive voltage that is equal to approximately one-half the peak voltage in value. This mode of operation is convenient for viewing a wide array of waveshapes with changing voltage amplitudes, without requiring constant adjustment of the LEVEL control.

While it is most unlikely that you could set the trigger LEVEL control outside the signal range of the auto mode, the auto sweep can be stymied by the absence of an input signal. To eliminate the inconvenience (and possible confusion) brought about by a blank screen, a timer is started after each sweep period prior to the triggering of the trigger generator. If another trigger is not found before the timer expires, the trigger generator takes it upon itself to force a sweep by manufacturing its own trigger signal. This causes a straight baseline to appear on the screen, a process that remains in effect until another trigger source is detected.

The normal mode (identified as NORM on the front panel), establishes an absolute trigger reference level for the sweep. In this mode, only those waveforms that contain the proper trigger requirements are displayed. The normal mode does not permit a trace to be drawn on the screen if there is no trigger. The normal mode also provides the widest range of triggering signals, ranging from dc to 60 MHz (full bandwidth) in the case of the Tektronix 2213A.

When servicing video monitors (including monitors for computer terminals and workstations) or televisions, the television triggering mode is most useful. Most oscilloscopes with this mode enable automatic triggering on video fields at sweeps of 100 μs/division and slower, and video scan lines of 50 μs/division or faster.

Probe Adjustment

Before an oscilloscope measurement can be made, the frequency compensation capacitor inside the scope probe must be adjusted. Probe adjustment is done using a square wave since it contains the largest range of low to high frequencies available in a single waveform. The lowest frequency determines the oscillation period, while the highest defines the sharpness of the corners. Most quality oscilloscopes provide an internal square-wave calibrator on the front panel for such purposes. On the Tektronix Model 2213A it is located between the BEAM FINDER and the FOCUS controls.

With the probe attached to the voltage PROBE ADJUST output, adjust the trimmer capacitor, located behind the screw hole in the head of the probe, until the waveform appears most square. What you are actually looking for is definition of the leading edge. If the probe is undercompensated — excessive attenuation of high frequencies — the corner appears rounded. Too much compensation results in signal overshoot, which may lead to ringing. Figure 5-9 illustrates the proper adjustment procedure.

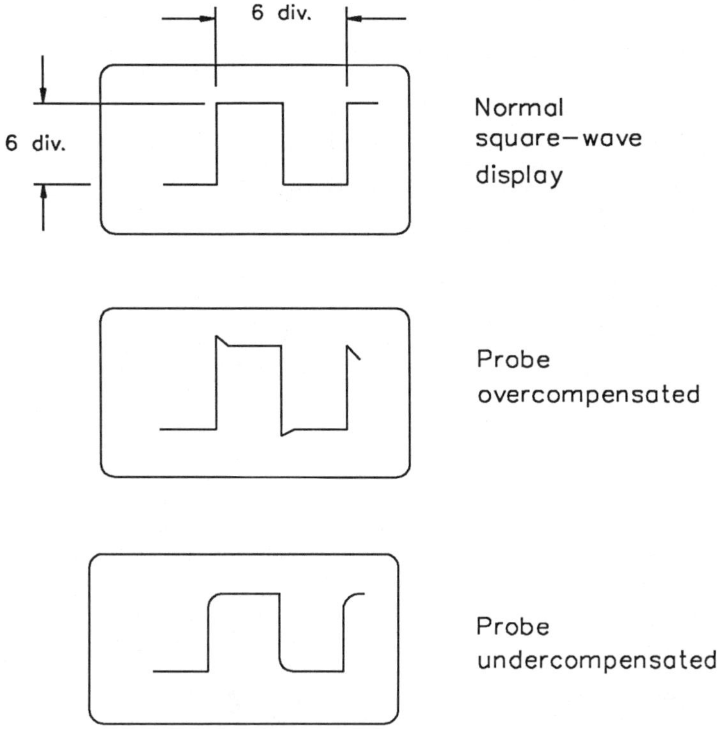

Figure 5-9 Test probe calibration

Ac Voltage Measurements

As can be expected from such a versatile instrument, using an oscilloscope to make voltage measurements is somewhat more complicated than using an analog or digital meter. There are no fixed numbers or pointers to use; all voltage values are interpreted from the screen according to size and distance. To assist in determining distances on the face of the screen, the CRT uses a graticule.

Because the CRT always displays the full waveform, *peak-to-peak* voltage measurements are the easiest to make. Peak-to-peak voltage is obtained by first stabilizing the waveform using the vertical, horizontal, and triggering front panel controls then measuring the vertical displacement of the electron beam.

To illustrate, consider the waveform in Figure 5-10. In this example, the screen displays a typical sinusoidal waveform with a peak-to-peak amplitude

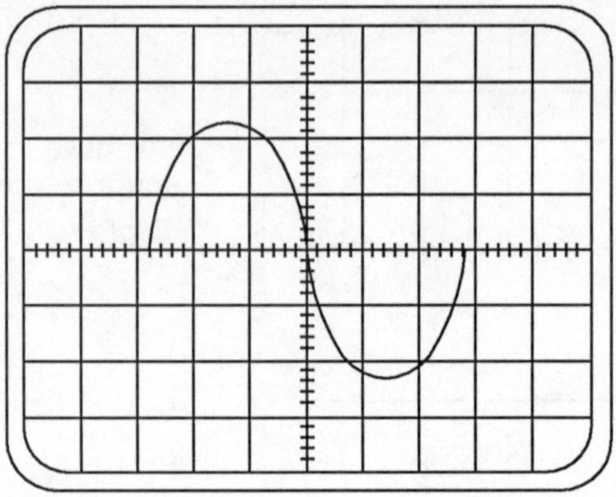

Figure 5-10 Sine-wave display

of 4.8 volts. Peak-to-peak amplitude is determined by counting the number of major and minor divisions the waveform spans top to bottom as the electron beam traces the waveform on the screen. Major divisions are counted as wholes, while minor divisions are parts of a whole. In the above example, the trace occupies four major divisions and four minor divisions (with each minor division representing 2 subdivisions), for a total count of 4.8.

The peak-to-peak voltage of the signal is the vertical amplifier sensitivity in VOLTS/DIV times the number of vertical divisions times the probe attenuation. With a vertical sensitivity of 0.1 volt per division and a probe attenuation of 10:1, the peak-to-peak voltage of the sinusoidal signal shown in Figure 5-10 is:

$$V_{p\text{-}p} = \text{No. of Div.} \times \text{VOLTS/DIV} \times \text{Probe Attenuation}$$

$$= (4.8)(0.1)(10)$$

$$= 4.8 \text{ volts (peak-to-peak)}$$

Bear in mind that this is a peak-to-peak voltage measurement.

DERIVED VOLTAGE MEASUREMENTS. Peak-to-peak values have little practical use, however. More practical values are root-mean-square (rms) and average voltage. While rms and average values cannot be displayed directly on an oscilloscope screen, they can be derived from formula. Derived voltage measurements are the result of calculations made after direct measurement.

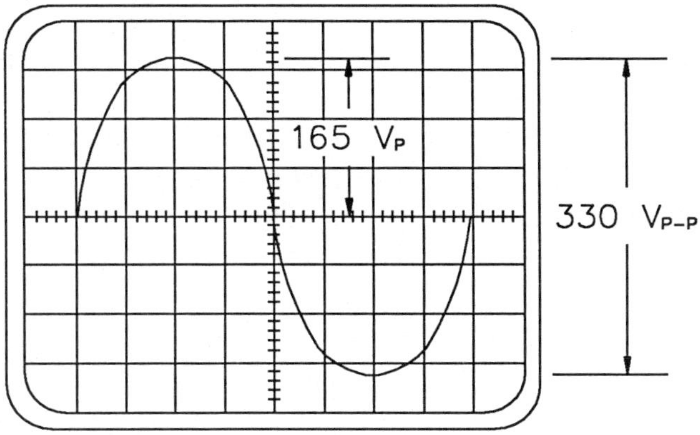

Figure 5-11 117-volt sine wave

Take the example in Figure 5-11, a typical sinusoidal waveform representative of 117-volt ac utility power. The waveform measures 6.6 divisions from peak to peak. With a vertical scale factor of 5 volts per division and a probe attenuation of 10X, the peak-to-peak voltage is:

$$V_{p\text{-}p} = \text{No. of Div.} \times \text{VOLTS/DIV} \times \text{Probe Attenuation}$$

$$= (6.6)(5)(10)$$

$$= 330 \text{ volts (peak-to-peak)}$$

The *peak voltage* of the waveform is half its peak-to-peak value, or:

$$V_p = V_{p\text{-}p} / 2$$

$$= 330 / 2$$

$$= 165 \text{ volts (peak)}$$

Peak voltage is also called the *maximum value* and is often important when determining peak inverse voltage (PIV) ratings for rectifier diodes.

The *average voltage* value is the total area under the peak voltage curve divided by the period in radians, or approximately:

$$V_{ave} = 0.318 \times V_p$$

$$= 0.318(165)$$

$$= 52.5 \text{ volts (average)}$$

When considering the sine wave as a whole, the average value is 0, because the positive and negative average voltage values are equal and therefore cancel.

The *root-mean-square* (rms) voltage is the effective value of ac voltage that has the same heating effect as a dc voltage. Rms voltage is equal to the peak value divided by the square root of 2:

$$V_{rms} = V_p / \sqrt{2}$$

$$= (165) / (1.414)$$

$$= 117 \text{ volts (rms)}$$

Rms may also be calculated directly from peak-to-peak voltage, using $V_{p\text{-}p}/2\sqrt{2}$.

Dc Voltage Measurements

Dc voltages can also be measured on an oscilloscope using the techniques already described. With the dc signal connected to the vertical amplifier input, set the input coupling switch to the GND position and center the trace on the screen. Move the input coupling switch to the dc position and note the trace deflection. If the trace deflects upward, the dc voltage is positive; if it moves downward, it is negative. The dc voltage is measured by multiplying the amount of vertical deflection by the VOLTS/DIV setting and the probe attenuation factor.

$$V_{dc} = \text{trace deflection (div.)} \times \text{VOLTS/DIV} \times \text{probe atten.}$$

It may be necessary to reset the input coupling switch to GND and reposition the baseline up or down (depending on signal polarity) so that the CRT screen can display the full trace displacement.

Frequently, the dc voltage will have an ac component riding on the dc level. A typical situation is the output of a power supply, a signal that normally consists of an ac ripple voltage superimposed on the dc output.

To measure signals with both ac and dc components, set the input coupling mode to AC and measure the peak-to-peak ac voltage. The vertical sensitivity should be set for the ac voltage to occupy a majority of the screen for this measurement. Next, set the input coupling switch to GND and position the baseline for standard dc voltage measurements (up or down, according to

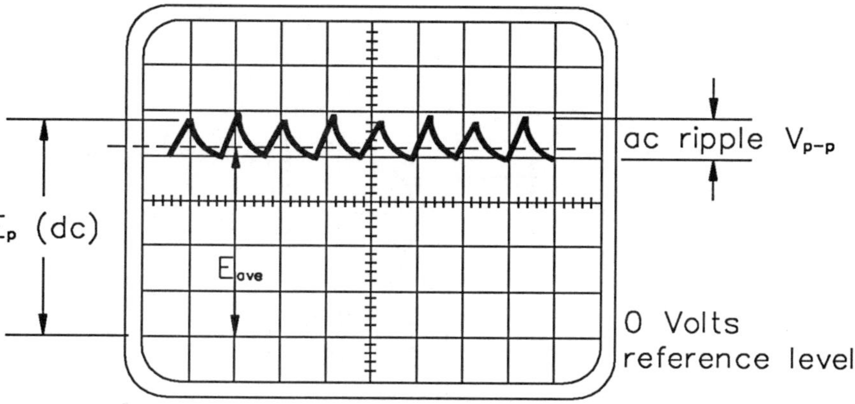

Figure 5-12 Ac ripple on dc output

dc input polarity). Set the input coupling switch to DC and adjust the vertical sensitivity to display the dc voltage trace on the screen. Measure the displacement from the base reference line and multiply by the scale factors.

Unlike meters, which are easily damaged by out-of-range input voltages, the oscilloscope is very forgiving. Unless the input signal exceeds the breakdown voltage of the vertical amplifier or test probe, an improper attenuator setting simply forces the trace off the screen with no damage done.

Time Period/Frequency Measurements

Time period and frequency measurements are made along the horizontal axis of the CRT screen. The time period of a signal is determined by measuring the time for one cycle of the waveform in horizontal divisions and multiplying by the setting of the horizontal SEC/DIV control. Signal frequency is then calculated as the inverse of the time period, or:

$$f = \frac{1}{t}$$

where f is in hertz (Hz) and t is in seconds.

The procedure is as follows. The signal to be measured is input to the vertical amplifier, and the front panel controls are adjusted to display at least one complete cycle covering a large part of the CRT screen. The time period for the waveform is found by counting the number of major and minor divisions

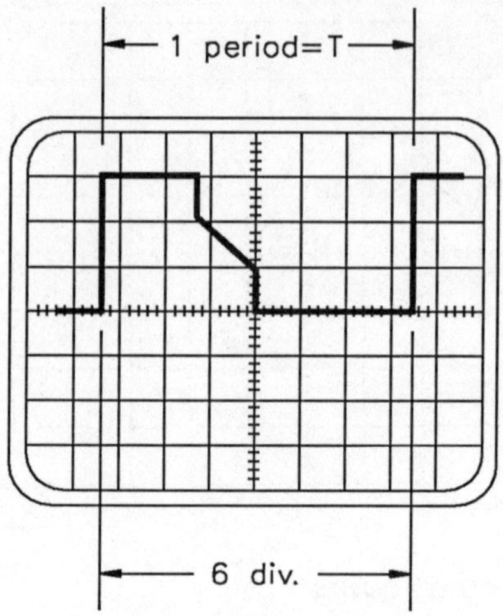

Figure 5-13 Measuring time/frequency

along the horizontal axis between the beginning and end of the cycle, and multiplying that value by the scale factor. The formula is:

$$t = \text{number of divisions / cycle} \times \text{SEC/DIV}$$

Inspect the waveform in Figure 5-13. In this example, the waveform is found to be 6 divisions long, with a sweep speed of 10 μs per division.

Substituting these values in the above equation yields:

$$t = (6)(10\ \mu s)$$

$$= 60\ \mu s$$

This is the time period of the waveform. To convert this quantity into frequency, take the reciprocal of t:

$$f = \frac{1}{t}$$

$$= \frac{1}{60 \times 10^{-6}}$$

$$= 16{,}667\ \text{Hz}$$

Frequency is, obviously, a derived measurement. Other derived measurements include duty cycle, duty factor, and repetition rate.

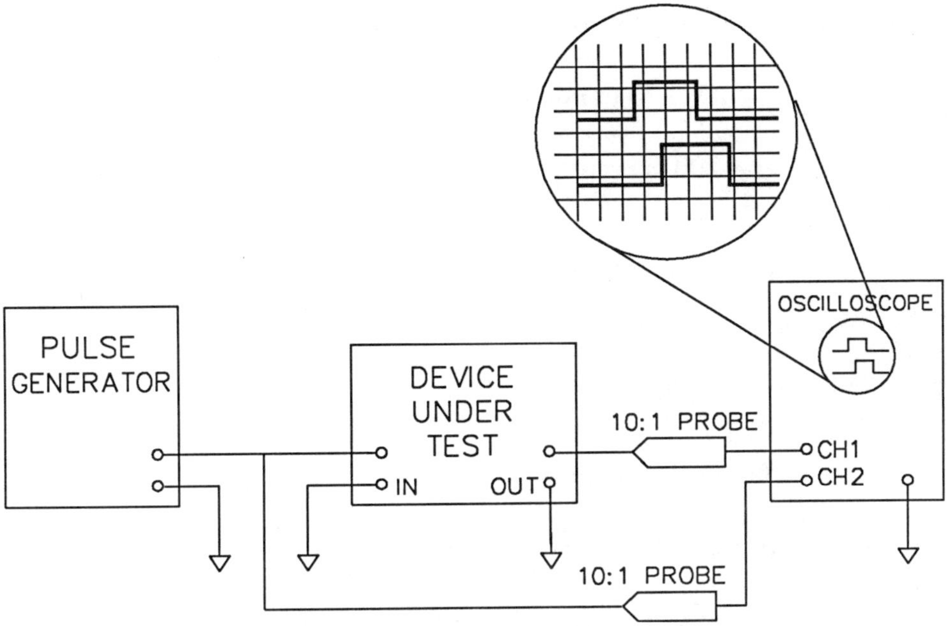

Figure 5-14 Test setup for measuring time/phase

Time/Phase Measurements

The time/phase relationship between two waveforms may be directly measured from the oscilloscope screen by viewing both waveforms on a dual-trace oscilloscope and determining the delay time between them. Measurements of this type are often encountered when working with digital clock frequencies, as illustrated in Figure 5-14.

In the above example, the reference clock has been placed on channel A and the clock under test on channel B (or channels CH1 and CH2, respectively). Using the front panel controls associated with each channel, the oscilloscope is set to alternate mode (for simultaneous viewing) and the two waveforms are adjusted to comfortably fill the screen. Since the situation requires triggering on the rising edge of the reference clock, the trigger generator is set to trigger on channel A only.

The time delay t_d between waveform A and waveform B is now read directly from the horizontal graticule. Returning to Figure 5-14, we find time period t to be:

$$t = \text{horiz. div. between waveform ends} \times \text{SEC / DIV}$$

$$= (10)(20 \text{ ns / div})$$

$$= 200 \text{ ns}$$

The delay between the two signals is found as:

$$t_d = \text{horiz. div. between A and B} \times \text{SEC / DIV}$$

$$= (2.5)(20 \text{ ns / div})$$

$$= 50 \text{ ns}$$

To convert time into phase angle, use the formula:

$$\text{phase delay } (\phi) = \frac{t_d}{t} \times 360$$

Substituting the values from Figure 5-14 into the equation, we calculate the phase shift to be:

$$\text{phase delay } (\phi) = \frac{50}{200} \times 360$$

$$\phi = 90 \text{ degrees}$$

SUM AND DIFFERENCE. An interesting aspect of dual inputs occurs when the two vertical signals are algebraically added together. In this mode, the outputs of the vertical preamplifiers are combined together in a *mixer circuit* instead of being selected individually by the electronic switch. The phase shift of the two signals within the mixer circuit is precisely controlled so that predictable results are obtained.

In the sum (**ADD**) mode, the two vertical input signals are mixed together as they emerge from the vertical preamplifiers. When two waveforms of identical phase are mixed in this manner, they combine to complement each other. Two sine waves of the exact same frequency and phase content, for example, reinforce to produce an output that has twice as much voltage swing as either has individually. (Refer to Figure 5-15.) The sum mode may be used to display vertical input signals otherwise too small for regular viewing by paralleling the two preamplifiers and using the "amplified" result to drive the deflection plates.

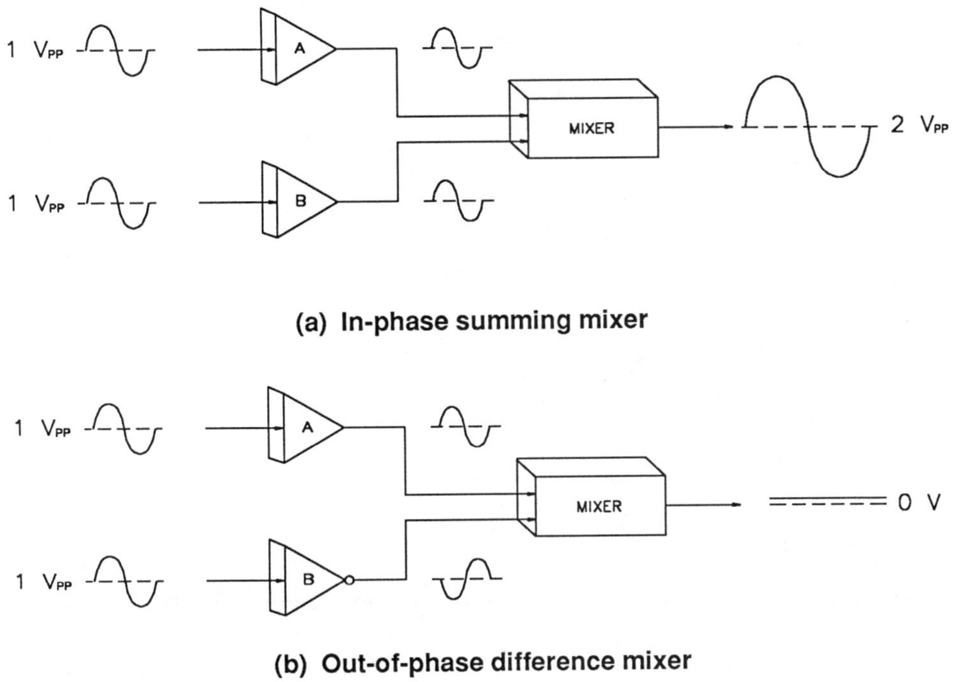

Figure 5-15 Summing and difference mixers

The difference (**SUB**) mode mixes the two vertical signals 180 degrees out of phase by first passing the channel B signal through an inverting amplifier. In this mode, identical voltages cancel, as Figure 5-15 shows. If identical sine waves are applied to the channel A and channel B inputs, the voltages oppose each other, the net result being a straight line across the center of the screen.

The difference mode is most useful for detecting small changes in a processed signal, as would be the case when comparing the input of an amplifier to its output. Because all in-phase components cancel themselves (after the appropriate gain adjustment), distortion and phase-shift anomalies become readily apparent and are the only images seen on the screen.

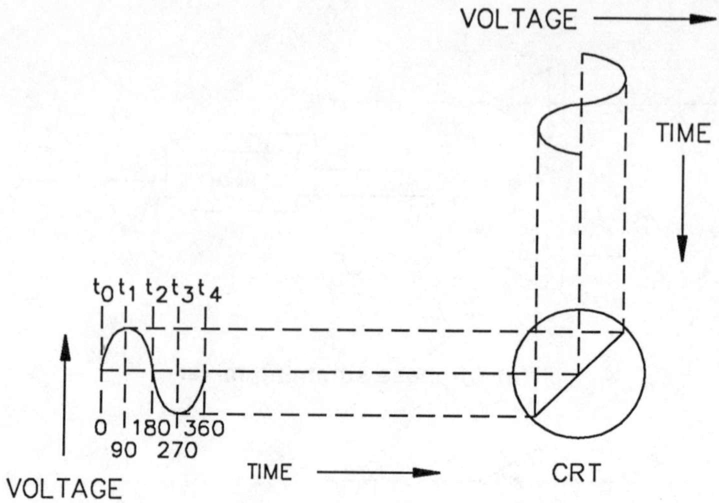

Figure 5-16 Nonlinear sweep

Lissajous Figures

A more accurate way to measure frequency and phase is to utilize the oscilloscope as a frequency comparator using a process that produces *Lissajous figures*. Up to now, the discussion has centered on linear sweeps: deflection voltages intended to move the electron beam across the face of the CRT at a uniform pace. While linear sweeps are most often used for oscilloscope measurements, nonlinear sweeps play an important role in oscilloscope flexibility.

The Lissajous figure is but one example of a nonlinear sweep. Unlike the linear sweep, which uses a sawtooth waveform input to the horizontal deflection plates, the Lissajous sweep is sinusoidal, a characteristic that causes the beam to travel at different speeds over different portions of the screen.

Nonlinear sweeps are obtained by setting the horizontal section controls of the oscilloscope to the X-Y mode and applying a sine wave to the CH2 input of a dual-trace scope or the EXT INPUT of a single-trace scope. In this mode, both the trigger system and horizontal system controls are disabled. Gain of the Y-axis input is controlled by the CH2 VOLTS/DIV selector switch.

When a sine wave is also presented to the vertical input, it produces a Lissajous pattern. Figure 5-16 shows the result of one sine wave sweeping another sine wave.

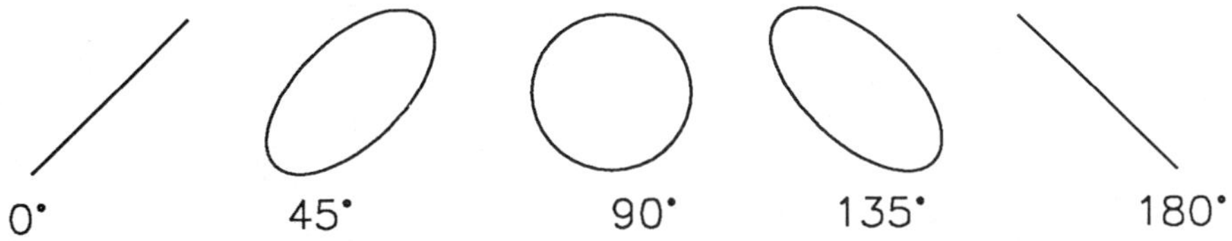

Figure 5-17 Phases of 1:1 Lissajous pattern

Starting at t_0 (the 0 degree point on the sweep cycle), the electron beam is deflected to the right of the screen as the signal voltage increases to its maximum value at t_1 (90 degrees). Simultaneously, the vertical input voltage is increasing to its peak value, forcing the electron beam upward. As the sweep progresses from t_1 (90 degrees) to t_2 (180 degrees), the amplitude of both waveforms decreases from their peak positive values to zero, retracing and repeating the values of the first quarter and returning the electron beam to the center of the screen. Continuing from t_2 (180 degrees) to t_3 (270 degrees) sees the waveforms approaching their peak negative values, with the subsequent deflection of the electron beam down and to the left. The final portion of the sweep, t_3 (270 degrees) to t_4 (0 degrees), retraces the path of the third quarter, bringing the electron beam back to its starting point.

Notice that throughout the sweep cycle the beam is always visible with no retrace period, thus forming a continuous trace. The oscilloscope displays it as a single line. As the phase angle between the two signals changes, though, the line opens up — first to an ellipse and finally into a circle. At this point, the two sine-wave signals are 90 degrees out of phase. Further phase advancement draws the circle back together again until, finally at 180 degrees, a single line (perpendicular to the first) appears.

This example is the simplest of the Lissajous figures, and it represents a frequency ratio of 1:1. Should the sweep rate be doubled and the input frequency maintained, only one-half the input waveform is traced. Because it is a closed figure, however, the resultant pattern contains two loops along the vertical axis and one along the horizontal axis, as demonstrated in Figure 5-18.

This configuration is expressed as a 1:2 ratio. Notice that the vertical frequency is recorded first and the horizontal frequency second. Similarly, a ratio of 1:3 produces three loops along the vertical axis and one along the horizontal, as shown in Figure 5-19. Ratios of 1:4 and 1:5 generate four and five loops, respectively.

When the sweep rate is less than the vertical input frequency, the loops are displayed along the horizontal axis, as shown in Figure 5-20. In this ex-

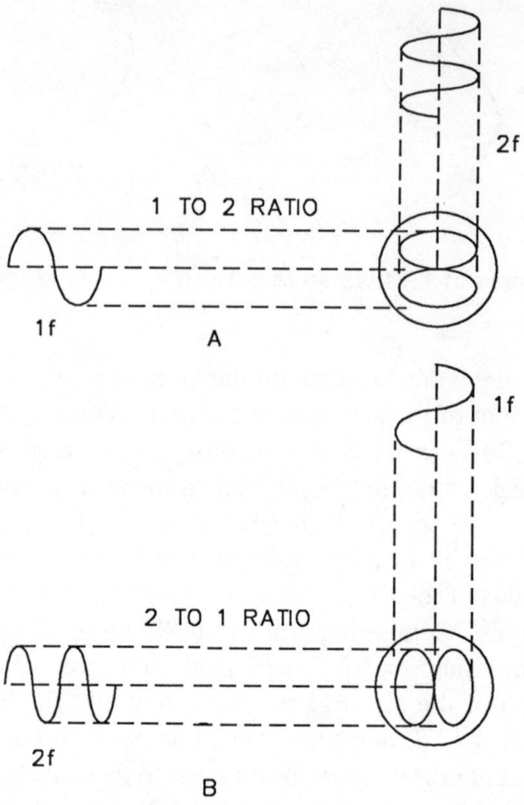

Figure 5-18 Simplex Lissajous figures

ample, the sweep rate is one-half the vertical input frequency, thus producing two full traces along the horizontal axis and one along the vertical axis. The ratio is expressed as 2:1.

The frequency ratio is determined by the number of loops in the Lissajous pattern. The representations of the Lissajous figures shown in Figures 5-19 and 5-20 should convey the general concept involved in comparing an unknown frequency to the reference frequency. Notice that when the frequency ratio exceeds unity, the number of horizontal loops exceeds the vertical loops. When the frequency ratio is less than unity, the vertical loops exceed the horizontal.

For example, it is easy enough to see that if there are five loops along the horizontal axis, the frequency ratio is 5:1. A simple ratio conversion is all that is necessary to derive the unknown frequency. By knowing that the horizontal frequency is 1 kHz, for example, one can deduce that the vertical input signal must be 5 kHz.

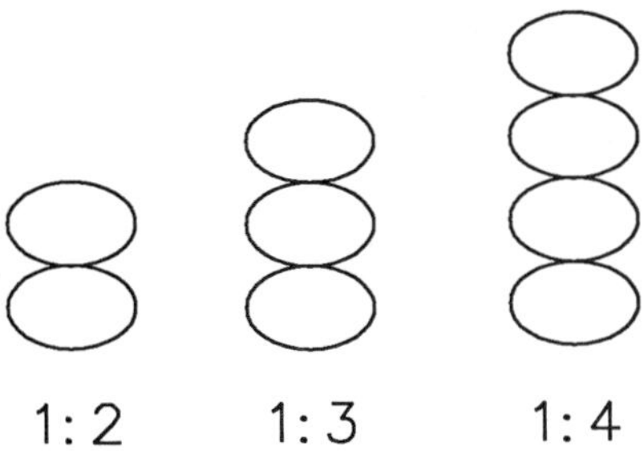

Figure 5-19 Multiple Lissajous figures

Because the 60-Hz line frequency is highly accurate and readily available, it is often used for Lissajous figure comparisons. Its only drawback is that the line frequency is so low that measurements above 1 kHz are hard to interpret.

FRACTIONAL RATIOS. While it is fairly easy to determine simple frequency ratios with Lissajous patterns, more often than not frequency ratios do not fall as exact increments of each other. It is more common to find ratios of 5:3 or even 19:20. That these ratios can also be displayed as Lissajous figures is in fact the most powerful feature of the Lissajous method. It is only a short step from simple Lissajous figures to more complex patterns.

As the frequency ratio becomes more complicated, the pattern grows nodes. Instead of the simple barrel-hoop patterns seen above, both axes contain loops that must now be counted. A frequency ratio of 5:3, for example, is displayed as five nodes along the vertical axis and three nodes on the horizontal axis. A ratio of 4:7 is four vertically and seven horizontally.

When determining the frequency ratio of complex Lissajous figures, simply count the number of nodes along each axis and put the numbers in ratio format. The ratio is now expressed as a fraction from which the unknown frequency can be derived. The first figure in Figure 5-21, for example, represents a 3:2 frequency ratio. If the horizontal frequency is 200 Hz, then the vertical input frequency is 300 Hz.

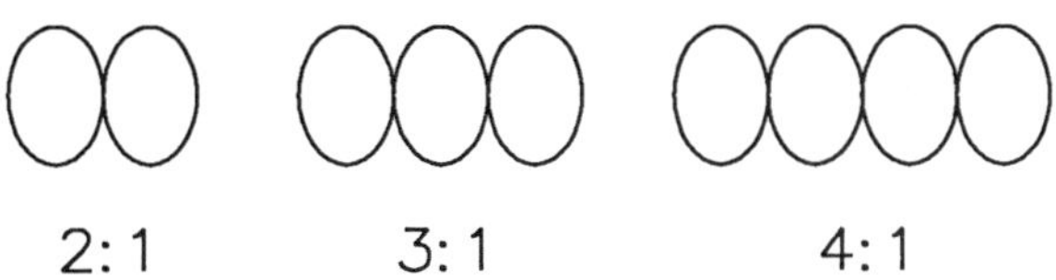

Figure 5-20 Multiple Lissajous figures

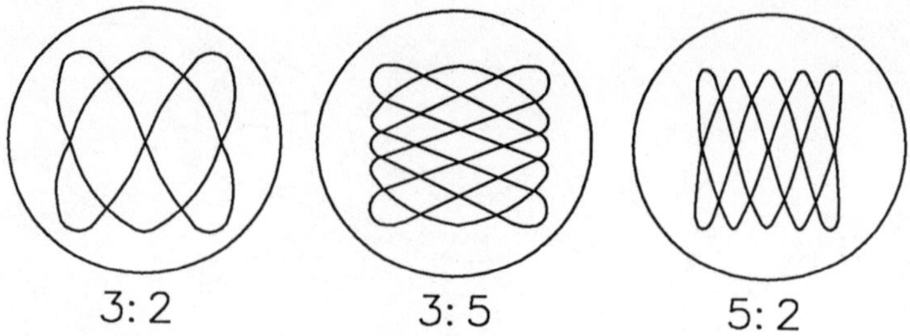

Figure 5-21 Complex Lissajous figures

When dealing with complex Lissajous patterns, be aware that the corner loops are counted in both the horizontal and vertical directions, seemingly serving double duty. An examination of the patterns in Figure 5-21 should quickly acquaint you with the procedure.

While the concept is simple enough, some skills must be developed in counting and interpreting Lissajous figures. High, low, or close ratios can be difficult to decipher. Misinterpretation of Lissajous figures is quite easy when there are many loops to identify. Not only is it possible to miscount the loops, but under unfavorable phase conditions, loops may not be fully displayed.

Sometimes it is better to deliberately rotate the pattern by purposely introducing a slight frequency misadjustment. Minor frequency differences are reflected as a slow gyration of the Lissajous pattern, making it easier to identify and count the hidden loops. Poor phase conditions can also hide nodes behind overlapping lines. If the scope has a phase control, adjusting it can often make the difference between an unreadable figure and an accurate one. Lissajous figures are easiest to interpret when the phase angle is 90 degrees.

Adjustment of the brightness, focus, and (when accessible) astigmatism controls can also be of help. The attainment of fine focus at the outer edges of a pattern, even at the expense of a fuzzy center, may provide the extra resolution needed to separate closely spaced loops.

Often a complicated figure can be simplified by an adjustment in the reference frequency. A ratio of 20:9, for example, is an extremely difficult pattern to deal with. A slight increase in the horizontal reference frequency, however, reduces this pattern to a 20:10 ratio, which the oscilloscope displays as 2:1.

The accuracy of the Lissajous method is limited only by the precision of the horizontal reference frequency, and accuracies to 0.001 percent are not uncommon. It should also be noted that Lissajous patterns are accurate even in the presence of noise. Noise spikes that would quickly render a frequency

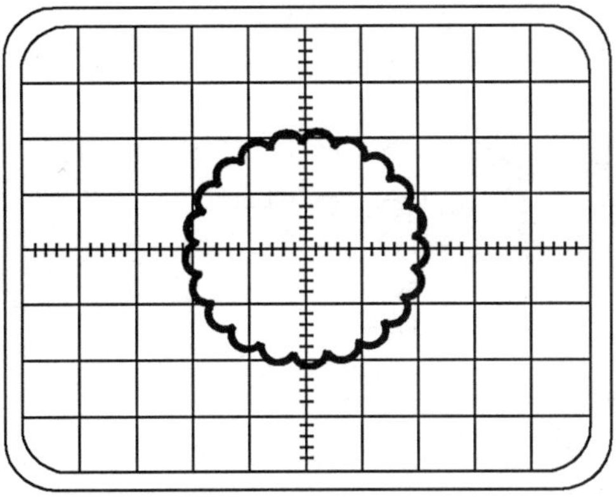

Figure 5-22 Gear-wheel pattern

counter useless appear as harmless barbs growing from the surface of the Lissajous curves.

GEAR-WHEEL PATTERN. A frequency determining method bearing resemblance to the Lissajous technique is the *gear-wheel* method. Like the Lissajous method, the gear-wheel process uses the oscilloscope as a comparator device by making external use of the horizontal input.

The technique takes its name from the characteristic mechanical gear-wheel pattern displayed on the screen as a result of mixing the two frequencies around a circular Lissajous figure, as shown in Figure 5-22. The basic circular pattern is created by comparing the reference frequency to itself on both the horizontal and vertical inputs with a 90-degree phase shift. The gear-wheel pattern is then superimposed on the Lissajous circle by modulating the horizontal input. The test setup is described in Figure 5-23.

Modulation is achieved by mixing the two frequencies in an isolation transformer. The isolation transformer prevents ground loops — the bane of electronic testing — from forming between the horizontal and vertical inputs. In situations where ground isolation is not necessary or desirable, the isolation transformer can be removed and the horizontal input modulated through direct signal injection (such as capacitive coupling).

Frequency is determined by counting the "teeth" on the gear and multiplying their number by the reference frequency. In order to carry off this task successfully, the pattern must be stationary. A stationary pattern can only exist when the unknown frequency is a whole-number multiple of the reference frequency. As it sometimes happens, a slight adjustment of the reference frequency is necessary to stop the gear from rotating before measurements can commence.

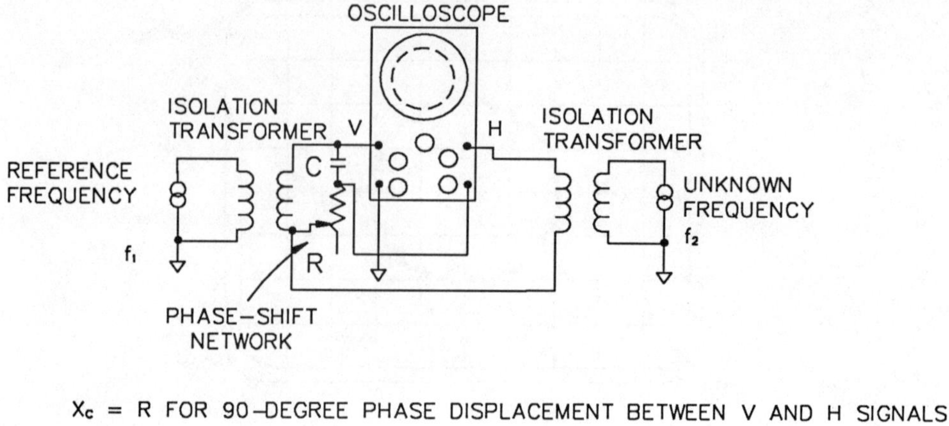

Figure 5-23 Gear-wheel test setup

When measuring frequencies that are lower than the reference frequency, the test setup must be modified to the configuration shown in Figure 5-24. In this setup, the horizontal input has been exchanged for the vertical input to produce the serrated pattern on the CRT screen. In order to use this modified scheme, the reference frequency would be read from the dial of a variable oscillator or with a digital frequency counter. The ratio equation is also reversed, making the unknown frequency the lesser component of the ratio.

The gear-tooth method of frequency determination is particularly useful when working with high ratios of known to unknown frequencies, as is the case when the 60-Hz line signal is the reference frequency. It is inherently easier to count a large number of bumps around a circle than it is to identify and correlate the maze of nodes present in a complex Lissajous figure.

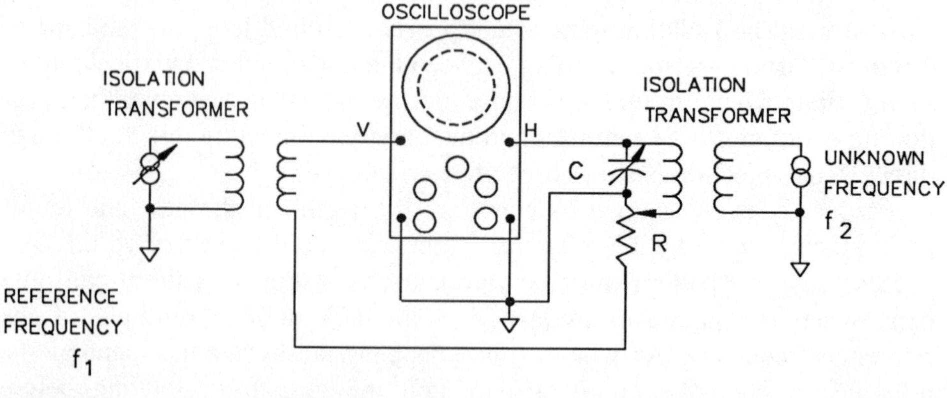

Figure 5-24 Modified gear-tooth test setup

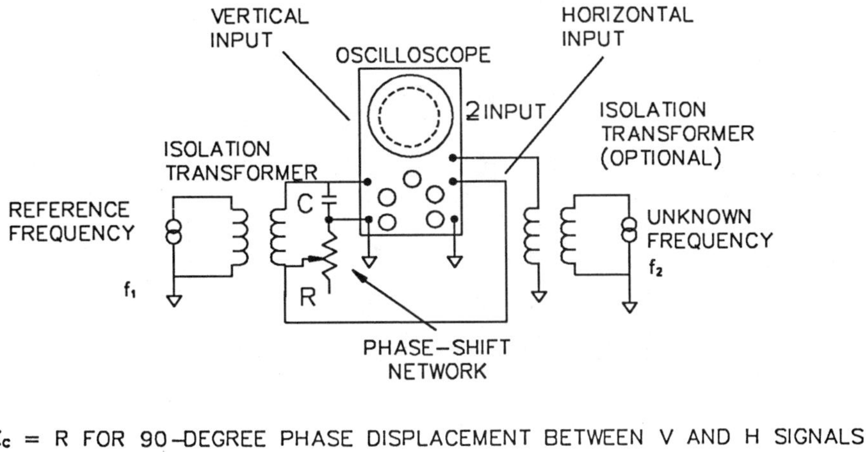

Figure 5-25 Slotted-wheel test setup

SLOTTED-WHEEL PATTERN. A somewhat more sophisticated version of the gear-wheel technique is the *slotted-wheel pattern*. The slotted-wheel method makes use of the Z-axis for its frequency measurements.

The Z-axis is no real mystery; it is simply a direct connection to the control grid of the CRT. Many scopes have this input located at the rear of the instrument rather than on the front panel. In normal operation, the control grid is used to adjust the intensity of the electron beam. Consequently, the brightness control uses the control grid to vary the intensity of the spot on the screen. When an external signal is applied to the control grid via the Z-axis input, the signal modulates the intensity of the display. In essence, this is the principle behind slotted-wheel frequency measurements.

The basic idea is to again establish a reference frequency on both the horizontal and vertical inputs, using a single reference source. The phase angle of the two input signals is adjusted, as before, to give a circular pattern on the screen.

Next, the unknown frequency is applied to the Z-axis input. When the Z-axis input amplitude is properly adjusted, blank spaces will occur around the ring. These spaces indicate places along the display at which the electron beam's intensity is fully cut off by the driving force of the Z-axis input.

The frequency ratio is determined by counting the number of times the circle is interrupted by the Z-axis signal. The interruptions appear as slots on the rim of the circular display, hence the name slotted-wheel pattern. Five slots, for example, indicates a frequency ratio of 5:1. Here, again, the unknown frequency must generate an integral number of interruptions within

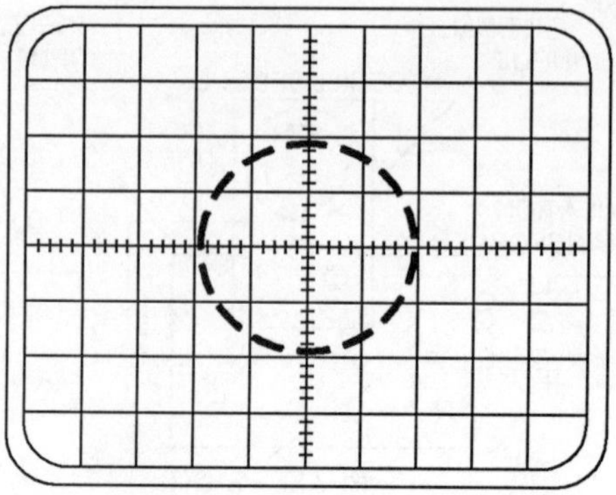

Figure 5-26 Slotted-wheel display

the reference circle; otherwise the pattern rotates like a spinning wheel. A slight adjustment in the reference frequency will correct the problem and bring the wheel to a halt.

Because it is relatively easy to count large numbers of slots accurately, this method is often better than either the Lissajous figure or gear-wheel method, under good conditions. While there can be confusion as to how many nodes or bumps exist on a particular pattern, there is no mistaking the whereabouts of empty slots in a stable slotted-wheel pattern.

The drawback to the slotted-wheel method is that the technique is not suitable for measuring frequencies lower than the reference frequency. Depending on circumstances, it may be possible to overcome this limitation by reversing the reference and test frequency inputs, sending the reference frequency signal to the Z-axis input and inputting the unknown frequency to the X-Y inputs.

PHASE MEASUREMENTS. Throughout the discussion of Lissajous figures and Lissajous-related techniques, emphasis is placed on the importance of the phase angle between the horizontal and vertical inputs. Phase angle is equally important in many electronic tests. Phase shift in an amplifier (be it audio or video), for example, produces distortion. Because of their high sensitivity to phase angle, Lissajous figures can be used for making phase measurements.

The basic test circuit for measuring phase angle using Lissajous figures is shown in Figure 5-27. With the oscilloscope's horizontal amplifier set to external mode, adjust the gain of the horizontal and vertical amplifiers so that each axis displays equal gain on the CRT screen. If the oscilloscope includes a vertical input phase adjustment, set the control for zero phase shift.

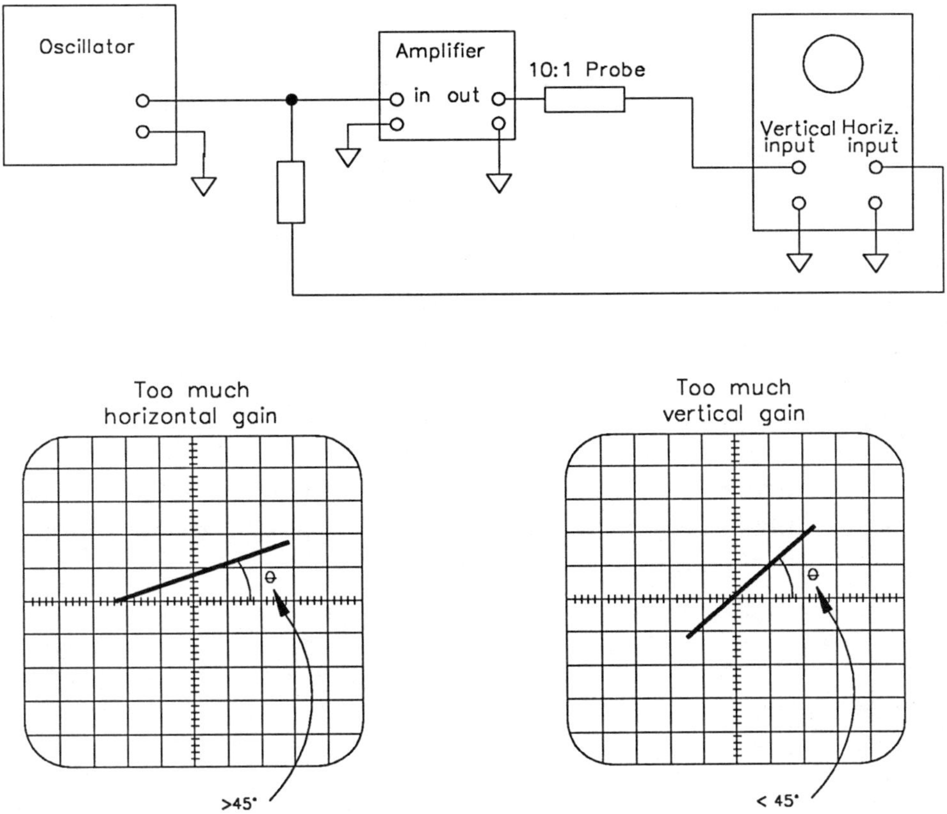

Figure 5-27 Test setup for measuring phase shift by Lissajous patterns

The amount of phase shift generated within the device under test is reflected by the shape of the Lissajous figure. When there is no phase difference between the two signals, the oscilloscope draws on the screen a straight line that is slanted 45 degrees to the right, as shown in Figure 5-28. As the phase angle between the two signals increases, the line opens up, first into an ellipse and finally into a circle. The circle occurs when the two signals are 90 degrees out of phase. Increasing the phase angle causes the Lissajous pattern to rotate to the left, again collapsing into the familiar elliptical figure. Finally, at 180 degrees phase shift, the display reduces itself to a single line slanting 45 degrees to the left.

The actual amount of phase shift displayed by the Lissajous pattern is determined by the ratio of vertical opening height to overall figure height. An imaginary line is drawn vertically through the figure, and two measurements are made using the graticule as a reference. They are shown in Figure 5-28 as

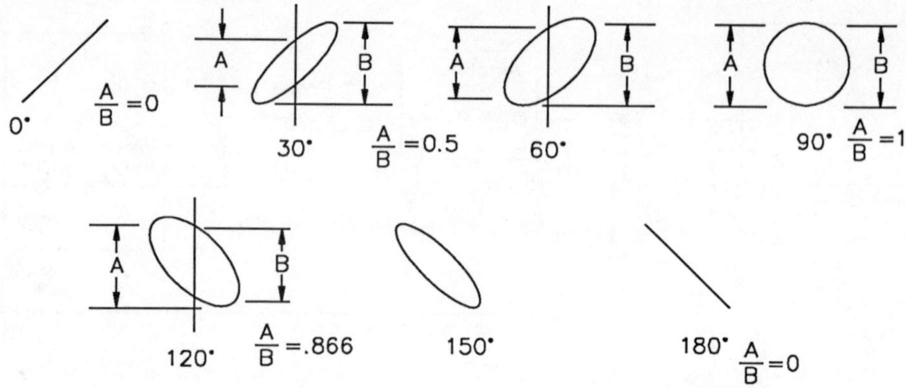

Figure 5-28 Phase-shift patterns

A and B. Measurement A is the size (in vertical divisions) of the elliptical opening, while B is the total vertical height of the Lissajous figure. The phase shift in degrees is then calculated from:

$$\text{phase shift } (\phi) = \sin^{-1}\left(\frac{A}{B}\right)$$

At both extremes, 0 degrees and 180 degrees, A is reduced to zero and the equation is not applicable.

Pulse Measurements

Pulse measurements are important when working with digital equipment and data communications devices. An ideal digital pulse contains five parameters. They are pulse rate, duty cycle, pulse width, rise time, and fall time.

The test setup for pulse testing a circuit is shown in Figure 5-29. While the setup shows the oscilloscope sweep being triggered from an external (EXT) input, similar results may be obtained using the channel A only trigger mode.

The *rise time* of a pulse is the time for the voltage to go from 10% to 90% of the pulse amplitude, as illustrated in Figure 5-30. Similarly, the *fall time* is the time required for the pulse voltage to drop from 90% to 10% of its amplitude. Most quality oscilloscopes include 10% and 90% marking on the graticule for convenient measurement of these vital parameters. As a rule, the amplitude of the pulse (which is generally defined by logic type) is not as im-

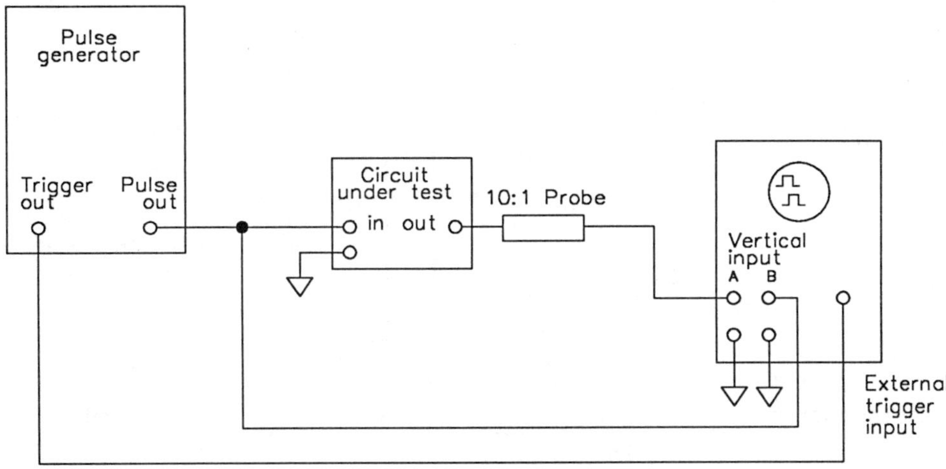

Figure 5-29 Test setup for pulse measurements

portant as rise and fall times, and the variable gain function of the vertical input amplifier is often brought into play to align the displayed pulse with these percentile marks for easier definition. Both rise and fall times are measured along the horizontal axis. With the SEC/DIV control set at 0.1 μs (100 ns) per division, the rise time and fall time values of Figure 5-30 are:

$$t_r \approx 0.7 \text{ divisions} \times 100 \text{ ns} = 70 \text{ ns}$$

$$t_f \approx 0.9 \text{ divisions} \times 100 \text{ ns} = 90 \text{ ns}$$

The pulse width (PW) is measured from the 50-percentile marks, while the time period (T) is the time required to complete one full cycle. From our example, these values are determined to be:

$$PW \approx 450 \text{ ns}$$

$$T = 800 \text{ ns}$$

thus yielding the pulse *repetition frequency* (PRF) of:

$$PRF = \frac{1}{T}$$

$$= \frac{1}{800 \times 10^{-9}}$$

$$= 1.25 \text{ MHz}$$

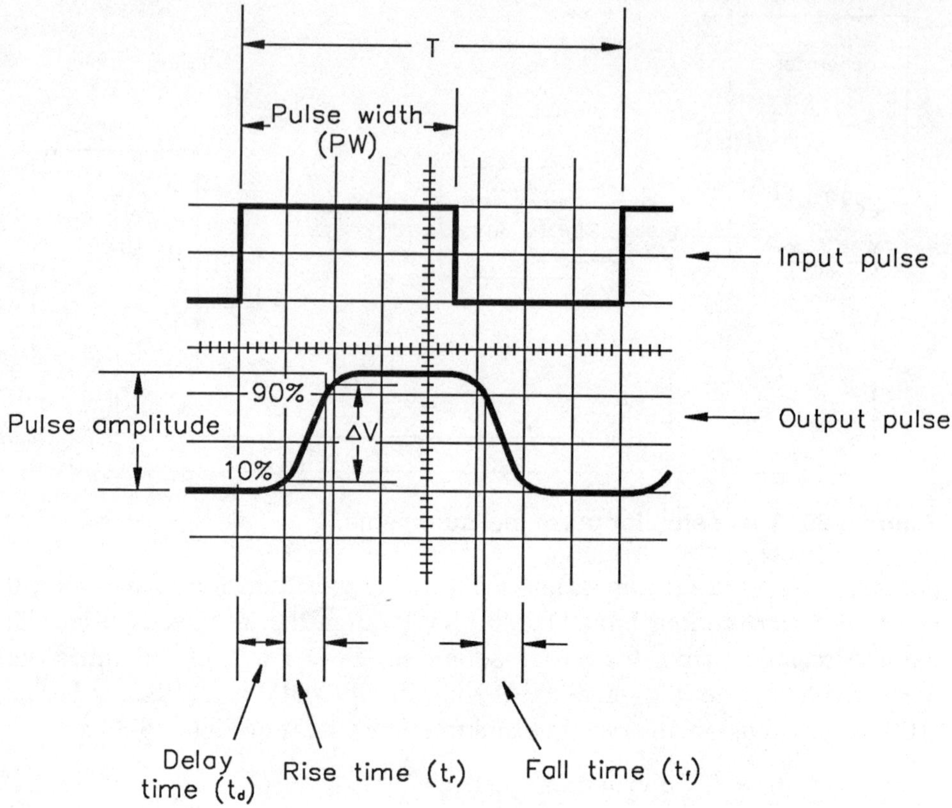

Figure 5-30 Parameters of a square wave

Duty cycle is calculated using the equations:

$$\text{duty cycle} = \frac{\text{PW}}{\text{T}}$$

or

$$\text{duty cycle} = \text{PW} \times \text{PRF}$$

Either equation produces a duty cycle of 56% when the test values of Figure 5-30 are inserted into the formula.

With many circuits under test having a rise time equal to that specified by the manufacturer for the oscilloscope, the displayed rise time may be different than the actual rise time. The error is a function of oscilloscope bandwidth, which is determined by the formula:

$$t_{r\,scope} = \frac{0.35}{BW}$$

If the bandwidth of the oscilloscope under question is 100 MHz, for example, then the rise time is

$$t_{r\,scope} = \frac{0.35}{1 \times 10^8}$$

$$= 3.5 \text{ ns}$$

The pulse generator rise time is measured by the oscilloscope, and the error is calculated using the following equation:

$$t_{r\,gen} = \sqrt{t_r^2{}_{measured} - t_r^2{}_{scope}}$$

To measure rise time accurately, it is normally necessary to stretch out the pulse so that only the leading edge is visible on the screen. Many times this dictates the use of the sweep magnifier.

Knowing these two factors, $t_{r\,gen}$ and $t_{r\,scope}$, the true rise time of the circuit under test can be derived from:

$$t_r = \sqrt{t_r^2{}_{measured} - t_r^2{}_{scope} - t_r^2{}_{gen}}$$

To illustrate, consider a test setup using a pulse generator with 2-ns rise time and an oscilloscope with a bandwidth of 100 MHz. If the measured rise time of the device under test is 10 ns, then the actual rise time is:

$$t_r = \sqrt{t_r^2{}_{measured} - t_r^2{}_{scope} - t_r^2{}_{gen}}$$

$$= \sqrt{10^2 - 3.5^2 - 2^2}$$

$$= \sqrt{100 - 12.25 - 4}$$

$$= 9.15 \times 10^{-9}$$

or 9.15 nanoseconds.

As a rule of thumb, measurements made using test equipment with better than a 3-to-1 performance ratio over the device under test can be assumed to be accurate to within 10 percent of the actual value.

6. Digital and Special-Purpose Oscilloscopes

Like the analog voltmeter, the analog oscilloscope has a digital counterpart. Within its bandwidth and sampling rate limitations, the digital storage oscilloscope in particular brings significant advantages to the user in that it easily displays single-shot or low-repetition rate signals, obtains pre-trigger as well as post-trigger information, and mathematically analyzes and manipulates captured waveform data.

To some, the term digital oscilloscope indicates an instrument which replaces every analog system and device with a digital unit. This simply is not the case. Like the digital voltmeter, the digital oscilloscope relies on analog-to-digital conversion and data sampling techniques for its operation.

Sampling Oscilloscopes

The forerunner of the modern digital oscilloscope was the analog *sampling oscilloscope*. Essentially, it was the technology pioneered by this instrument that made digital conversion and display on a CRT feasible.

Most analog oscilloscopes have an upper frequency limit of 50 MHz to 100 MHz. Analog frequency limits are largely due to the brightness of the trace and not the electronics of the system. Higher input frequencies cause the electron beam to sweep across the face of the CRT so fast that little of the

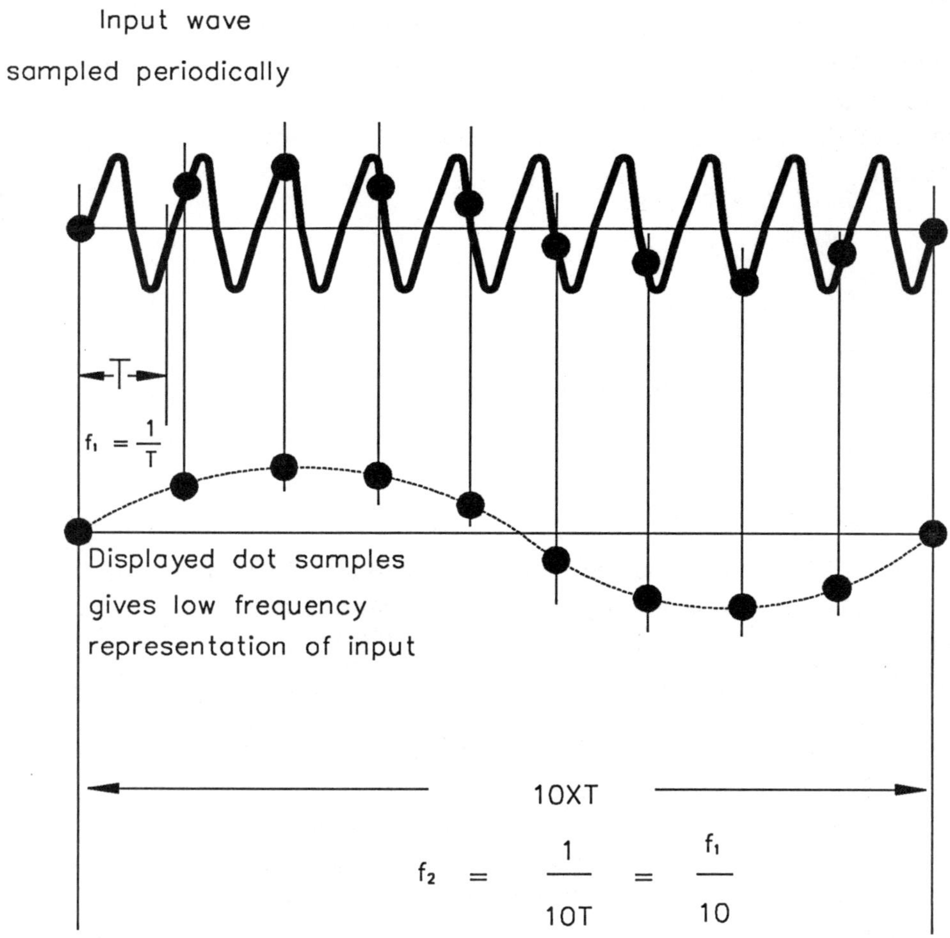

Figure 6-1 Sampling display

phosphor is excited. A point is eventually reached where the trace becomes so faint as to be unusable, thus limiting the input frequency.

The sampling oscilloscope overcomes this difficulty by displaying a low-frequency dot representation of the signal. Each dot represents an amplitude sample of the input signal, with each sample taken not from a single waveform, but from different points along consecutive input signal cycles, as illustrated in Figure 6-1. While the sampling circuit must be capable of operating at very high frequencies, the CRT and its associated circuitry may be relatively low-frequency devices.

Consider the example in Figure 6-1, which shows ten cycles of a high-frequency waveform expanded to a single cycle for a display using sampling techniques. Notice that ten samples are taken of the wave train, each at successively later times in each cycle. The result is an approximation representative of the original waveform at a lower frequency. The technique is called *equivalent-time sampling*.

Suppose the input signal has a time period of 0.01 microseconds. Its frequency is:

$$f_1 = \frac{1}{0.01\ \mu s}$$

$$= 100\ \text{MHz}$$

The dot display has a time period of $10 \times f_1$, or:

$$f_2 = f_1 \times 10$$

$$= 10\ \text{MHz}$$

Thus, the 100-MHz signal has been converted into a 10-MHz waveform for display. If one cycle of a dot display is created by sampling 100 cycles of a signal, the frequency of the displayed wave is 1/100 of the signal frequency, and so forth.

DESCRIPTION OF CIRCUIT FUNCTIONS. A simplified block diagram of a sampling oscilloscope is shown in Figure 6-2. Critical to the performance of a sampling oscilloscope is the *sampling gate*. When triggered, the sampling gate opens, allowing the instantaneous amplitude of the high-frequency input waveform to enter the vertical deflection amplifier.

Control over the opening and closing of the sampling gate is done using a voltage comparator and a *staircase generator*. The output of the staircase generator is a series of increasing voltage steps — or staircase — which rise from a negative voltage value to a positive value.

The stairsteps of the staircase generator are created using a pulse generator and a Miller integrator. An equivalent block diagram of a staircase generator is shown in Figure 6-3. Each negative-going output pulse from the pulse generator turns transistor Q1 on for a brief period of time. During this time, a current flows into capacitor C1, thus increasing its charge and the voltage across it. Every current pulse charges C1 to a new voltage level, which remains constant until the next current pulse arrives, and thus is created the stairsteps.

When the capacitor voltage has reached and exceeds an upper limit set by the Schmitt trigger, Q2 is turned on, discharging C1 to its starting level. The number of steps required to trigger the Schmitt circuit is dependent on the

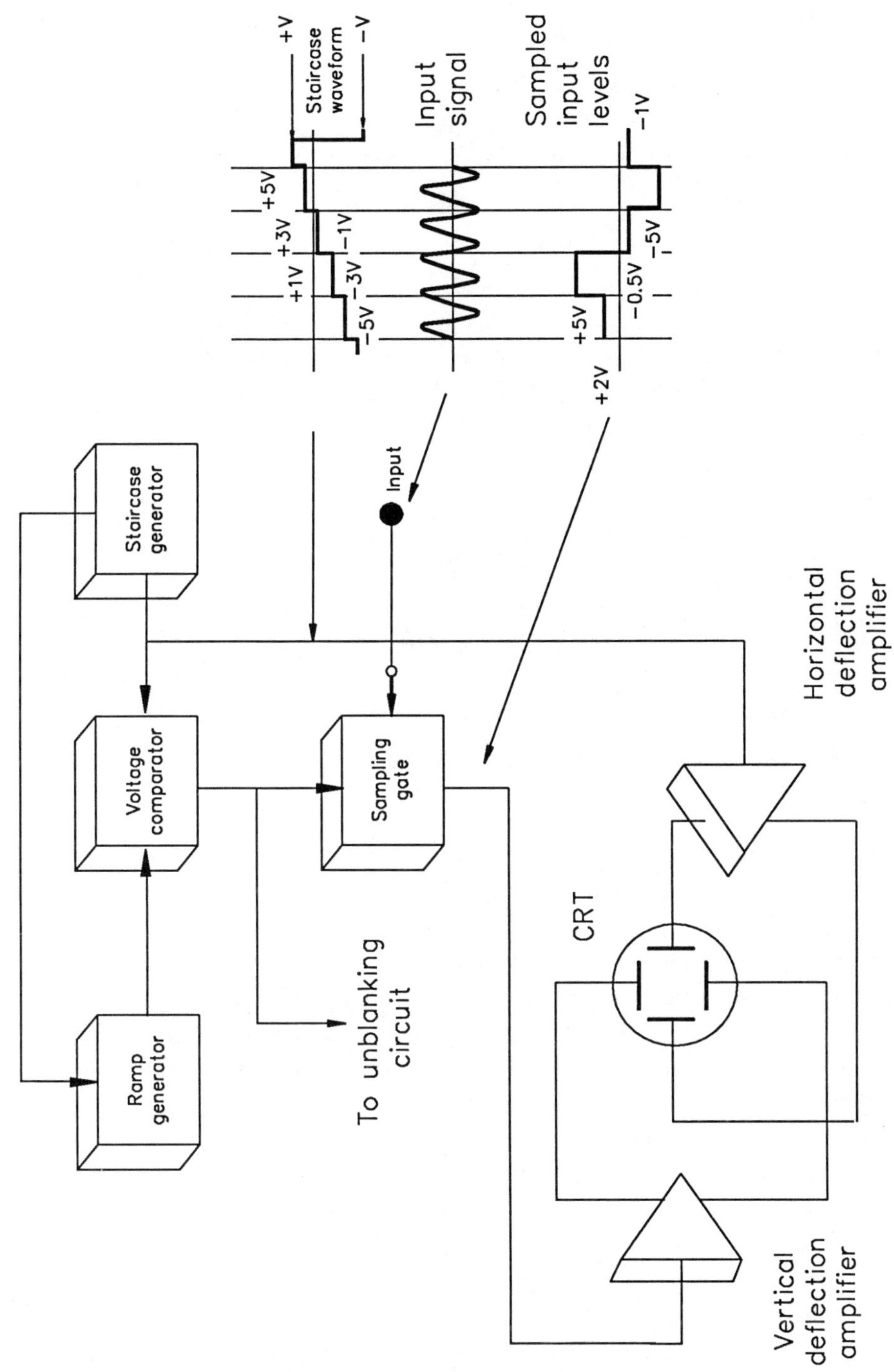

Figure 6-2 Sampling scope block diagram

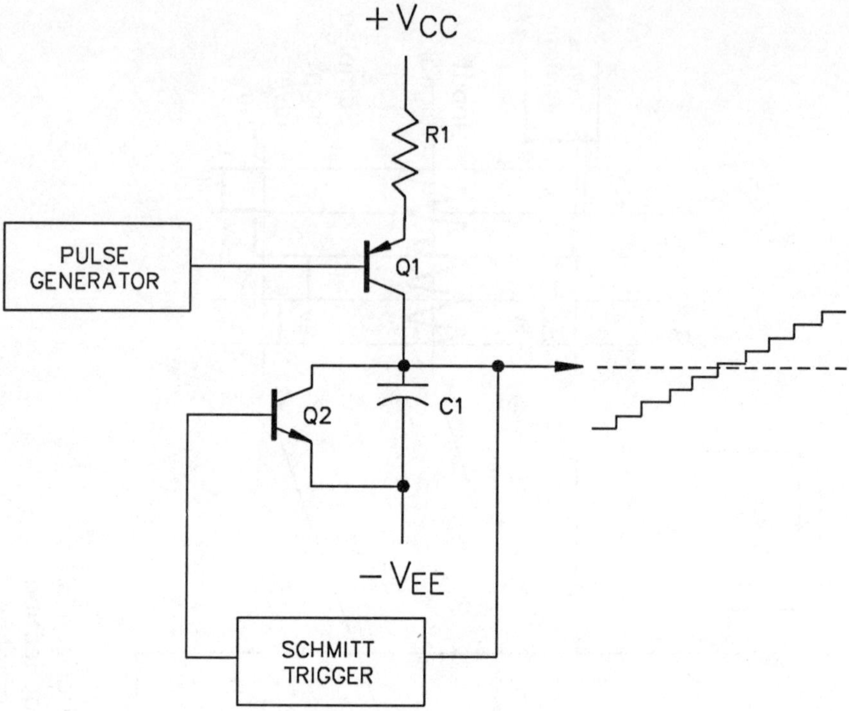

Figure 6-3 Staircase generator block diagram

width of the current pulse, its amplitude, and the capacitance value of C1. Different stepping values may be easily obtained by substituting different values for C1 using a simple switching arrangement.

The staircase waveform is then applied to a voltage comparator. Compared to the staircase voltage is the output voltage of a ramp generator. The ramp generator is triggered into action by the pulse generator inside the staircase circuit. Each time a new stairstep is created, the ramp generator begins a positive-going slope. When the output voltage of the ramp generator equals the staircase output voltage, the sampling gate is triggered into conduction.

The sampling gate is typically a sample-and-hold circuit. The sample-and-hold circuit consists of an input preamplifier, a sampling diode bridge, a holding capacitor, and an intermediate buffer amplifier, as illustrated in Figure 6-4.

In this circuit, the input signal is buffered and conditioned by the preamplifier before sampling by the sampling bridge. When a differential sampling pulse from the comparator is applied to the bridge, the four diodes conduct in the forward direction. That action momentarily connects the input

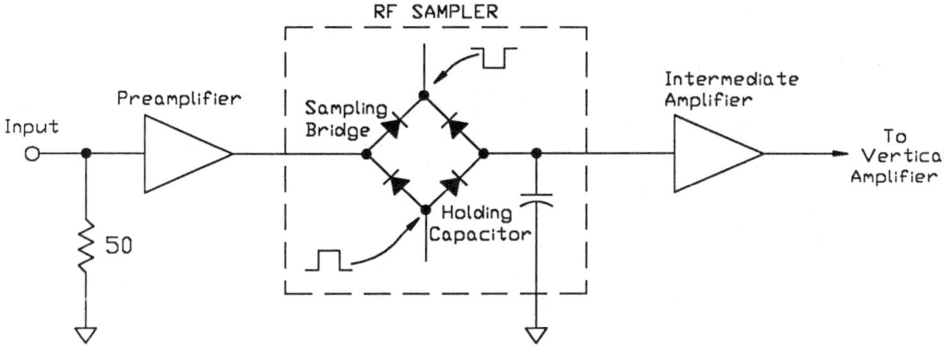

Figure 6-4 Sampling gate

of the sampling bridge to the holding capacitor. The sampling capacitor is then charged to the level of the input voltage. When the sampling bridge is turned off, the sampling capacitor retains its voltage at the level of the last sample. The sample is then buffered by a high-impedance intermediate amplifier to prevent discharge of the holding capacitor, and applied to the vertical deflection amplifier.

The input frequency bandwidth of the instrument is limited only by the input preamplifier and appariture time of the sampler, and is not affected by the buffer amplifier or any subsequent circuits unless they fail to operate at the division ratio created by the sampling rate. Because of the bandwidths involved, the bridge circuit often uses high-speed Schottky diodes built on a microwave hybrid chip or similar microstrip construction.

Because the staircase output voltage increments itself following each sample, the comparator has a new voltage reference which to compare to the ramp generator output each time — a reference voltage that becomes higher with every step. Consequently, it takes the ramp generator more time to achieve that voltage, thus delaying the triggering time of the next sample by an amount equal to the slope of the ramp and the stairstep voltage increment.

The output of the staircase generator is also input to the horizontal deflection amplifier. When viewed from afar, one can readily see that the staircase waveform is essentially a digitized version of a ramp voltage. When applied to the vertical deflection plates, the staircase waveform causes the electron beam to scan across the face of the tube in the same manner as does the sweep generator of a regular oscilloscope. When this sweeping action is integrated with the vertical dots of the sampling gate on the face of the CRT, our eyes perceive a continuous waveform representative of the higher input frequency signal.

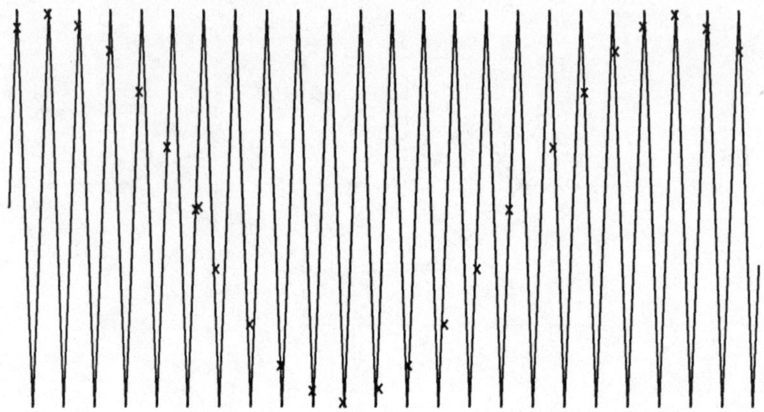

Figure 6-5 Perceptual aliasing in the figure above causes the display to appear at 1/19 the actual frequency.

ALIASING. A problem with using sampling techniques to construct a waveform is a phenomenon known as *perceptual aliasing*. Perceptual aliasing is inherent to dot displays and is essentially a kind of optical illusion.

Aliasing occurs when a signal is sampled less often than it should be. The result is a display that may be an alias representation of the waveform at a much lower frequency, as demonstrated in Figure 6-5.

The danger in aliasing lies in the fact that one might not even know it is occurring. In the example of Figure 6-5, the signal is sampled only once a cycle, where the ratio of sample frequency to input frequency is 1:1. Because of the sample points, however, the display appears as a signal with 1/19 the actual input frequency, making accurate frequency measurement from the screen image impossible. For more complex waveforms, the danger is even greater.

To avoid perceptual aliasing error, more samples per time period are required. Sampling theory according to the *Nyquist criterion* states that to completely acquire a waveform, sampling must occur at a frequency greater than twice the highest frequency component. Therefore, it might be reasonably assumed that an oscilloscope that samples at 100 MHz could be used to view signals up to 50 MHz. That assumption, however, would probably be wrong.

Data acquisition is only part of the aliasing problem. The waveform must also be reconstructed on the screen using the dots acquired. The normal method of reconstruction involves passing the input samples through a low-pass anti-aliasing filter to remove any unwanted high-frequency components which may interfere with the display and forwarding the processed information to the CRT for display.

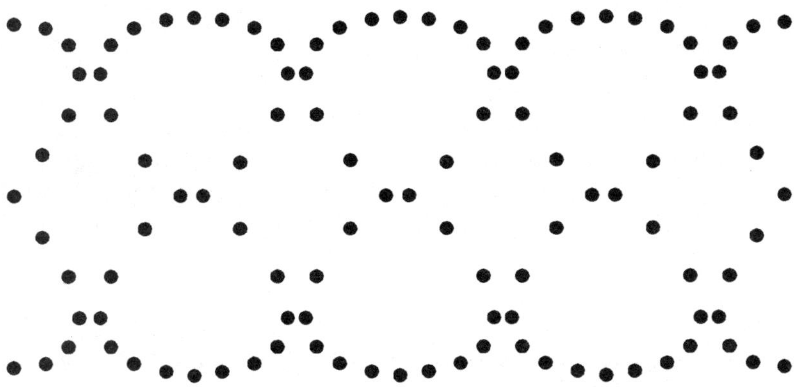

Figure 6-6 Even when the Nyquist criterion is satisfied, perceptual aliasing can occur. What appears to be many untriggered sine waves in the display above is really the same high-frequency waveform seen in Figure 6-5.

But the concept of waveform reconstruction is totally dependent on the mind's ability to form continuous waveforms by connecting each dot with its neighbors when viewing a dot display. The next closest dot in space, however, may not be the next sample dot of the waveform. The screen in Figure 6-6 demonstrates the point.

In this example, the Nyquist limit of two samples per cycle has been exceeded. But what seems to be many untriggered sine waves is actually only one waveform at a much higher frequency than you perceive. In fact, it is the same original waveform shown in Figure 6-5. Your eye — not the oscilloscope — makes the mistake. To be reasonably sure that what you see is what you get, it takes about 20 to 25 dots per cycle of screen reconstruction to present a recognizable image.

Storage Oscilloscopes

Another disadvantage of the equivalent-time sampling method is that only purely repetitive signal waveforms can be investigated. If the waveform changes over several cycles, the dot pattern will be in error. A transient waveform cannot be displayed using this sampling technique.

One way around the problem is to capture the fleeting image using a *storage oscilloscope.* Basically, there are two types of storage oscilloscopes:

analog and digital. The more sophisticated of the two is the digital storage oscilloscope, with its ability to display transient events as well as provide pre-trigger and post-trigger information.

It is appropriate, however, to mention conventional analog storage methods in order to understand the chain of events which led up to digital storage. The following discussion will also serve to highlight the advantages and disadvantages of the two systems, both of which are in common usage.

Analog Storage Oscilloscopes

BISTABLE PHOSPHOR CRT. The first storage oscilloscope to appear on the scene was the *bistable* (phosphor) CRT storage oscilloscope. Bistable CRT storage uses a special phosphor with two stable states for image display: either the waveform is displayed or not displayed. There is no control over the intensity of the displayed image.

The bistable CRT screen has a *storage layer* of phosphor material that is capable of emitting secondary electrons, but which retains a very high insulation resistance between particles. (As you recall from Chapter 4, secondary electron emission occurs when high-energy electrons strike a surface and cause other electrons to be emitted from the surface. The generation of X-rays, for instance, is totally a secondary electron emission process.) A transparent metal film is deposited between the glass of the CRT and the storage layer to encourage the generation of secondary electrons, just the opposite of the construction methods used to dispel secondary electron emission in a conventional CRT. Another metal film layer, called the *collimator*, is deposited around the neck of the tube.

Bathing the fluorescent screen with a constant flood of electrons are two *flood guns*. The two flood guns are simply cathodes gently heated to generate low-energy electrons. Located within the neck of the tube is a standard CRT electron gun assembly — including electron lens, acceleration anodes, and horizontal and vertical deflection plates — called the *write gun*.

In normal operation, the two flood-gun cathodes are at ground potential, as is the collimator ring. The metal film deposited on the face of the tube is at a potential of +1 to +3 volts with respect to ground. The clouds of electrons emitted by the flood guns are attracted by this positive potential so that they flood the fluorescent screen with slow electrons, as illustrated in Figure 6-7. If the electron beam from the write gun has not been activated, the phosphor layer is not affected by the low-energy electrons from the flood guns. These

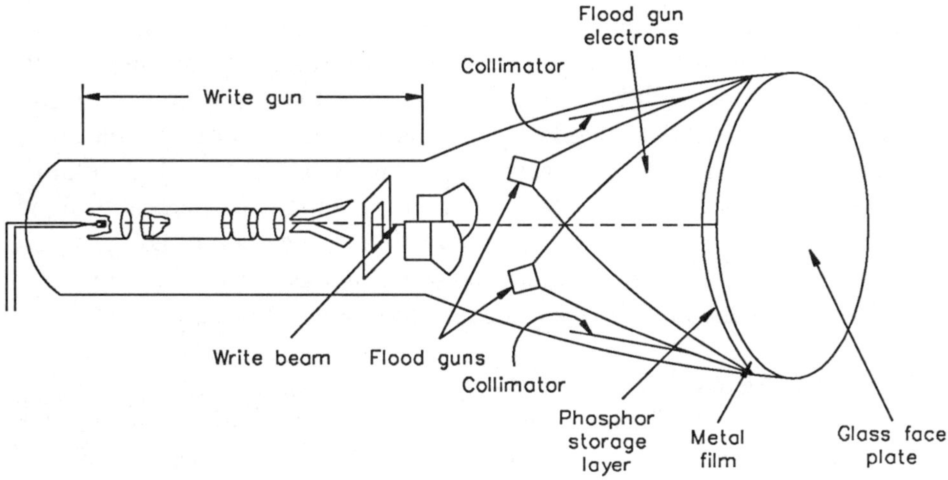

Figure 6-7 Bistable-type storage CRT

electrons do not penetrate the phosphor, rather, they recoil as they strike, and are collected by the collimator. No display occurs in this condition.

When the write gun is energized, a beam of high-velocity electrons impinges on the screen. These electrons strike the phosphorus layer with sufficient energy to produce secondary emission at the point where the high-energy beam strikes. These secondary electrons are attracted to and collected by the collimator. Due to the high insulating properties of the phosphor, however, every point where secondary emission occurs becomes positively charged because the electrons are lost from that point. The low-energy electrons from the flood guns are now attracted to the positive charge areas, and they pass through it to the more-positive metal film. In passing through the storage layer, the flood-gun electrons cause the phosphor to glow.

When activation of the write gun is inhibited until an event of significant interest occurs at the input of the oscilloscope, the bistable CRT is able to capture transient signals that elude normal oscilloscopes and viewing conditions. Once activated, the phosphor may continue to glow for hours, giving the operator plenty of time to make screen photographs, signal analysis, or compare signals. More than one trace may be stored on the display using sequential triggering of the write gun.

Erasure of the screen is effected by forcing the faceplate metal film negative. This action repels the flood-gun electrons back into the storage layer, where they accumulate and return the written area to the same potential as the surrounding unwritten areas.

Bistable storage is the easiest to use and least expensive CRT storage, but with the slowest writing speed. The principal applications for bistable storage are in mechanical applications, signal comparisons, and data recording. Most bistable CRTs have split screen capabilities with the ability to store signals on one half the screen independently of what is on the other half. This unique advantage of bistable storage provides the capability of displaying a known waveform while viewing other waveforms for comparison. Screen resolution is also greater than that of other storage methods. The main disadvantages of bistable storage are the inability to reposition a waveform after capture for graticule alignment and the lack of control over screen brightness.

VARIABLE PERSISTENCE CRT. While not a distinct disadvantage, the bistable phosphor CRT cannot erase itself. Screen erasure must be done manually. There are times, however, when a self-cleaning screen would be handy. Such CRT designs are called *variable persistence* CRTs.

Variable persistence CRTs have the ability to vary the persistence of the signal or vary the amount of time that the signal is displayed on the CRT. The decay time is adjustable so that the writing of one signal coincides with the fading of another.

A storage CRT capable of variable persistence operation is shown in Figure 6-8. The major difference between the variable persistence CRT and the bistable CRT is that a fine-wire *storage mesh* and *collector mesh* are now included inside the glass envelope. Unlike the bistable design, the screen is the normal type of CRT screen with a low-persistence phosphor layer backed by an aluminum film. The storage layer is deposited on the inside surface of the storage mesh. One additional change is that control grids are now positioned in front of each flood gun.

Normally, the collector mesh is biased with a positive potential of approximately 100 volts, and the storage mesh biased with a voltage variable between 0 and –10 volts. In operation, the high-energy electron beam from the write gun produces secondary emission at the storage layer and creates a positive charge path in the shape of the input waveform. Low-energy flood gun electrons are repelled from all parts of the storage layer except where the write beam has produced the positive trace path. It is within the trace area that the flood-gun electrons pass through the phosphor and generate a trace on the screen. The secondary-emitted electrons and the flood-gun electrons that do not pass through the storage layer are attracted back to the highly positive collector mesh. Although the screen phosphor has low persistence, the flood-gun electrons passing through the positive trace on the storage layer continue to maintain a display on the screen.

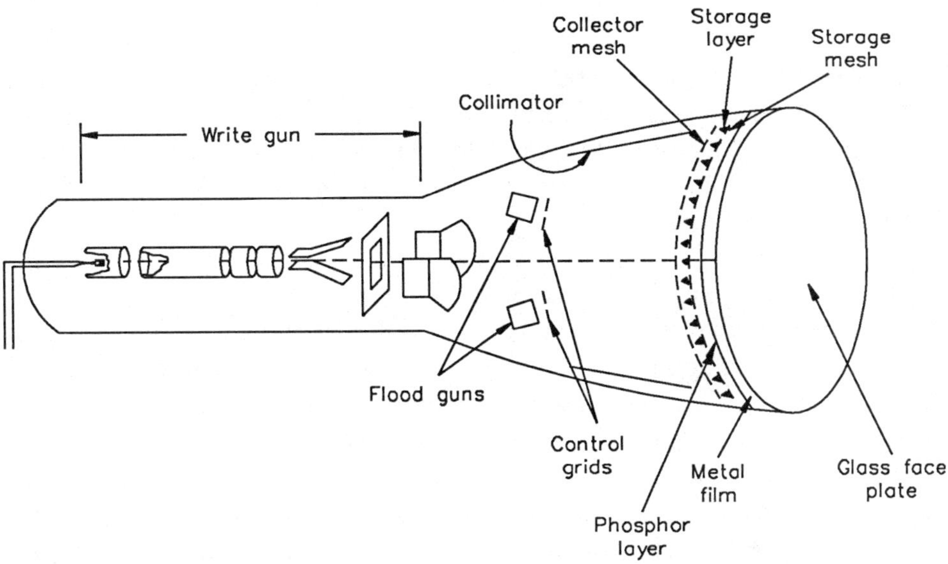

Figure 6-8 Variable persistence CRT

Erasure of the stored waveform is accomplished by connecting the storage mesh to the same high positive voltage as the collector mesh for a brief period of time. In this case, the flood electrons produce secondary emission all over the storage layer so that the written waveform is wiped from the screen. Returning the storage mesh to normal bias sets the stage for another screen write.

Erasure can also be performed by adjusting the display persistence control. Variation of the display persistence is obtained by slowly erasing the waveform using variable-duty pulses to the storage mesh. A repetitive negative pulse of –4 volts to –11 volts is applied to the storage mesh to discharge the positively charged area. The repetition rate is typically 1.2 kHz and the duty cycle is continuously variable between 0 and 100 percent. The narrowest pulses give the longest display persistence, and the wider pulses shorten the persistence by discharging the storage layer more rapidly.

The brightness of the displayed waveform depends on the number and energy level of the flood gun electrons that strike the screen. A bias voltage applied to the control grids of the flood guns controls the number of electrons that pass to the screen. A high negative bias cuts off the flow of electrons, thus negating the display, while a zero bias voltage permits maximum current flow for maximum brightness. Between these two extremes are continuously variable levels of excitement and screen brightness.

Digital Storage Oscilloscopes

The fundamental difference between the digital storage oscilloscope and its analog counterparts is the form of storage. Analog oscilloscopes use mechanical means to store the waveform. Digital oscilloscopes, on the other hand, sample the input waveform and convert it into binary data that is stored in semiconductor memory for recall on the screen.

Digital storage requires both digitizing and reconstruction of the input signal. Digitizing consists of sampling and quantizing. Using the sampling techniques discussed earlier in this chapter, the input signal is first sampled and then each analog sample converted into a digital sample of each using an *analog-to-digital* (ADC) converter. The digitized sampling data is then stored in semiconductor memory. To display the waveform, the digitized data must be converted to analog signals using a *digital-to-analog* converter (DAC). The reconstructed analog sampling voltages are then synchronized with the horizontal deflection circuits and displayed on the screen in dot-pattern format.

A block diagram of a digital sampling and storage oscilloscope is illustrated in Figure 6-9. Central to the operation of the digital oscilloscope is a *clock timebase*. The clock timebase generates a pulse waveform at the desired sampling frequency that drives the ramp generator, staircase generator, and horizontal sweep. This clock gives discrete points in time to reference the quantized values of the input signal. Each tick of the timebase opens the sampling gate for a brief moment and allows a sample to be taken. The rate of the ticks is called the *digitizing rate* and is usually specified in mega-samples per second (MS/s) or points per second.

ANALOG-TO-DIGITAL CONVERTER. The sample voltage is then buffered and converted into digital format using an analog-to-digital converter. There are several methods of digitizing or quantifying the different voltage levels of a waveform. The three most popular are: successive approximation converters, flash converters, and scan converters.

Successive approximation converters compare the input voltage to the output of a digital-to-analog converter. If the input is higher than the digital-to-analog output, the most significant bit (MSB) is set high, if it is lower, the MSB is set low. The converter than progresses to the next significant bit and makes the same comparison and decision. Unfortunately, it is a slow, tedious process that requires at least one clock pulse for each digit quantized, and is little used in modern digital sampling oscilloscopes. The process of successive approximation conversion is detailed in Chapter 7.

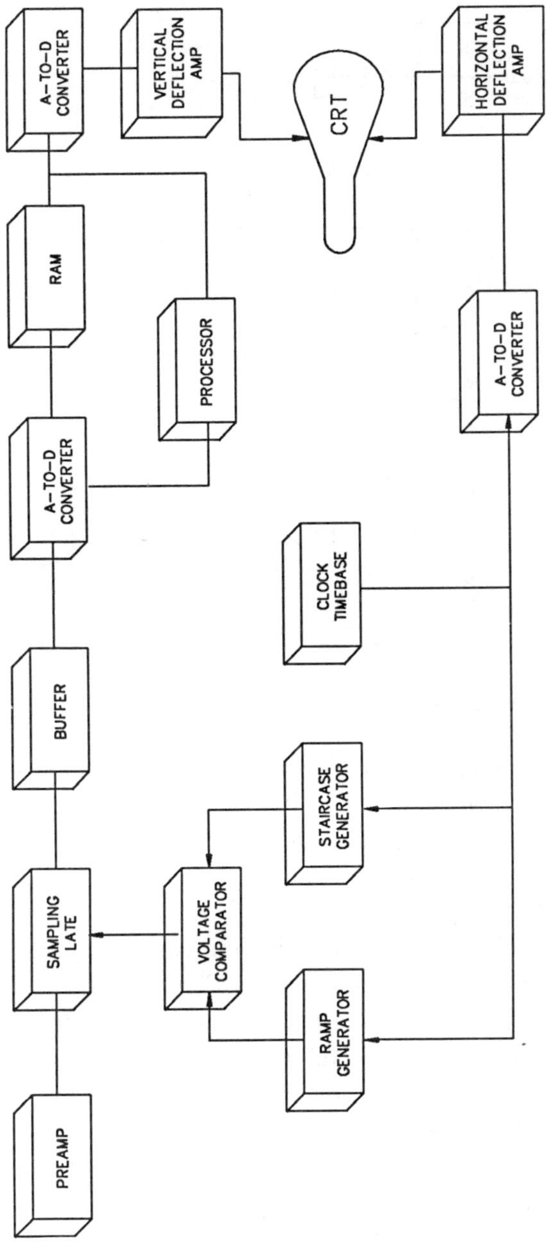

Figure 6-9 Digital storage oscilloscope block diagram

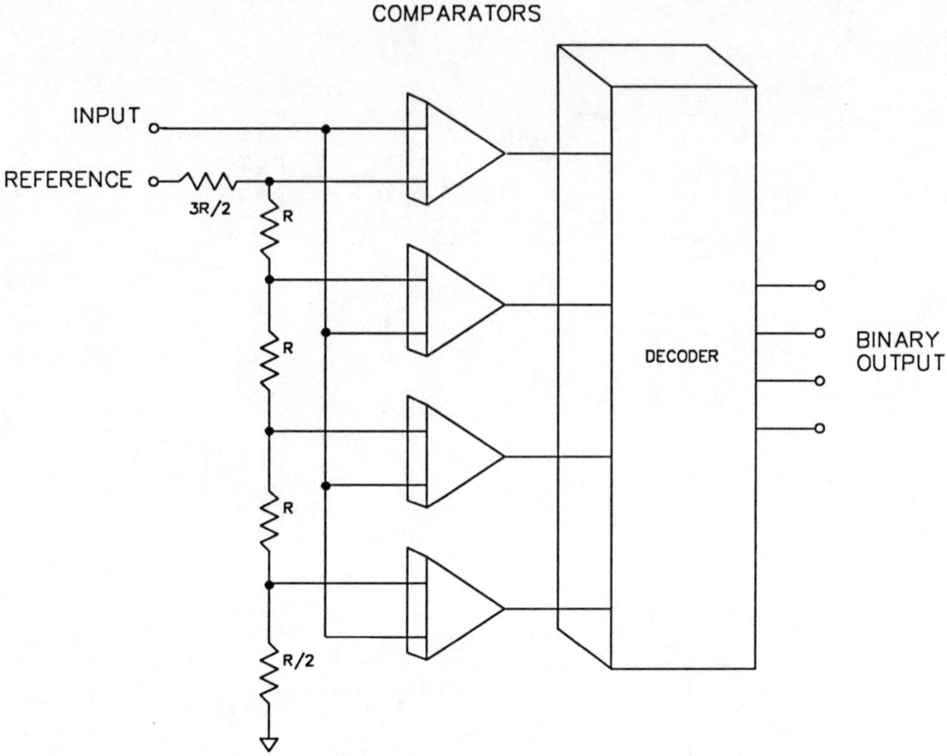

Figure 6-10 Flash converter

Flash converters, on the other hand, convert an analog input into a binary number in a single clock pulse. Whereas a successive approximation converter requires eight clock pulses to convert a 7-bit number, the flash converter can do it in one, thus increasing the sampling rate and permitting a greater bandwidth.

In flash conversion, analog-to-digital conversion is accomplished through a parallel array of voltage comparators. The inputs of the voltage comparators are connected to a weighted resistor divider that scales the analog value. Depending on the magnitude of the analog input voltage, a comparator may or may not turn on. The number of comparators that do activate is proportional to the analog input voltage. A digital decoder converts this number into binary format for further processing. An in-depth analysis of flash conversion is detailed on page 254 of Chapter 7.

The *scan conversion* is a mechanical method of converting analog information into digital format. It involves storing the information on a target in a specially constructed CRT and reading this data using a separate scanning beam on the other side of the target. A scan converter contains a double

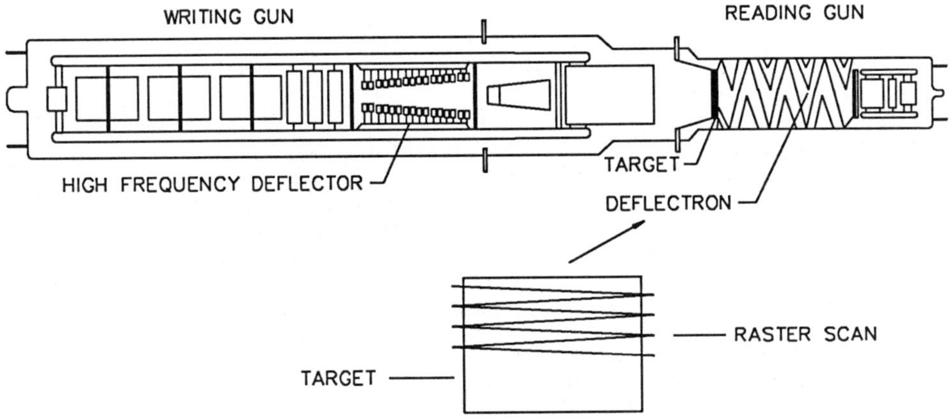

Figure 6-11 Scan converter

ended CRT. The writing beam is like that in an oscilloscope CRT, except that it uses a very small and very special target, one made up of a silicon diode array.

The target is then read by a slower reading beam that scans across the target in a raster similar to that used to build up a picture on a television screen. Very high sampling bandwidths, up to 500 MHz, are possible using this technique. The scan converter device is, however, quite expensive and found in only the most sophisticated digital storage oscilloscopes.

MEMORY. Once the input signal has been sampled and digitized, the binary number is stored in memory. Many different types of memory, called by many different names, are associated with digital storage, but typically fall into two categories: semiconductor and magnetic. RAM (random access memory) chips are classified in the former, while bubble memory, tape, and disk are in the latter.

Semiconductor RAM is the faster of the two, with typical access times of 100 nanoseconds and less. A RAM consists of two functional blocks: the memory cell array and the peripheral interface circuitry.

The memory cell array is the heart of the memory unit. It is usually a square matrix of individual one-bit memory elements arranged in rows and columns. Because the memory cell arrays are for the most part square, the capacity tends to be an even power of 2. There are many 256, 1K, 4K, 16K and 64K RAM chips, but very few 512, 2K, and 8K RAMs.

There are many approaches to the design of the individual memory cells. The most common are static flip-flop cells and transistor-base-charge dynamic cells.

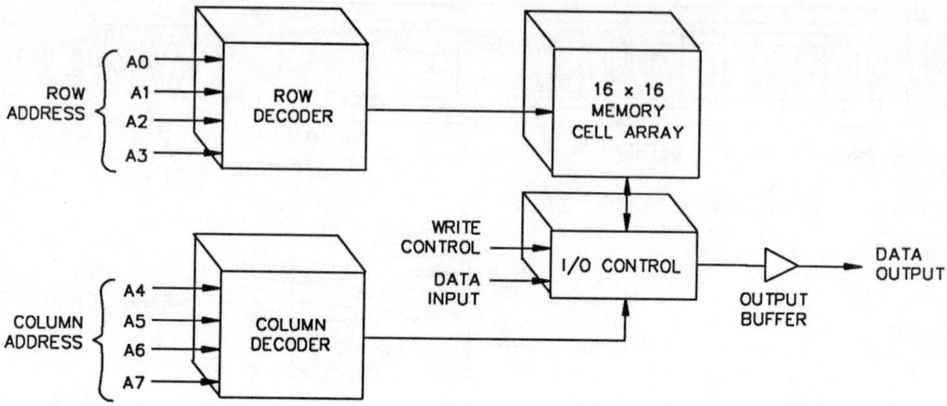

Figure 6-12 Memory peripheral circuitry and cells

The *static memory cell* is a two-transistor flip-flop bistable multivibrator. By activating either the right or left transistor, a 0 or a 1 can be stored in the cell. This data is retained until an external current changes the flip-flop's state or until power is lost. The cell is called static because it can hold information indefinitely as long is it has power; there is no apparent change in the cell once it has been set unless acted upon by an external force.

The semiconductor *dynamic memory cell* consists of a single transistor. A current into the base of the transistor turns the stage on by building up a charge in the transistor base region. An on state represents a 1 while an off state represents a 0. While it may seem that a single-transistor dynamic cell is better than the static two-transistor cell because it uses less parts, therefore consuming less power, the dynamic cell cannot hold this state indefinitely. It must be refreshed. The term dynamic refers to the gradual change of state as time goes on. A transistor can only hold a base charge for a few milliseconds before it bleeds away. If the charge leaks away without being replenished, the stored data is lost. To accomplish the replenishing or refreshing, refresh circuitry must be built into every dynamic RAM.

As Figure 6-12 illustrates, a RAM must take in an address through a set of address lines and select the appropriate row and column corresponding to the addressed cell. If data is being written into the cell array, the input data control unit must turn the addressed memory cell on or off to indicate a 0 or 1. This is a write operation. When a cell is examined by the input/output (I/O) control circuitry for disclosure of its logical state, it is a read operation.

The most common method to refresh dynamic cells is to recharge every memory cell in a given row of a cell array whenever any element in that row is read. Most dynamic RAMs, therefore, require a read operation at least every 2 milliseconds. If this comes naturally in a machine cycle, so much the better; if not, a time interval must be set aside to refresh the RAM by sequentially reading its contents every few milliseconds.

Magnetic memory is usually external to the instrument. Typical external devices are floppy disk, hard disk, magnetic tape, and, in some cases, bubble memory. The magnetic media is usually used as a form of mass storage and is typically accessed via GPIB or RS-232 through a controller. Details of this technology are presented in Chapters 7 and 8.

MEMORY AND RECORD LENGTH. The installed memory is then divided into records. A record represents one waveform. A particular instrument, for example, may have the ability to acquire 1024 points per waveform; this would be the record length. The same instrument, however, may also be able to store several such records in its 32K memory — up to 32, in this example.

It may seem that almost any length of memory is possible to achieve internally with today's memory technology. However, there are other factors that enter into the design of the digital storage oscilloscope that places restrictions on how much memory is feasible.

Specifically, one factor is the update rate. The larger the memory, the more time it takes to acquire a record and the longer it takes to refresh the display. With a record length of 1024 bits, the resolution of the waveform is nominally better than most CRT storage oscilloscopes.

DIGITAL-TO-ANALOG CONVERTER. To display the digitized data, either from memory or in real time, the digital signal must be converted to an analog voltage for display on the CRT. Digital-to-analog converters, when used for reconstructing the digital data, do not require the performance characteristics that the analog-to-digital converter needs since the conversion rate is so much slower. The digital-to-analog converter's main purpose it to take the quantized data and convert it to an analog voltage.

Most digital-to-analog converters use a parallel input circuit. This design accepts a parallel binary input code and outputs an analog voltage representative of that code. The input code is converted to an analog voltage using binary weighted switches that react simultaneously to the binary input data. Figure 6-13 shows an example of a typical digital-to-analog converter for use with a digital storage oscilloscope.

In this example, it is assumed that the analog-to-digital converter has a resolution of 8 bits, while the digital-to-analog converter has a resolution of 10 bits. Even though the analog-to-digital converter is an 8-bit device, the

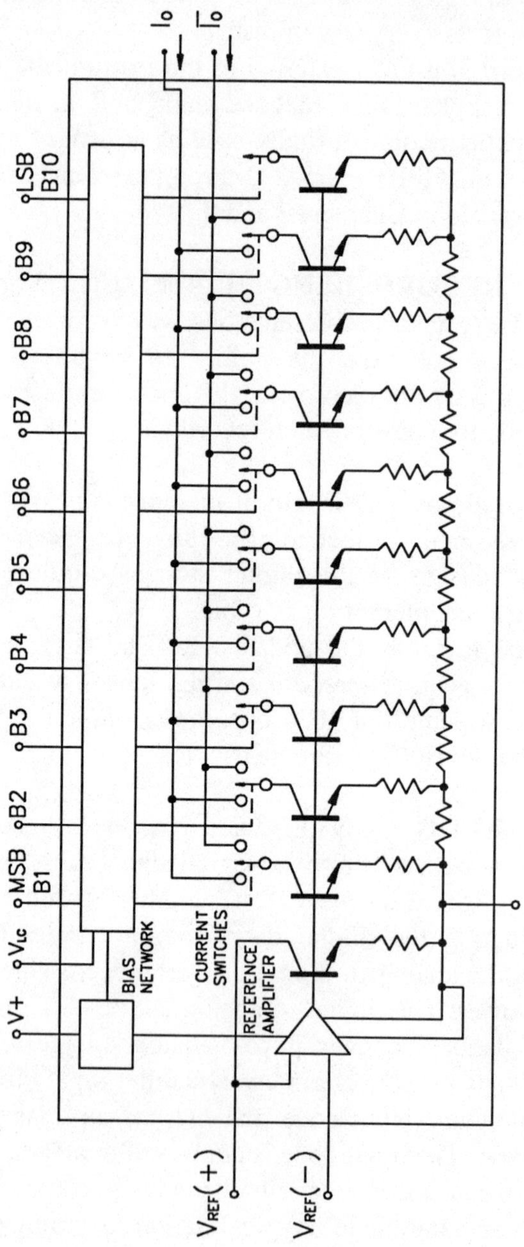

Figure 6-13 Digital-to-analog converter

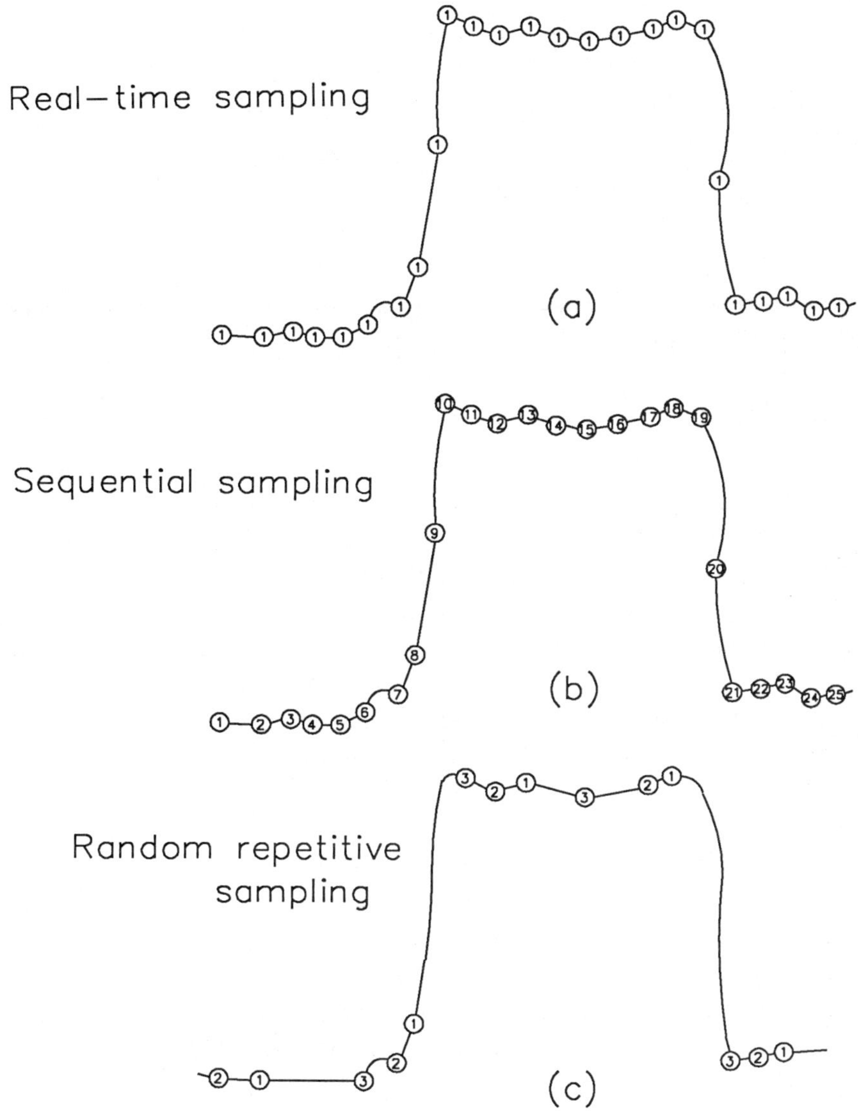

Figure 6-14 Digital sampling techniques

digital-to-analog converter takes advantage of electronic averaging to display 10 bits of vertical resolution.

REAL-TIME SAMPLING. One of the primary reasons for using storage is to display events that happen only one time. Unfortunately, the equivalent-time sampling technique requires several cycles to build its waveform. To capture one-time events for digital processing and storage, a sampling technique called real-time sampling must be used.

In real-time sampling, the input signal is sampled and digitized during the first cycle, as shown in Figure 6-14; the pass numbers are shown in the small circles. This technique requires that the sampling rate be at least 10 times faster than the highest frequency component of the input signal.

Digital Storage Oscilloscope Performance

Digital storage oscilloscopes can be used to display sine, pulse, and transient waveforms. As with analog sampling oscilloscopes, the quality of the displayed waveform depends upon the number of samples per cycle. The more samples taken, the greater the resolution of the image. The ratio of sampling rate to useful display rate is the bandwidth of the storage oscilloscope.

In analog CRT storage, input bandwidth is specified as the write speed. It is the amount of charge that can be deposited on the target that determines the upper limit of the instrument's storage capabilities, and not usually the frequency response of the instrument's amplifiers. Writing speed is directly related to the frequency of the signal that the instrument can store, as illustrated in Figure 6-15.

USEFUL STORAGE BANDWIDTH. Because the digital storage oscilloscope captures the waveform electronically, and not mechanically, writing speed has little meaning. Instead, the terms *useful storage bandwidth* and *useful storage rise time* are used as figures of merit. Useful storage bandwidth is dependent on the digitizing rate and on the waveform type. For a full-scale, sinusoidal signal, the useful storage bandwidth (USB) is defined as:

$$\text{USB(MHz)} = \frac{\text{Maximum digitizing rate (MHz)}}{25}$$

where the denominator represents the 25 dots required for waveform recognition. Note that the number of samples per cycle necessary to recognize the input signal on a dot display changes with the amplitude of the trace. If the trace is smaller, then the data points on the screen are closer together and perceptual aliasing is reduced.

To obtain a more recognizable display with less samples, digital oscilloscopes normally have a facility for drawing lines between the dots in the dis-

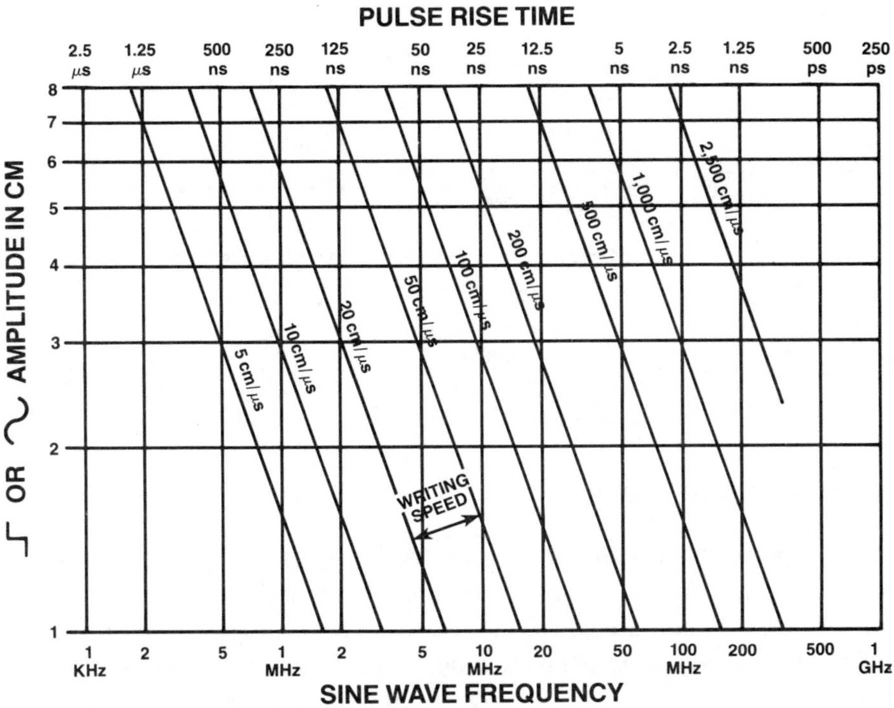

Figure 6-15 Analog writing speed

play. This is referred to as *interpolation*. Interpolators that draw straight lines from dot to dot are called *linear* or *pulse interpolators*. While this method greatly reduces perceptual aliasing by adding vectors to the display as shown in Figure 6-16, an envelope error persists because the vectors are only straight lines joining the data points. The signal could still lie outside the waveform traced by the vectors.

Some digital storage oscilloscopes use a vector generator which draws lines between the data points on the screen. This interpolation then is used to display a sine wave; perceptual aliasing is eliminated and only 10 samples per cycle of the sine wave are necessary to reconstruct a recognizable display. Pulse interpolators also make glitches more visible by connecting every data point and preventing a single dot far from the rest from being overlooked. The digital storage bandwidth for a linear interpolator is:

$$\text{USB} = \frac{\text{Maximum sampling rate}}{10}$$

As long as the vectors on the screen are short, an accurate representation of the input sine wave is possible. But with long vectors, the displayed waveform

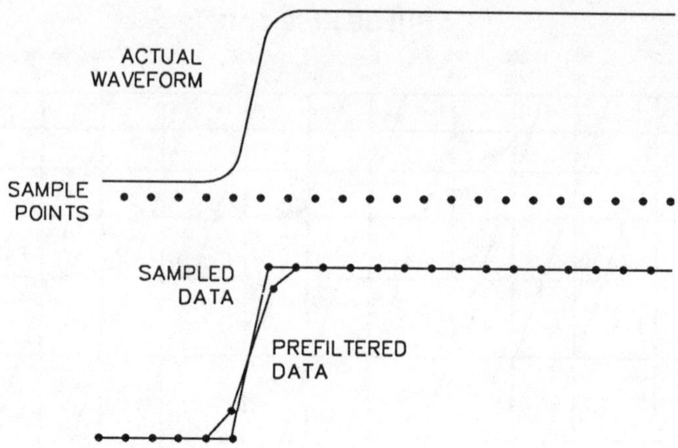

Figure 6-16 Linear interpolator image

and the input signal might not coincide. A linear interpolation of a minimally sampled signal can indicate a different shape for the waveform anytime the samples taken do not fall exactly on the peaks of the signal.

Sine-wave interpolators, on the other hand, introduce a sinusoidal function between the dots which gives a smooth flow to the vector line. As long as no aliasing took place when the original data words were gathered, the sine-wave interpolator will not introduce errors when displaying sine waves. With a sine-wave interpolator, only 2.5 samples per cycle are required for a satisfactory display, which changes our usable storage bandwidth equation to:

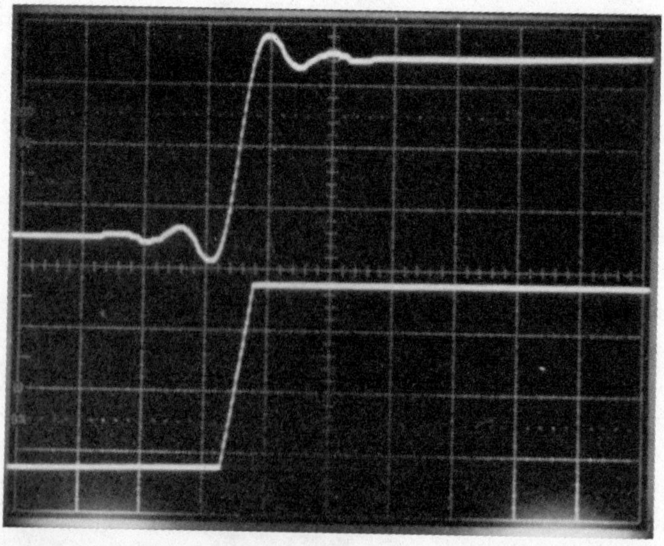

Figure 6-17 Sine-wave interpolator error

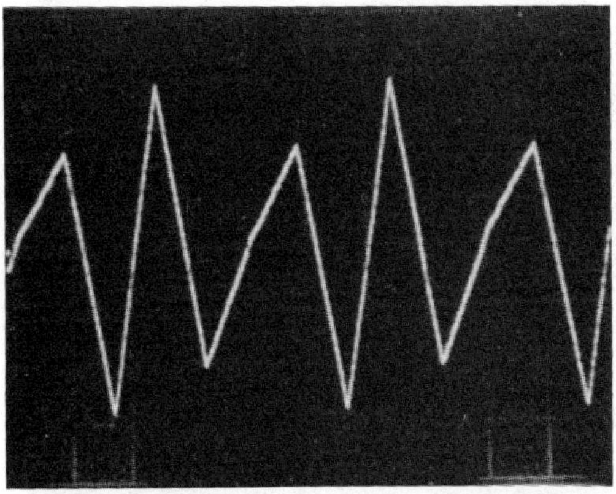

Interpreted waveform

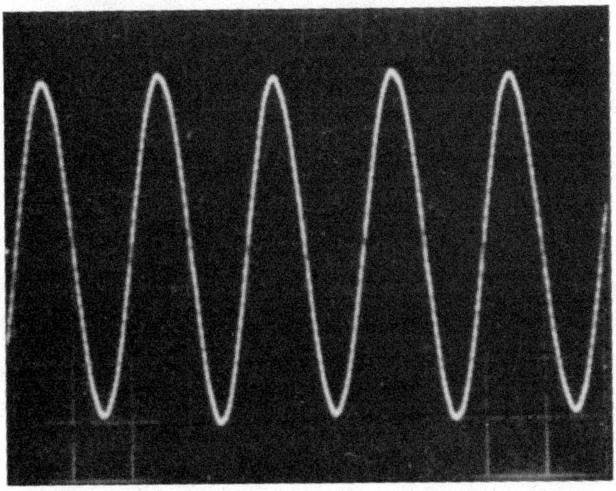

Actual waveform

Figure 6-18 Modified sine interpolation

$$\text{USB} = \frac{\text{Maximum sampling rate}}{2.5}$$

The same interpolator that makes sine waves more accurate, however, introduces what appears as preshooting and overshooting on pulses (see Figure 6-17). One method of alleviating this problem is to use a digital pre-filter. The digital filter coupled with the sine interpolator permits reconstructing the waveform to a better representation of the actual waveform, as indicated in Figure 6-18.

Note that the useful storage bandwidth divisor above is dependent on the particular interpolator used and the waveform displayed. As a rule of thumb, most waveforms can be interpolated correctly using a sampling factor of 10 with the proper interpolator.

USEFUL RISE TIME. One of the most important parameters of pulse waveform investigation is rise time. For a satisfactory dot display, approximately 10 samples of the leading edge are required. Since the time period of the sampling rate is 1/(sampling rate), the minimum rise time that can be investigated is:

$$t_{r(min)} = \frac{10}{\text{sampling rate}}$$

For a sampling rate of 50 MHz:

$$t_{r(min)} = \frac{10}{50\text{ MHz}}$$

$$= 200\text{ nanoseconds}$$

Pulse waveforms with shorter rise times than this can be investigated, but the edge of the pulse cannot be studied with any detail.

As it turns out, the maximum positive rise-time errors produced by a pulse interpolated display closely follow the form of an analog oscilloscope's error (Figure 6-19) when the analog system has an equivalent rise time of 1.6 sample intervals. Maximum negative rise-time errors are much smaller. Consequently, the most limiting measurement errors that occur when using 1.6 sample intervals as the nominal sampling rate, and the useful rise time can be defined as:

$$Ut_{r(min)} = \text{Minimum sample interval} \times 1.6$$

If, for example, you wanted to calculate the useful rise time of a digital oscilloscope with a maximum digitizing rate of 10 MHz, by multiplying the result would be:

$$Ut_r = 0.1\text{ ms} \times 1.6$$

$$= 160\ \mu\text{s}$$

Notice that the useful rise time is based on a pulse interpolator. Dot displays have additional error possibilities because of the truncation of resolution; with the same number of samples on the input step, the dots simply don't trace the shape of the signal. Interpolators designed for displaying sine waves show rise times that are faster than the input signal because of the preshooting and overshooting introduced when there are only a few samples on the input step.

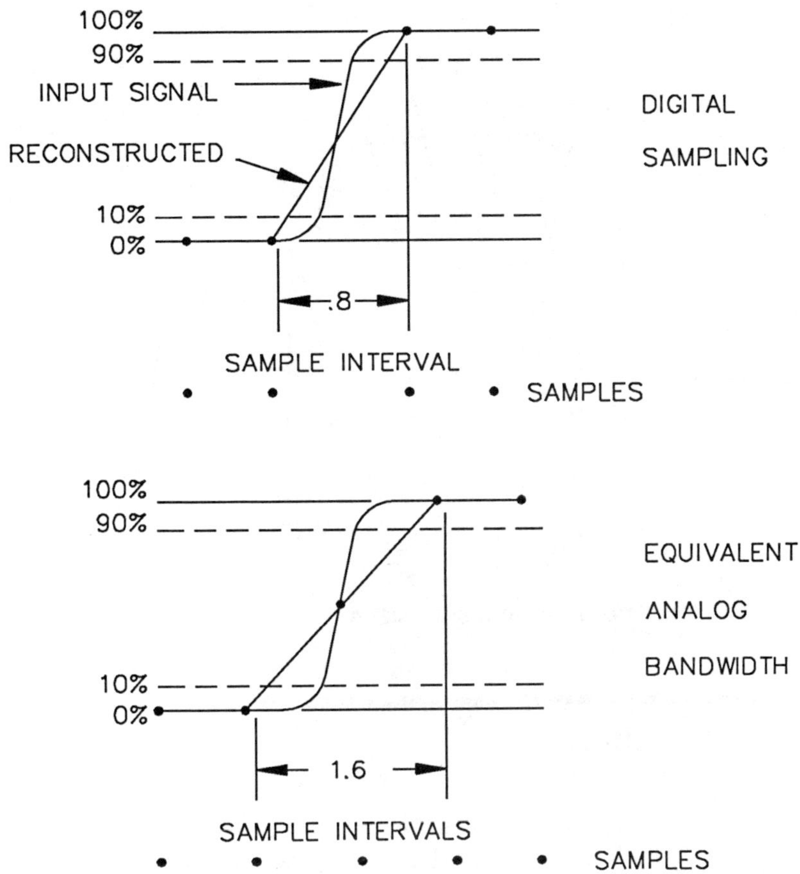

Figure 6-19 Sampling rise-time measurements

It is important to remember that, unlike the rise time of an analog oscilloscope, you cannot use useful rise time and the measured rise time to work back to the signal's rise time. The root-squared equation presented in Chapter 4 does not hold water. The useful rise time is based on the worst-case condition where errors are minimized. The actual error in any given measurement can vary between the maximum negative and maximum positive values, depending on the placement of the samples on the input signal.

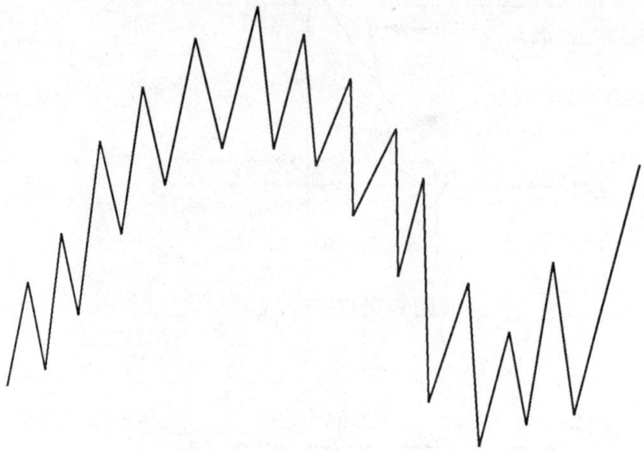

Figure 6-20 Noise riding atop a sine wave

Signal Processing

Signal processing is another possible feature offered by digital storage oscilloscopes. It includes the translating of raw data to determine average value, rms voltage, frequency, etc. of the stored waveform.

Signal averaging is a most useful feature because it can eliminate glitches and noise riding on the input waveform. Processing a signal with averaging can increase both the resolution and the accuracy of the measurement by comparing identical samples over a longer period of time and displaying only the average value. Here is where the memory size, and not the record length, becomes important. The larger the digital memory, the more accurate the result.

Signal processing can increase the resolution of a signal so that the smallest unit can be distinguished. Consider, for example, a random noise element equal to one-third the signal level riding atop a sine wave, as shown in Figure 6-20. Under most circumstances, the best resolution possible could be as poor as 1-1/2 divisions and as good as 1/2 division, depending on the instantaneous noise level at the time. After averaging ten consecutive waveforms, however, the resolution would be closer to 1/5 or 1/10 of a division.

With increased resolution, the accuracy of the measurement is also increased. The reason digital averaging has this effect is because the signal to be measured is related to time in a very specific manner and uses the same trigger each time the oscilloscope reads it. The noise surrounding the signal, on the other hand, is not related to time and does not depend on a trigger. Because the noise is random, its arithmetic mean value is zero. As the oscilloscope averages the waveforms, the noise is stripped from the signal. For uncorrelated noise, the signal-to-noise ratio (S/N) is improved by a factor of of the square root of n:

$$S/N = \text{noise (percentage)} \times \sqrt{n}$$

where n is the number of waveforms averaged. For example, if the measurements vary by 10 percent because of noise, the signal-to-noise ratio is 10:1. Taking an average of four signals increases the signal-to-noise ratio to:

$$S/N = 10\sqrt{4}$$

$$= 20{:}1$$

or 5 percent.

There are several ways in which averaging can be accomplished. One method is by obtaining just a simple average:

$$\frac{\Sigma \text{ wfm data}}{n}$$

where wfm = waveform n data, and n = number of averages.

Another method that can be used is by normalized averaging in which the last waveform acquired would have a binary weight assigned to it, such as 1, 2, 4, 8, 16...256. The algorithm for this average function is:

$$Avr_S = Avr_{S-1} + ((I_S - Avr_{S-1}) / 2^n)$$

where:

s = number of waveforms averaged

Avr_S = average after s waveforms

Avr_{S-1} = previous average

I_S = current waveform data

n = positive integer ($0 \le n \le 8$), slaved to elapsed sweep

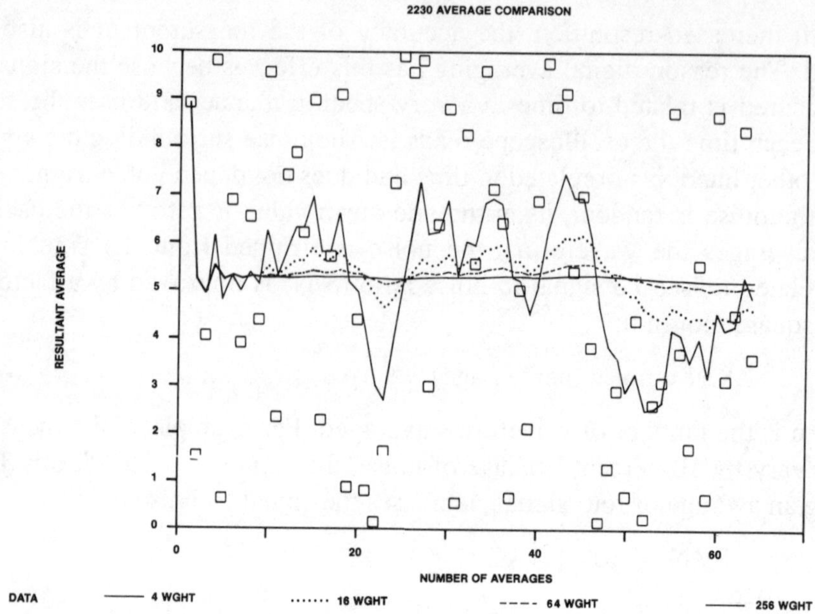

Figure 6-21 Effect of weighting

Figure 6-21 indicates the difference the weighting algorithm has on a series of randomly generated numbers.

The above examples are certainly not the only signal processing capabilities that digital storage offers. Two different signals can be digitally added, subtracted, multiplied, or converted to an X-Y display. Some instruments can integrate or differentiate a waveform at the touch of a button. In other words, just about any math function is possible given the right instrument. Several digital storage oscilloscopes also contain a GPIB connector for data transfer to and from a computer-based system for even higher-level processing of the signal. For more information about these oscilloscopes, refer to Chapters 7 and 8.

Applications

Digital technology in a storage oscilloscope offers abundant advantages. Compared to analog storage instruments, digital oscilloscopes are easier to use and have more capabilities. This section contains a description of how to use digital oscilloscopes.

As with any storage oscilloscope, waveforms may be stored or saved for further examination. In both cases, the digital storage oscilloscope substitutes push-button controls for several interactive potentiometers on an analog oscilloscope. In fact, many digital storage oscilloscopes are programmable, permitting the setting of display parameters from a predefined software program.

Digital storage oscilloscopes may have two or more channels of data acquisition and a time base that behaves just like nonstorage oscilloscopes, Some will have delayed sweep capabilities, or plug-in modules for special-purpose applications. Generally, if you know how to use a nonstorage oscilloscope, you can operate a digital storage oscilloscope with little, if any, additional effort.

ROLL MODE. Some digital oscilloscopes offer a display feature called *roll mode*. In this mode, new information is acquired by the oscilloscope and used to constantly update the screen. This involves continuous scanning of the scored waveforms, with the display moving from right to left on the screen. The effect is like viewing a strip chart recorder.

PRE-TRIGGER. With digital storage oscilloscopes, the point at which the trigger occurs does not have to be the first. Data can be stored in any proportion before and after the trigger point. If you are operating the oscilloscope so that the display begins before the trigger point, you have selected pre-trigger viewing and can see what happened prior to the displayed trace.

Pre-triggering is possible because a digital storage scope is constantly translating the input voltage into binary words. The trigger does not actually start the recording; it is merely a reference point.

A typical application of pre-triggering is catching glitches on the run. Glitches are exasperating creatures. Sometimes they are there; sometimes they are not. This disappearing act makes them hard to catch, particularly if the glitch is the trigger of a more severe malady.

Glitches, by their very nature, are fast. This, obviously, places restrictions on their recovery and display. What makes them particularly aggravating is that they occur when they want to — not when it is desirable. Using the pre-triggering mode, it is possible to review the events which lead up to a particular occurrence. Such a study is not possible with analog storage devices.

7. COMPUTER-BASED TEST INSTRUMENTS

Personal-computer-based test instruments have fast become the quick and easy way for companies of all sizes to realize the advantages of computer-aided testing (CAT) without incurring the enormous expense of program development and mainframe time. Such instruments are able to meet the goals of reduced labor cost, increased productivity, and the elimination of human error in reading and processing measurements, by utilizing the power of software to perform many of the functions traditionally done by workbenches loaded with hardware.

The general design of these instruments is around a desktop personal computer, such as the IBM PC or the IBM PC AT. PC instruments, as they have come to be known, are categorized as being either internal or external.

Internal Adapters

PC instruments designed for internal use are fabricated on one or more computer adapter boards. These adapter boards are physically identical to the video and I/O adapter cards required by the computer for normal operation. To install the instrument into the computer, the card is simply plugged into an available expansion slot on the motherboard, and the system is powered up as usual. Use of the test instrument is then controlled by software from the computer's keyboard. There are no knobs, switches, or indicators available to

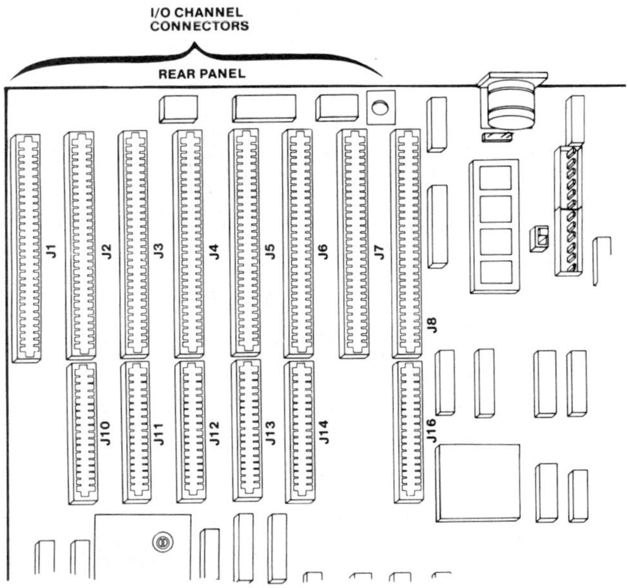

Figure 7-1 AT motherboard (courtesy IBM)

the user. The entire operation of the instrument is through the computer and its interfaces.

An example of an internal-adapter PC-test device is an analytical oscilloscope from R.C. Electronics called the COMPUTERSCOPE-IND IS-16. The IS-16 Data Acquisition package consists of a 16-channel analog-to-digital conversion board, an external instrument interface box, and appropriate software.

In operation, the IS-16 offers a 1-MHz aggregate sampling rate capability on 16 individual input channels with 12-bit resolution at input voltages within the range of –10 to +10 volts. Fully automated keystroke commands (which programmers call hotkeys) provide the user with control over all features of the instrument, including channel selection, trigger control (internal to any channel, external, +/– level or slope), sampling rate, and memory buffer size from 1K to 64K. In effect, the hotkeys are designated keys on the computer keyboard that act like the knobs and levers found on an oscilloscope's front panel.

Beyond the simulated mechanical aspects of an oscilloscope, the IS-16 employs a ring buffer that allows the capture of data in pretrigger intervals of virtually any length. Software commands permit timebase expansion and contraction, left and right scrolling, independent vertical gain adjustment, and waveform storage and retrieval.

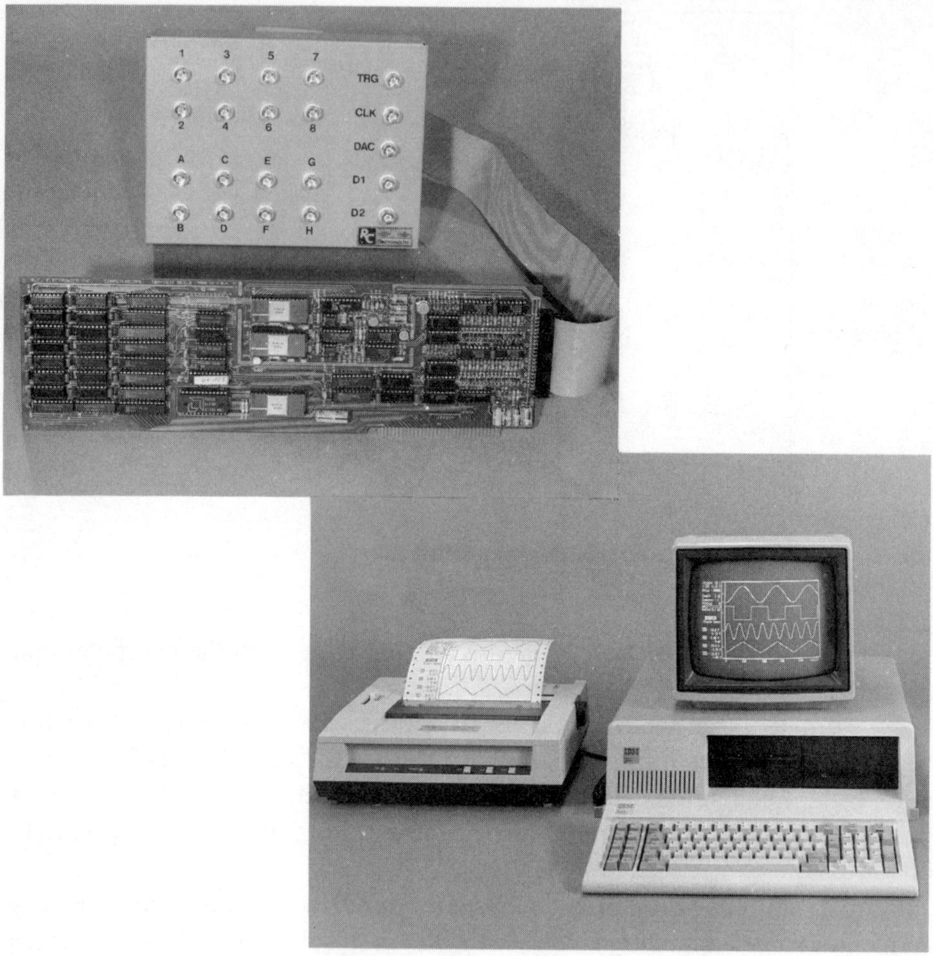

Figure 7-2 The IS-16 adapter card plugs into any slot of an IBM PC personal computer, transforming the system into an analytical oscilloscope (courtesy R.C. Electronics, Inc.)

It is this last feature, waveform storage and retrieval, that sets the PC-based IS-16 oscilloscope apart from its mechanical counterpart. Input measurements can be stored in computer files or temporary computer memory (RAM) for archival purposes or further processing. The fact that the

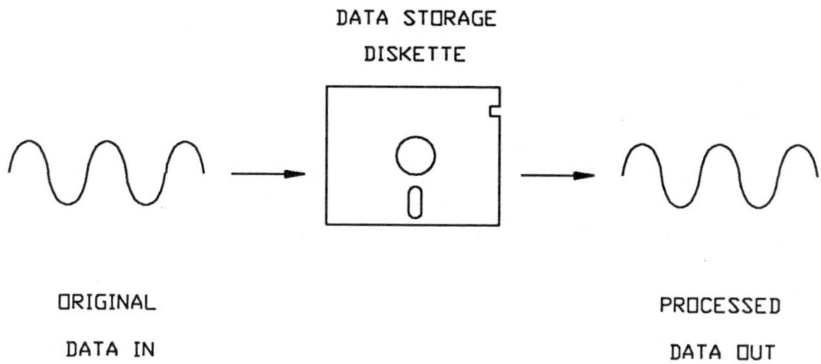

Figure 7-3 Computer diskettes save measured raw data for subsequent processing

dynamic input becomes static in nature provides the user with the ability to modify the contents of an existing file in many useful ways, so that subsequent analysis of the signal may be effectively performed. The entire operation is done in software, following the acquisition of the input signal.

The program begins by creating a verbatim copy of the input signal in a computer data file, using strings of binary words to represent instantaneous events and values in real time. Once saved, the entire measurement procedure can be duplicated down to the last detail by simply playing back the file record into the computer, in much the same way music is forever captured on a record disk or magnetic tape and played back upon demand. Each channel of data may be saved and processed separately using this method.

The saved data may now be processed by the computer's software to derive measurements not initially performed by the PC instrument. Table 7-1 gives a brief summary of the procedures that may be performed on the raw data. When one realizes that all these measurements can be made after the fact, the implications are staggering. One-time events that are elusive to conventional measurement can be done routinely in the leisure of an office setting by simply running the recorded data through the computer one more time using the appropriate software selection.

Furthermore, the results can build one on another. For example, in the first pass, noise and amplification factors may be adjusted, resulting in a data file that can now be averaged, integrated, or differentiated to see the effects of each, without ever having to change the physical setup of the test instrument or device under test.

GLOSSARY

Signal averaging: Provides greater resolution by averaging the results over a user-specified period of time.

Full-wave rectification: Data points with voltage values below a user-specified threshold are converted to voltage values above the threshold, relative to the initial difference in absolute magnitude.

Data inversion: The voltage value for each recorded data point is inverted in its relationship to a user-specified threshold.

Gain adjustment: Data points may be reassigned voltage values according to a new gain factor.

Integration: Channels may be integrated individually with a user-specified time constant, the results of which are adjusted in their temporal relationship so that no time lag is introduced.

Differentiation: The analog waveform is converted to its first derivative.

DC offset adjustment: The value for each recorded data point is increased or reduced by a user-specified constant, thus adjusting baseline voltage.

Algebraic functions: Each waveform may be added, subtracted, multiplied, or divided by a user-specified factor.

Digital filtering: Five differential digital frequency filtering functions — low pass, high pass, band pass, band stop, and notch — permit the user to determine the cutoff frequencies and roll-off efficiency independently for each input channel.

Three-point symmetric smoothing: A three-point symmetric smoothing function with double weighting on the piddle point provides the ability to attenuate high-frequency noise in the data.

Table 7-1 Summary of Computerscope IS-16 computer-aided measurements

External PC Instruments

Such performance is not limited to internal adapter devices or PC-based oscilloscopes. Similar results can be achieved using externally connected PC instruments.

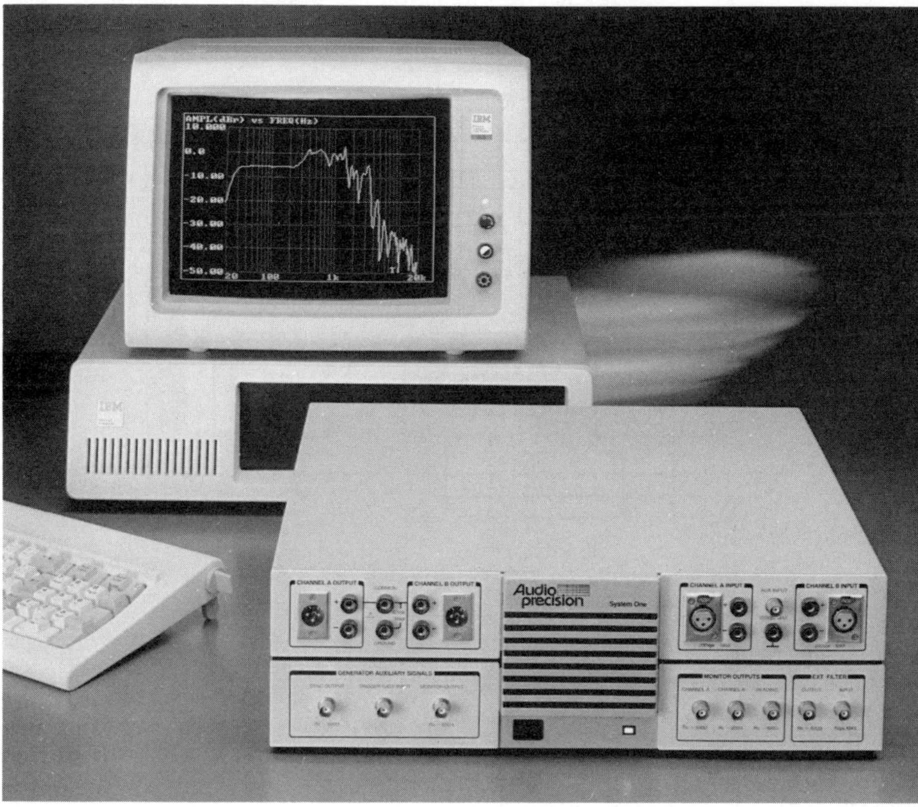

Figure 7-4 System One audio analyzer (courtesy of Audio Precision)

In an externally connected PC instrument, the test instrument is housed in a conventional cabinet external to the computer's cabinet. Installed in the cabinet may be input-output connectors, selector switches, and maybe an LED or two. There is also a connector cable coming from the external instrument that plugs into the PC.

The connection between the external PC test instrument and the computer is in the form of an umbilical cord that feeds voltage, data, and control signals between the two devices. Without this connection, the test instrument is little more than a stand-alone device with features less than comparable to the instruments examined earlier in this book. When connected to the computer, however, it becomes a virtual powerhouse.

Inside the computer is an interface card that translates the signals coming from the external measuring instrument into digital pulses with voltage levels and timing requirements compatible with those on the computer bus.

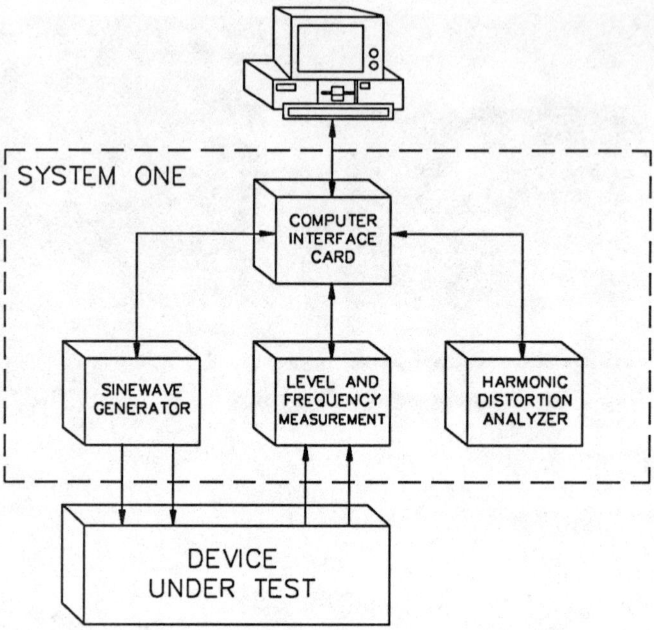

Figure 7-5 System One block diagram

An example of an external PC-test instrument is the System One audio analyzer from Audio Precision. The System One is designed as a computer-interfaced audio test station capable of performing more than 36 standard performance tests on audio amplifiers, magnetic tape systems, and related audio components.

The System One unit contains of an ultra-low-distortion (0.0008% at mid-band) sine-wave generator, a level-detector and frequency-measurement amplifier, a harmonic distortion analyzer, and a computer interface card.

The System One has no controls or displays other than power on/off. Located on the front panel of the instrument are the audio connectors required to interface the System One with the device under test. All user-related functions of control, such as selection of test type, level adjustment, frequency output, and display are accomplished via the keyboard and display of an IBM PC or PC-compatible personal computer.

Like the COMPUTERSCOPE-IND IS-16, all equipment and test functions are controlled by predefined keystrokes on the PC keyboard. Because most audio measurements are sets of measurements rather than an individual spot measurement, the keyboard is programmed to execute an entire series of measurements at the touch of a key, when requested.

Amplitude
Phase
Noise
Common-mode rejection ratio
Crosstalk
Equalization curve
Frequency response
Frequency drift
Phase jitter
Tape noise
Tape frequency response
Head azimuth alignment
WOW & Flutter
THD amplitude
THD frequency
THD versus power output

Table 7-2 System One test functions

Generally, audio tests are performed using sweep frequencies that span beyond the range of the device under test. Some of the tests available are listed in Table 7-2.

The type of test performed, such as frequency response, is selected from a menu of standard tests. In these tests, the sweep range, rate, and amplitude are preset, according to SMPTE, DIN, CCIR, EIA, IHF, NAB, and other established audio standards. Each of these test patterns is recorded within the software of the System One application package for easy access through a screen menu.

The user may also change any of these parameters before initiating the sweep by calling up a control panel screen on the computer. The control panel is essentially a screen version of a mechanical control panel, with the difference that values are entered directly in numerical quantities from the keyboard. There is no twisting of the knob or changing of the ranges. A frequency change is as simple as locating the line displaying the current frequency and entering the new value. While conventional hardware panel instruments are limited to the units of measurement for which the designer has the panel room and the meter scale space, the choices of the System One panel values can be changed to include volts, dBV, dBu, watts, dBm, or percentage for any specified impedance or function. If desired, the new parameters may be saved to a computer file for future use.

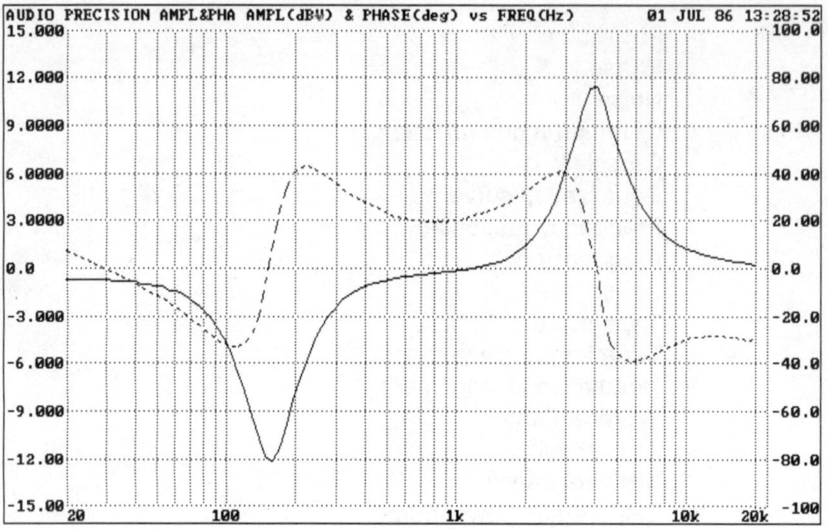

Figure 7-6 Typical System One screen display

Once the testing procedure starts, the System One software begins its measurement of the unit under test. During the course of testing, the measurements are displayed on the computer's video monitor. The result is an X-versus-Y graph representation of the test(s) according to the response of the unit under test. A typical screen display is shown in Figure 7-6. Multiple line charts may be displayed by entering the proper values into the opening menus.

Test measurements are also saved to a computer file. As with the PC oscilloscope described above, these files may be played back for re-enactment of the tests performed or verification of test results. Unlike the COMPUTERSCOPE, the System One does its testing in real time. Because of the nature of the tests involved, measurements are more accurate when programmed to be performed on the unit under test at the time it is connected to the test equipment than are measurements derived from calculated values. That is not to say that new measurements and conclusions cannot be derived from the data in the files created by the System One, because they can. Therein lies the advantage of the PC-based test instrument over other types of testing devices.

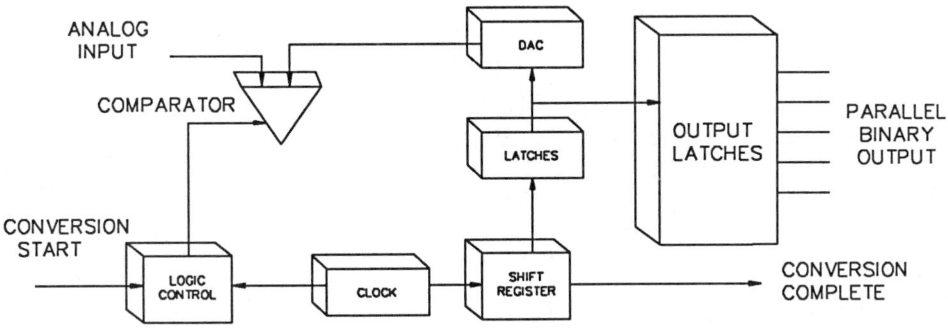

Figure 7-7 Successive approximation register ADC block diagram

Analog-to-Digital Conversion

Central to the performance of any PC-based test instrument is an *analog-to-digital* converter, or ADC. The ADC converts an analog input voltage into a representative binary word. For example, an input of 2.00 volts may be represented by the binary word 0011001000, whereas 2.01 volts would be converted to 0011001001. The accuracy of the conversion is totally dependent on the number of bits in the binary word. In the above examples, the values are accurate to 10 places, or 10 bits.

Basically, there are two analog-to-digital conversion methods used for PC-based test instruments. They are the successive approximation register type and the flash converter.

SUCCESSIVE APPROXIMATION REGISTER CONVERSION. Until the advent of monolithic electronics, the *successive approximation register* (SAR) method was considered the black sheep of the ADC family because of its slow conversion rate. But with today's technology — ultra-fast comparators, fast current switching, and ECL logic — very useful performance is obtainable from successive approximation converters.

An attractive feature of the successive approximation technique is that it requires very little hardware. Consequently, it is possible to integrate an ADC of this type into a single chip with 12-bit or higher resolution at very little cost.

The successive approximation converter, a block diagram of which is shown in Figure 7-7, uses a digital-to-analog converter (DAC) in a feedback

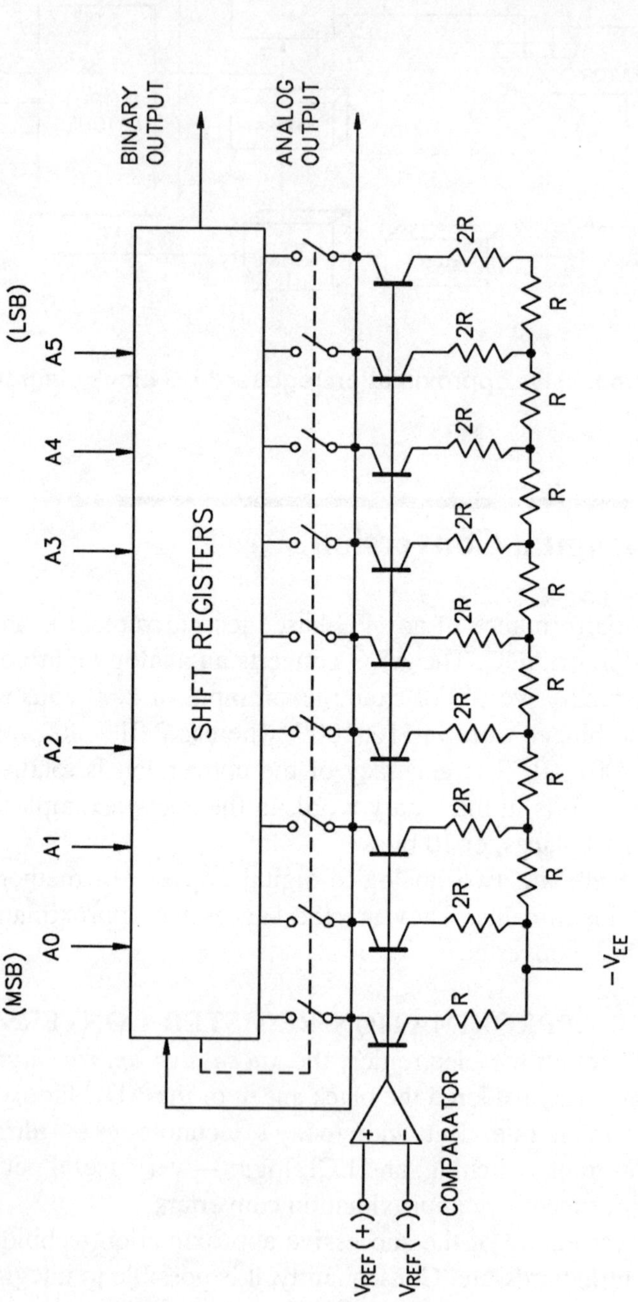

Figure 7-8 Block diagram digital-to-analog converter

loop. The digital-to-analog converter is generally a resistive network that generates an output voltage proportional to a digital code, using precision resistors in a ladder array. With a constant current passing through the resistor array, the output voltage is proportional to the total resistance, which is dependent on the number and value of the resistors inserted in the circuit by a binary code.

In operation, the shift register sets a one (1) in the MSB latch of the approximation register and resets all other latches to zero (0). The digital converter's (DAC) analog output is then compared with the analog input voltage to determine which value is higher. If the analog input voltage is less than the output voltage from the DAC, then the MSB latch is cleared to zero (0) and a one (1) is placed in the next-most-significant bit register. All this takes place within a single clock cycle.

During the succeeding clock cycle, the new DAC voltage is compared to the analog input voltage, and a determination is made as to which is greater. For the sake of discussion, let's assume that the analog input signal is greater than the digitally simulated output voltage from the DAC. When this is the case, the control logic retains the one value in the current latching register and places a one (1) into the next latch in line. On the next clock cycle, this new binary value, 0110000..., is compared to the input analog voltage and a decision made about whether to keep the newly inserted one or discard it.

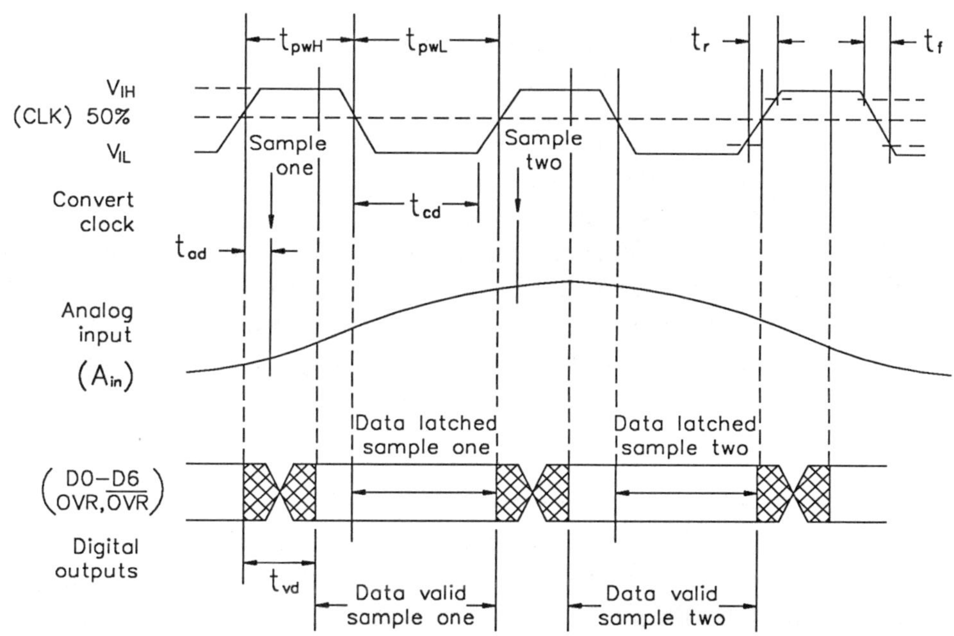

Figure 7-9 Flash converter timing

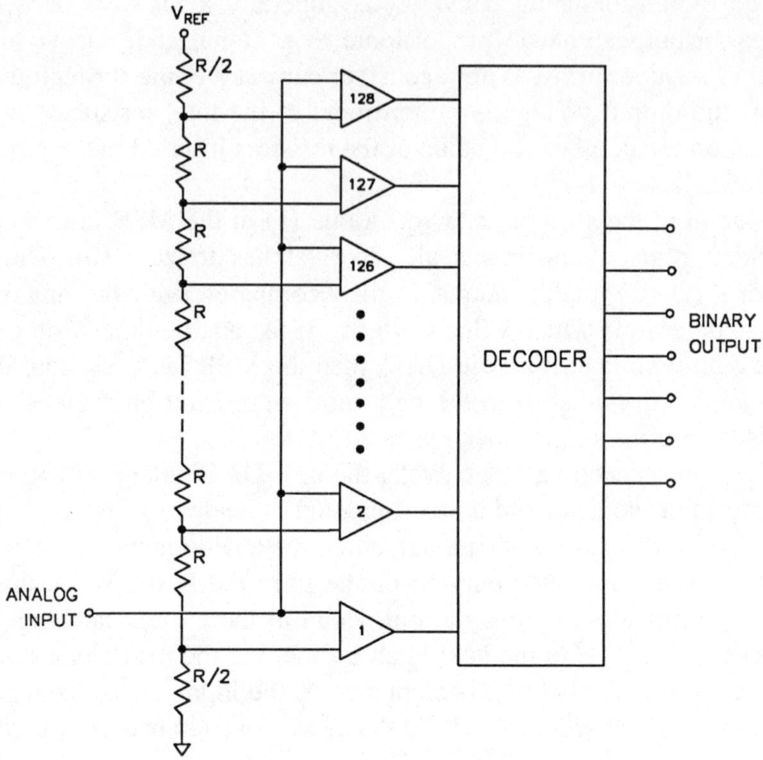

Figure 7-10 Flash-converter block diagram

During successive clock periods, this process is repeated over and over with bits of diminishing significance until the register is exhausted. The DAC output therefore becomes a progressively more accurate approximation of the analog input voltage, taking N clock periods to achieve N-bit resolution. At the end of the conversion period, the binary word is read into the computer as an input value and another conversion started. While conversion rates as high as 15 MHz are possible using current design techniques, most successive approximation converters for use with PC instruments operate within the 1-MHz to 5-MHz range.

FLASH CONVERTERS. As economically attractive as the successive approximation converter may be, it cannot possibly address the issues of video frequency or digital sampling applications. For such applications, another type of ADC called the *flash converter* is used.

The flash converter performs a conversion and outputs data within a single clock cycle. Referring to Figure 7-9, note that the clock input is sensitive to the rising and falling clock edges. All significant operations are referenced to these edges.

Analog-to-digital conversion is accomplished through a parallel array of voltage comparators. In the block diagram of Figure 7-10, 128 parallel comparators are used to achieve a resolution of 7 bits.

The reference voltage to each of the 128 comparators is derived from a uniformly spaced voltage divider that spans both the positive and negative range of the input signal. All 128 sampling inputs, on the other hand, are commonly tied together and connected to the analog input voltage.

As the analog input voltage changes from one value to another, comparators are turned on or off according to the value of the analog input to the reference voltage. At an absolute value of zero, no comparator generates an output signal; at V/2, 64 of the 128 comparators indicate an output signal. These outputs are analyzed by a logic circuit, called a 127-of-7 encoder, and converted into a binary number representing the number of comparator gates turned on and the number turned off. On the falling edge of the clock pulse, this value is locked into the output register and is input to the computer for evaluation.

Computer Interface

Thus far we have indicated that the digital data bits from the PC instrument are input to the computer for processing, but little has been said about the method of transfer. In the case of the internal PC instrument, the data bits are transferred directly to the computer using the computer's internal data bus. The only interface to the outside world by the computer is through the test instrument's input/output ports.

In the design of external PC-test instruments, provisions must be made to interface the external box to the internal computer bus. In the case of the System One audio test station, the connection is made through a proprietary communications link to a digital interface card inserted into a computer expansion slot. The interface card then translates the digital input pulses into signals that the computer bus understands. Many external PC-based test equipment manufacturers have taken this route, using an enormous variety of proprietary interface schemes.

In some cases, the external test instrument manufacturer attempts to standardize its interface, making it compatible with a wide range of computers and not limiting it to the IBM PC family. One such approach is the use

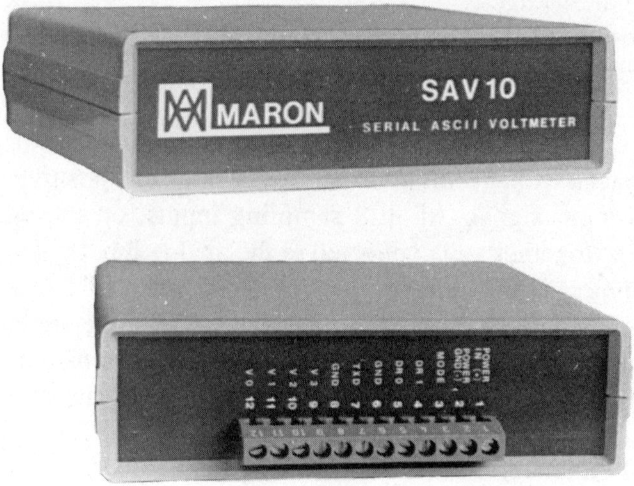

Figure 7-11 SAV 10 Serial ASCII Voltmeter, front and back views (courtesy of Maron Production, Inc.)

of the RS-232 serial interface port available on every computer and many other types of devices.

An example of a typical RS-232 interfaced PC instrument is the SAV 10 Serial ASCII Voltmeter from Maron Pròduction Inc. The SAV 10 is a multi-channel ASCII voltmeter that monitors four analog voltage inputs, converts the information to digital values, and transmits them as ASCII encoded messages in a format compatible with standard data display terminals and computer RS-232 serial ports. The unit requires no control messages from a host computer and operates independently of the computer system. If effect, the SAV 10 is a satellite data-gathering device that inputs raw data into the computer for subsequent storage or processing.

GENERAL-PURPOSE INTERFACE BUS. As the need for electronic measurements has grown in number and complexity, more and more instrument users have come to realize that their traditional design and test procedures are inadequate. The modern technician and engineer have new needs that can be satisfied only by assembling individual test instruments into interactive, automated systems. With more than one million personal computers in the hands of engineers and scientist, it was inevitable that instrument measurements and automated testing would find their way into the realm of desktop computing.

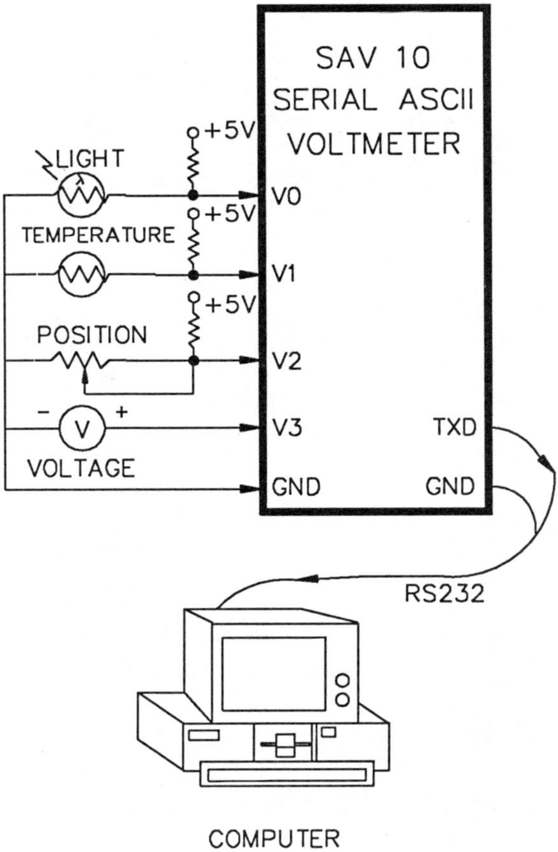

Figure 7-12 Typical SAV 10 test setup and computer interface

The first major step toward this goal was taken by Hewlett-Packard in the early 1970's with their introduction of the Hewlett-Packard Interface-Bus (HP-IB). The HP-IB is essentially a communications link that allows one instrument to talk to another over a standard electrical bus composed of many wires. Before HP-IB, connecting programmable test instruments to a computer desktop calculator or local area network (LAN) was a major undertaking. Different standards among instrument manufacturers created proprietary interfaces that tied the user to one manufacturer and denied use of test instruments not available from that supplier. The HP-IB standard once and for all defined an interface that made it possible to put together computer-controlled test systems.

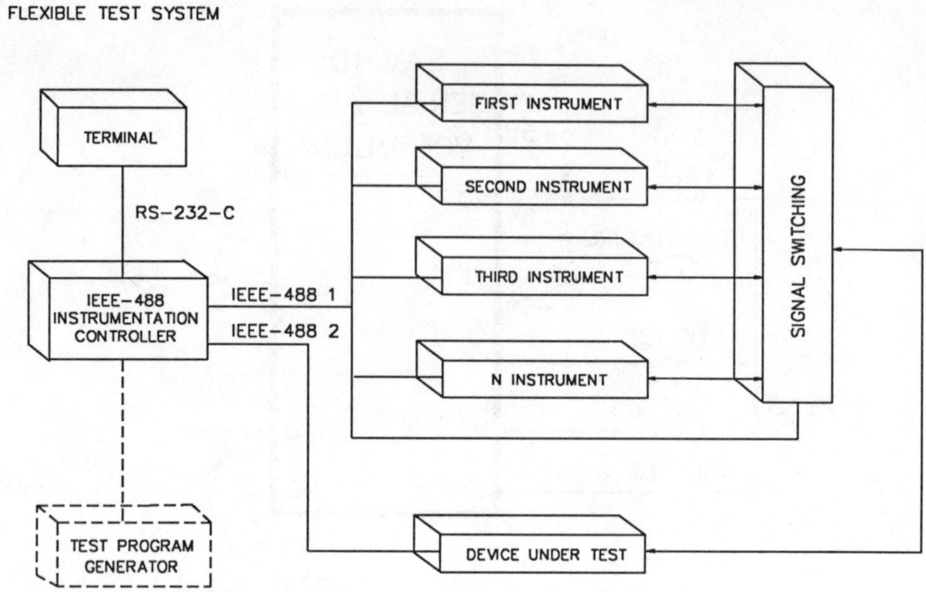

Figure 7-13 Mix and match instrument modules using the GPIB

In 1975, the HP-IB standard was adopted by the IEEE committee as the IEEE-488 standard. The IEEE-488 standard was updated in 1978 to IEEE-488-1978 and officially labeled the General Purpose Interface Bus, or *GPIB*.

The IEEE-488 interface is an asynchronous, handshaking-oriented parallel computer bus that uses communication protocols similar to Digital Equipment's LSI-11 bus (which is based on DEC's extremely popular PDP-11 minicomputer system). Central to the operation of any GPIB system is the *controller*. The controller is analogous to a telephone operator in that it receives, directs, and routes messages across the network. The GPIB uses talkers and listeners to communicate data over the IEEE 488 bus. Talkers send data to listeners via the controller. GPIB devices or instruments can assume one of three communications roles: talker, listener, or talker/listener.

A talker, which is normally referred to as a *talk-only* device, can only send data over the bus; it cannot receive data or commands. A common example of a talk-only device is a voltmeter that only sends measurements. Talkers are generally not desirable for IEEE-488 test systems because the controller cannot turn them off or on. Such devices, like the voltmeter, must be actuated by a separate trigger line or front-panel pushbutton. A system containing a talk-only device is generally a very simple setup with only two components, such as a voltmeter and a data printer. In this arrangement, the printer would periodically print out voltage readings as they were generated

Figure 7-14 Typical GPIB address selector

by the voltmeter. When a talk-only instrument is used in GPIB systems with more than one other device, however, the talk-only device could end up monopolizing the bus and interfering with commands issued by the controller and/or data on the bus coming from other instruments.

An instrument that is a listener, or a *listen-only* device, can only receive data over the bus and cannot transmit. Although listen-only devices are not common in IEEE-488 test configurations, they are practical for some non-measurement devices. A good example of a listen-only device is a power supply whose function is to receive voltage and current programming information over the bus rather than send information. Even in this case, however, it is beneficial for the power supply to have talker capabilities in the event of a failure. An out-of-spec warning signal transmitted over the GPIB bus by the power supply can give the controller an opportunity to disable the power source before the unit under test is damaged.

The preferred instruments for an IEEE-488 test system are *talker-listeners*, which can both send and receive information over the bus. Talker-listener capability can make the tasks of programming, troubleshooting, and maintaining IEEE-488 test systems much easier. Most new GPIB instruments are of the talker-listener variety. When the application requires stricter control, these instruments can usually be configured for talk-only or listen-only operation, using either a software command or a hardware switch.

By combining the various features of the GPIB with available GPIB test instruments, one can achieve a well-conceived test system capable of performing many measurements simultaneously, with data transfer between the GPIB devices for storage, print out, and evaluation. An in-depth analysis of the GPIB, its applications, and implementation are taken up in Chapter 8.

PC-based Workstation

There are, of course, applications for which the owner of a PC may want higher performance than that offered by the GPIB alone, with its odd assortment of devices. Such systems are generally referred to as *PC-based workstations*. In a PC-based workstation, instrument modules with matched performance and compatibility are clustered around a PC computer. The PC-based workstation allows data from multiple instruments to be rapidly gathered, stored, graphed, analyzed, and incorporated into report form on a single PC. Systems of this type are generally put together by one manufacturer to enhance performance and cooperation between the instrument modules involved.

An example of a PC-based workstation is the PC Instruments system introduced by Hewlett-Packard. PC Instruments is a series of nine dedicated instrument modules designed to operate over a dedicated interface bus by an HP Touchscreen II personal computer, an IBM PC/XT/AT, an AT&T PC 6300, or an HP Vectra PC. Among the instrument modules available for PC Instruments are a 50-MHz digitizing oscilloscope, a 12-bit dual-channel digital-to-analog converter, a 100-MHz universal counter, a 5-MHz function generator, and a digital multimeter.

Test data gathered from the instrument modules are fed directly to the PC for storage, display, and analysis. The instrument modules themselves have no meters or displays of their own. All outputs, including waveform displays, are through the PC screen, and up to eight different instrument outputs may be displayed simultaneously.

PC Instruments communicates with the PC via the PC Instruments Bus, or PCIB. PCIB is implemented with a single interface card that plugs into one expansion slot of the PC motherboard. Each PCIB interface card supports up to eight instrument modules. Installing a second PCIB card into the computer permits support for an eight additional instrument modules.

Figure 7-15 PC Instruments (courtesy of Hewlett-Packard)

The PCIB is unique in that it supports both parallel and serial communications. At first glance, the idea of having two communications channels in a single bus system appears to be redundant. On closer analysis, however, a cost saving is realized and the redundancy is reduced.

In a typical GPIB system there are instruments, such as voltage and current meters, that require the inputs and/or outputs to be isolated from earth ground and the computer so that floating measurements can be made. The isolation is typically provided by optoisolators communicating the data from the GPIB side to the instrument in serial fashion using a parallel-to-serial converter. Data coming from the instrument likewise must pass to the GPIB through an optoisolator using a serial-to-parallel converter. Consequently, every device that requires isolation also requires its own pair of optical isolators and respective converters.

In PC Instruments, the test instrument may choose between an isolated serial communications channel or the faster, nonisolated parallel communications bus. The parallel-to-serial and serial-to-parallel converters for the serial channel are placed on the interface card, thus reducing system cost while improving performance.

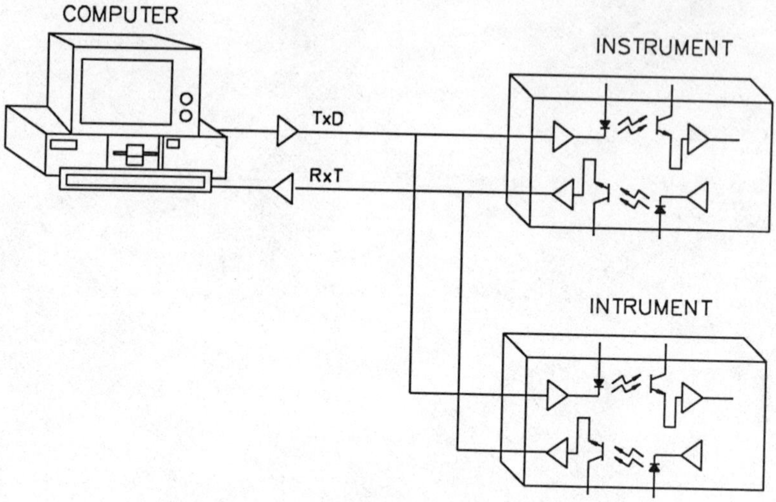

Figure 7-16 PCIB serial communications channel

The PCIB parallel communications channel (Figure 7-17) offers a high-speed data path for instruments that do not need isolation. Data can be transmitted at rates up to 100K bytes/second, subject to limitations of the host personal computer. The parallel communications channel consists of an 8-bit data path, two control lines, two handshake lines, and an interrupt request line (SRQ). The 13 signal lines, with appropriate ground returns, are part of the 26-conductor ribbon cable that connects the host personal computer to the modular instruments. The PCIB parallel communications channel is also upwardly compatible with the GPIB, making PC Instruments a most versatile system.

One of the most impressive aspects of PC Instruments is the soft front-panel system software, which provides an interactive graphics mechanism for the user to control instruments manually. The control panel can also be invoked directly from either DOS, BASIC, or any other language that produces the correct response, thus lending the system to computer automated testing using customized software routines.

The front-panel software emulates a benchtop stacked with traditional instruments, allowing the user to control one module while viewing them all. It is a simple matter to select a new instrument using simple touchscreen, mouse, or keyboard inputs.

The software for each installed instrument module is contained within the host computer at all times. The performance of each test module has been optimized for both individual performance and interactive cooperation. The

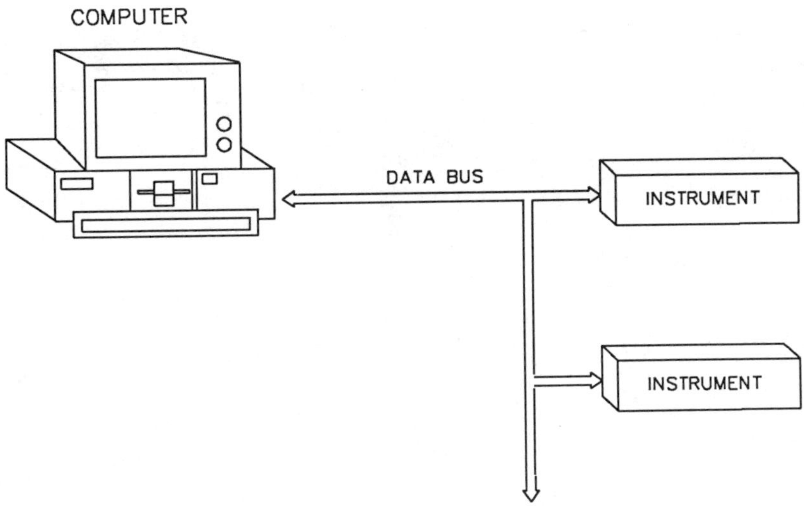

Figure 7-17 PCIB parallel communications channel

software for the digitizing oscilloscope, for example, is the most complex in the current PC Instruments line, and is representative of what the system is capable of doing. The digitizing oscilloscope software package is broken into five major functions: hardware setup, data acquisition and display, user interface, measurements, and program control.

The hardware setup sets the vertical and horizontal sensitivities, delay time, coupling input, acquisition mode, and trigger configuration for the oscilloscope module. When invoked, the setup instructions are sent over the PCIB to the oscilloscope module for execution.

Data acquisition consists of reading a 251-byte data string from the oscilloscope module following a measurement procedure. The data is scaled according to the hardware settings described in the procedure above. The settings may be modified by the user using the front-panel graphics (see Figure 7-18) that are available on every screen. Consequently, changes in the display of the data can be made on the fly.

The measurement package includes control of voltage and time markers and waveform analysis for automatic parametric measurements of rise time, fall time, period, frequency, pulse width, overshoot, preshoot, and peak-to-peak voltage. These measurements are based on a statistical analysis of the 251-byte data string from the oscilloscope module, using a histogram to determine the absolute and relative maximum, minimum, and percentage points, as well as the location of rising and falling edges. Because the data is displayed statistically, not physically, it is also subject to software processing

Figure 7-18 PC Instruments (courtesy of Hewlett-Packard)

in much the same way that signal processing was done with the R.C. Electronics IS-16 COMPUTERSCOPE described at the beginning of this chapter.

The PC Instruments software also allows for data exchange between instruments via the computer. For example, changes seen on the oscilloscope display can be adjusted using software feedback loops to the function generator, relay actuator, etc. One application may be in the distortion measurement of a waveform, in which amplitude may be expected to vary with frequency. By monitoring the output voltage of the device under test, using the oscilloscope's data string and software processing, adjustment to the output level of the function generator can be made to automatically compensate for signal fluctuations.

AUTOMATIC TEST EQUIPMENT. The PC-based workstation, whether purchased as a system like the Hewlett-Packard PC Instruments or assembled from GPIB components, also lends itself well to automated testing environments. Automated testing systems, assembled from *automatic test equipment* (ATE), are a result of the computer being interfaced with digitally controlled stimulus and measurement instrumentation.

ATE is quickly replacing the benchtop setups of instrumentation that were manually controlled to make measurements. Once started, ATE can continue its operation with no outside intervention and can test thousands of parameters in seconds. ATE does not make mistakes when recording test data, does not compromise a test, does not forget tests, and does not get tired. Consequently, ATE dramatically improves throughput and measurement accuracy. It can even correct its own errors by comparing its measurement results to a known standard and compensating for the difference.

ATE test systems may be divided into three classes: benchtop, dedicated, and general-purpose. For manufacturers of semiconductors and large electronics systems, dedicated automatic test equipment is a necessary, albeit hefty, investment. But for thousands of small systems and subsystems manufacturers, the less costly benchtop PC-based workstation is the better solution.

A flexible ATE system usually consists of off-the-shelf instruments joined by an IEEE-488 or RS-232 bus to relieve the incompatibility problem that can occur in linking parts from different vendors. A major advantage of this approach is that it can easily be reconfigured for a variety of test applications by simply swapping one instrument for another.

Flexible ATE is generally fully programmable for applications in which testing is to be done with no operator intervention. Nonetheless, its components should also allow for operator intervention or manual operation when the situation so dictates. For example, manual operation is desirable when an instrument has to function off-line for maintenance or troubleshooting purposes, for minor testing tasks, or for writing and debugging the system test software.

A drawback to ATE is the expense incurred in writing the software. Because writing software can be the largest single task when configuring an ATE system, care must be exercised in the programming language selected and the structuring of the program. In some estimates, the cost of software development can run 90 percent of the cost of the system. Generally, an ATE system will cost four to seven times more than equivalent manual instrument.

The merits of automated testing are not to be debated here. That issue is best left to another treatment. What needs to be said is that, when considering PC-based measurement systems, consider the long-term investment and do not dismiss the advantages of ATE, should that possibility loom in the future. It is much easier to upgrade to ATE status than it is to start from scratch.

8. IEEE-488: THE GENERAL PURPOSE INTERFACE BUS

Although there are several methods by which a personal computer may be interfaced with a PC-based test instrument, the GPIB interface is the most universally accepted. Originally introduced as the Hewlett-Packard Interface-Bus (HP-IB), the HP-IB standard was adopted by the IEEE committee in 1975 as the IEEE-488 standard. The IEEE-488 standard was updated in 1978 to IEEE-488-1978 and officially labeled the General Purpose Interface Bus, or GPIB.

While the concept of the GPIB bus evolved from computer needs, a computer controller is not essential to its operation. The GPIB standard only defines the physical medium by which the devices are interconnected. It is up to the system designer to decide whether the system warrants a computer controller or not. Many systems can suffice with a simple board-level controller.

The GPIB Interface

The IEEE-488 cable contains 16 wires that terminate into a 24-pin GPIB connector. Eight of the lines are reserved for control signals, five of which constitute interface management lines, and eight are dedicated to data signals. Each GPIB instrument or peripheral contains a female 24-pin GPIB connector, typically mounted on the rear panel, to which the GPIB cable con-

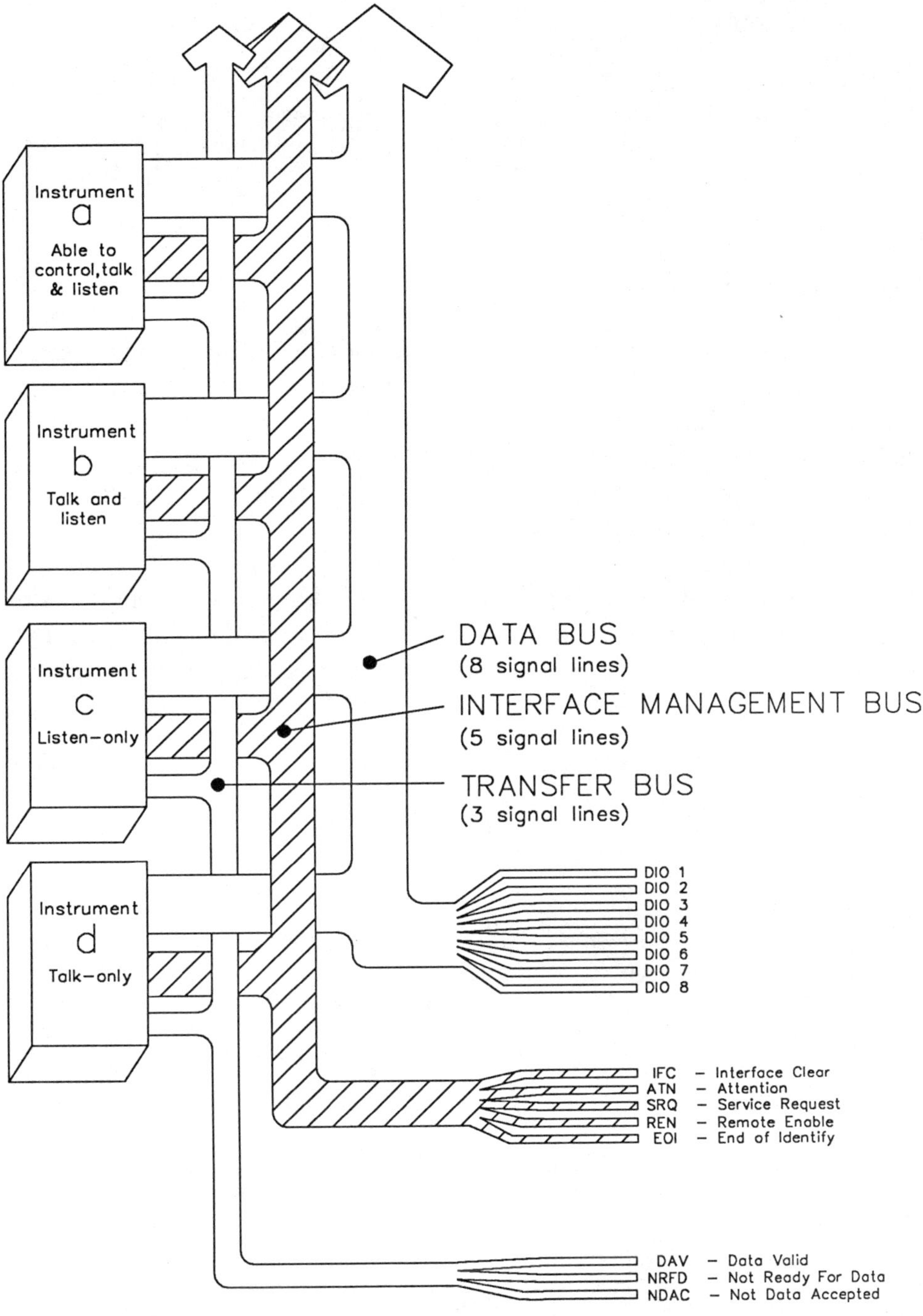

Figure 8-1 General Purpose Interface Bus (GPIB)

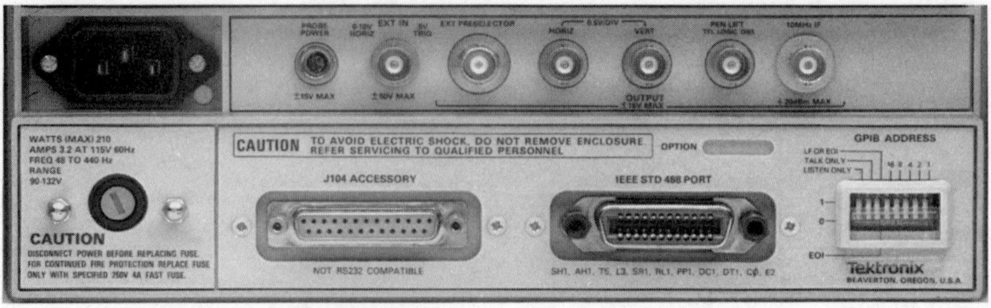

Figure 8-2 GPIB connector

nects. Each end of the cable is molded with both a male and female connector, one above the other, to allow two or more cables to be stacked in a single outlet, similar to the arrangement afforded by multiple strings of Christmas lights. This arrangement permits the GPIB network to be fashioned in either a star or daisy-chain configuration, with mixing of the two configurations permitted within a system.

GPIB COMMUNICATIONS. When the GPIB bus talks, it does so using binary bits organized into words called bytes. The order of the bits in the byte defines its meaning, which the GPIB device interprets as data or a command.

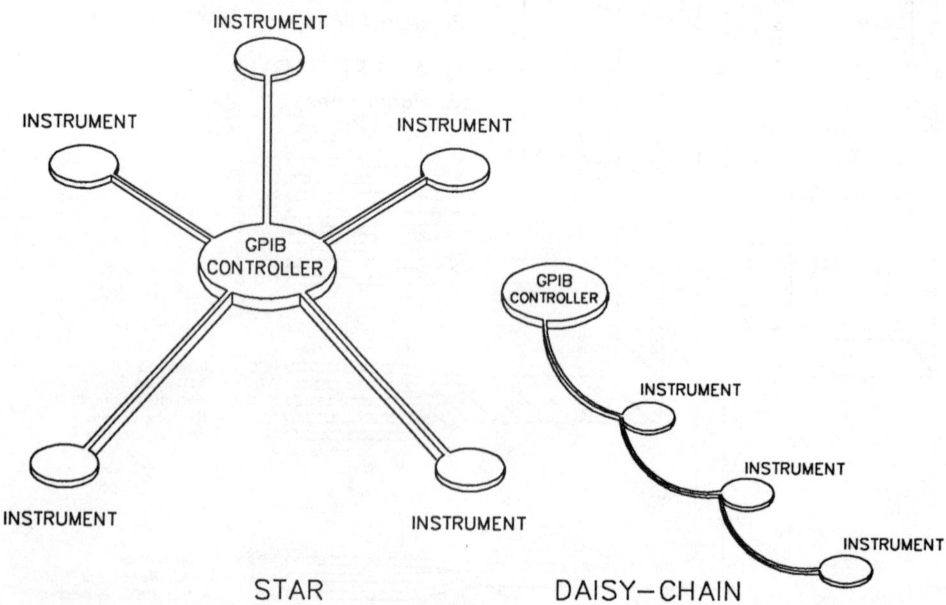

Figure 8-3 GPIB interconnect configurations

The object of the eight data-bus lines is to expediently transport these bytes from instrument to instrument without losing their meaning.

For minimum error rate, a GPIB bus must be properly loaded by adding a GPIB device for every two meters of cable length. A cable run of six meters, for example, requires that three GPIB devices (or their equivalent load) be attached to it. The devices do not have to be equally spaced along the length of the cable, however, and can actually be placed at the ends of the cable, if necessary. Maximum cable length for any single run is limited to 20 meters. Because of noise limitations, however, it is recommended that, whenever possible, a load be placed every four meters. The 20-meter limit on cable length may be extended by devices called IEEE-488 extenders. Basically, an extender is a low-impedance line driver that reduces the effects of induced EMI/RFI and permits cable lengths up to several hundred meters.

The IEEE-488 standard specifies that a maximum of 15 devices — including instruments, controllers, and peripherals — can be connected to the bus at any one time. In a typical system, this means a system controller with up to 14 instruments or peripherals. More than one bus may be used in a system, however, if they are physically separated. Data transfer from one bus to the other is done in software at the controller level.

GPIB ADDRESSING. To keep track of who's who on the GPIB bus, each device or instrument is assigned a unique address called the *primary address.* The system controller uses these primary addresses to route information and commands over the bus to the proper instrument or device. The range of primary addresses spans from 0 to 31, with primary address 31 reserved for control purposes. Primary addresses are generally selected via a binary code associated with a set of rocker switches, rotary switches, or jumper straps located on the back panel or somewhere inside the instrument. There is no particular sequence or order to the numbering of the GPIB devices except that no primary address can be duplicated within the bus. An instrument may be assigned any available address you wish. The exception is when a single device requires two primary addresses, in which case the two addresses are assigned consecutively.

When a controller supports more than one GPIB bus, each bus can have primary addresses identical to those used by another bus, as long as the addresses are not duplicated within the same bus. To differentiate between duplicate addresses on different buses, the controller interface card for each bus has a hardware-selectable *interface number.* When working with multiple controllers and buses, the software address must include both the interface number and the primary address for proper routing of the message.

Looking at the other end of the address spectrum, one finds instruments that use a *secondary address* in addition to the primary address. In such instruments, the primary address provides access to the device, while the secondary address takes us a step further into the device itself. Such an instrument could be a graphics plotter, for example, which might have, in addition to a primary address, two fixed secondary addresses that control the lifting and lowering of the plotting pen. The primary address gets us inside the plotter; the two secondary addresses control the lifting and lowering functions. Secondary addresses also serve in such equipment as a transient digitizer with a programmable mainframe and programmable plug-in modules for support of more than one input channel. For example, the digitizer, as a GPIB device, has both a primary and secondary address. The primary address accesses the digitizer and the secondary address routes the message to the specific plug-in module.

Like primary addresses, each instrument may have up to 31 secondary addresses in the range of 0 to 31. Secondary addresses within a device cannot be duplicated, but secondary addresses between devices may be. If, for example, the system contains two plotters, each must be assigned a different primary address, for obvious reasons. The secondary addresses controlling the pen, however, can remain identical in both units — a technique that is recommended as an easy way to alleviate the burden of remembering which secondary address controls what function.

Instrument addressing begins at the top of the address path with the interface number, which selects the desired GPIB bus. From this follows the primary address, which selects one of the 14 possible GPIB devices on that bus. The final leg of the path leads to one of 31 possible secondary addresses, which controls a function within the GPIB device itself. Those familiar with the structuring of files on a personal computer will recognize the topology as that of an inverted tree, with the root at the top and pathways branching downward. As with computer files, the complete address path, from interface number to secondary address, must be stated exactly and in the correct syntax to arrive at the desired destination.

IEEE-488 Protocol

To make more efficient use of the bus, communications over the GPIB takes place in packets called *message units*. A message unit represents a burst of

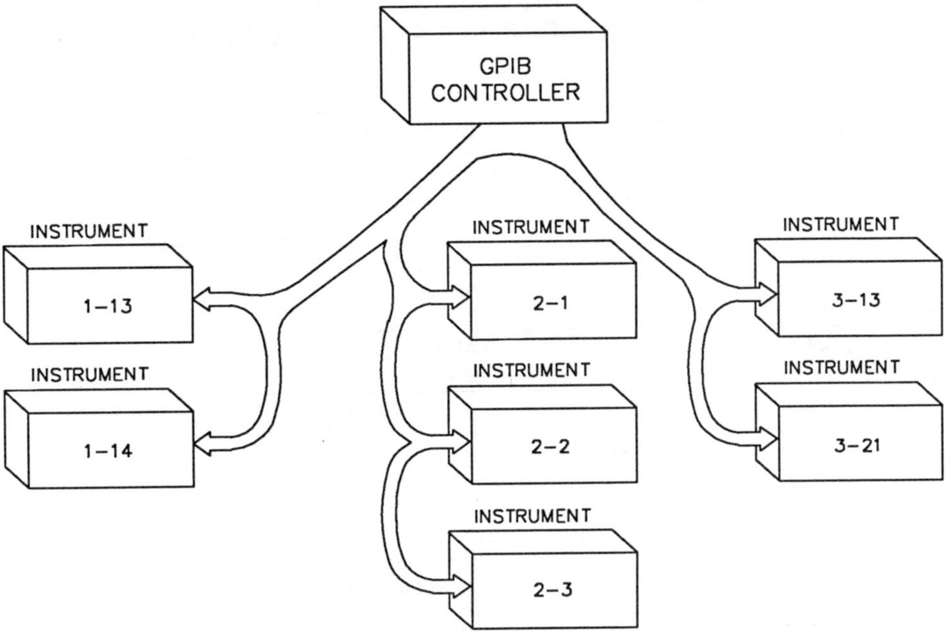

Figure 8-4 GPIB address structure

information that has a very defined beginning and end. On the GPIB bus, there are two types of message units.

The first is defined as an *interface message* unit. Interface messages are commands issued by the system controller to maintain order and initiate proceedings within the GPIB. Altogether, there are ten IEEE-488 commands, called *interface functions*, used for GPIB control and communications. The Talker (T) command, for example, instructs an instrument to commence sending data, Listener (L) enables an instrument's receiver function, Source Handshake (SH) is the synchronizing protocol command, and so forth.

The controller can send two types of interface messages: unaddressed and addressed. Every interface message begins with the controller issuing an *attention* signal. The attention signal consists of a voltage applied to one of the control wires in the GPIB cable (see Figure 8-1). When present, this attention voltage alerts all devices connected to the bus that an interface message is pending, thus forcing all devices with listening capability to receive the next transmission. In the unaddressed mode, the command is received by all devices on the bus that have listener capability. An example of an unaddressed command is the Device Trigger (DT) function, which can be used to synchronize the actions of several instruments.

**INTERFACE FUNCTIONS
DEFINED BY IEEE STANDARD 488-1978**

Function	Description
Source Handshake (SH)	Synchronizes message transmission
Acceptor Handshake (AH)	Synchronizes message reception
Talker (T)	Allows instrument to send data
Listener (L)	Allows instrument to receive data
Service Request (SR)	Requests service from controller
Remote Local (RL)	Allows instrument to select between GPIB interface and front-panel programming
Remote Poll (PP)	Allows up to eight instruments to simultaneously return a status bit to the controller
Device Clear (DC)	Puts instrument in known state
Device Trigger (DT)	Starts some basic operation of the instrument
Controller (C)	Sends device addresses and other interface messages

Table 8-1 GPIB interface functions

In the addressed mode, the command is received by only a single device, that which is addressed. When sending addressed commands, the interface message unit consists of at least two parts. The first part of the message unit contains the primary address of the intended receiver. The second part of the message unit is the interface function itself.

Attention Signal	Primary Address	Interface Function

To enable a voltmeter to transmit data, for example, the controller would begin with the attention signal, after which it would send the primary address of the instrument followed by the Talker (T) interface command. The controller then backs off and waits for the procedure to run its course.

Depending on the instrument involved, it is possible for the interface message to contain a third element: a *device-dependent command*, or, as it is sometimes called, an instrument command. The device-dependent command is used by the system controller to program an instrument for a specific function or situation. This situation may occur in such a piece of equipment as a waveform digitizer, in which more than one parameter is involved. If, for example, one requires a frequency measurement, as opposed to rise time, the interface message could read:

Attention Signal	Primary Address	Interface Function	Read Frequency

This procedure allows devices with similar functions, such as power supplies, to respond individually to different programming procedures. After the address is sent, the commands follow a handshaking procedure in which the instrument acknowledges receipt of the command message character by character.

DEVICE-DEPENDENT MESSAGE UNITS. The information transmitted by the voltmeter or digitizer is the second type of interface message unit. It is called a *device-dependent message unit.* Device-dependent message units contain no commands, only data usually in the form of an ASCII code that relates a value. Here is where things get a bit sticky in GPIB communications.

There is no set format for device-dependent message communications. While most instrument makers define the message unit in ASCII, others may prefer pure binary or even BCD. Even when two manufacturers agree on a format, there is still the question of bit organization; that is, whether the byte is organized least-significant bit (LSB) or most-significant bit (MSB) first. There is also the issue of the header format used to identify the type of data (volts, amps, etc.) contained within the message unit.

The answer to the dilemma lies in a proposed IEEE-P981 amendment to the IEEE-488-1978 standard. The IEEE-P981 standard sets forth a definitive set of communications rules that:

- Establish a common message structure
- Describe communication elements and how they may be combined
- Define control protocol procedures

Taken together, this complete set of rules allows system developers to write simpler system software, with a defined language syntax, while facilitating program modification.

TEKTRONIX CODES AND FORMATS FOR GPIB INSTRUMENTS. The Tektronix Codes and Formats for GPIB instruments, which Tektronix has been incorporating in the design of their test instruments for many years, is one among many device-dependent formats similar in content to the proposed IEEE-P981 standard.

The Tektronix device-dependent message is composed of one or more message units separated by message-unit delimiters. This means that a device-dependent message may consist of a single value measurement, such as a voltage or a current; a group of values related to a single type of measurement, such as a list of voltage measurements taken around a test board; or a string of unrelated values generated by a GPIB device in a mixture of measurements.

The first two cases are rather straightforward and clear. When a GPIB device is asked to report a measurement, the value is placed on the bus and transmitted to the listener(s) by request of the controller. An example of a voltmeter report might simply read:

1.4999

Should the device be required to report a group of measurements, as would be the case with a digitizer reporting the position coordinates of a waveform, the numbers are sent sequentially, one after the other. Separating the values is a delimiting comma. A typical measurement may be sent as:

.732, 1.52

While both of the methods described above appear simple and clear to the user, they may be confusing to the system processor. Consider the situation in which an instrument is required to make and report on two different types of measurements, as would be the case with a waveform digitizer. Among its many other duties, the instrument may be required to give results on frequency and phase measurements. Unless the processor knows which value is related to what parameter, however, things can become a bit tangled. The Tektronix Codes and Formats standard solves the dilemma by requiring a "header" to proceed the value. A header is an ASCII label that identifies the quantity coming up next. When the digitizer wants to speak of frequency, for instance, the transmission might appear as:

FREQ 1.234

while phase angle, on the other hand, would be sent as:

PHASE 56

It is now clear to the system controller and any other device listening in on the conversation what value is related to which quantity. When a series of values of differing quantities are required from a single instrument, the message units are separated by a semicolon, thus:

FREQ 1.234; PHASE 56

Series values within a message unit separated by commas are also permissible, using the following format:

VOLTS 0.795, 2.04; FREQ 1.234; PHASE 56

This format is defined as a mixed data message unit, and its construction can best be illustrated using the framework structure:

header	space	*argument*	*comma*	*argument*	*comma*
character argument		optional; any type		optional; any type	

The first frame entry is the header, which is defined by a *character argument*. A character argument is an ASCII label that identifies the quantity directly following it. It can be used to identify quantities (VOLTS, FREQ, etc.) or commands (SET). The header is always followed by a space.

The remainder of the frame contains optional arguments expressed as either a character argument or *noncharacter argument*. Noncharacter arguments are of many forms, and include numbers, statements, and control characters. Noncharacter arguments are defined in Table 8-2. There is no limit to the number of arguments a single frame can contain, as long as they are separated by a comma. Mixed data message unit frames can be concatenated if the message units are separated by semicolons.

Numbers transmitted in Codes and Formats message units are defined in ANSI X3.42 format. In effect, this format states that there are three types of numbers: integers (NR1), real (NR2), and real with exponents (NR3), as illustrated in Table 8-3: ANSI X3.42 numbers are always sent most-significant character first.

ARGUMENT TYPES AND EXAMPLES	DEFINITION	PURPOSE
Character Argument Trigger	One alphabetic ASCII character optionally followed by any number of ASCII characters excluding space, comma, semicolon, question mark, the control characters and rubout.	Used to transfer alphanumeric data such as message headers, labels, commands, etc.
Non-Character Arguments		
Number –12.3	A numeric value in any of the formats shown in Figure 8-5.	Used to pass numeric values in an ASCII format.
String Argument "Remove Probe"	Opening delimiter (single or double quote) followed by a series of any ASCII characters except for the opening delimiter and a closing delimiter identical to the opening delimiter.	Provides a means for transmitting ASCII text to an output device.
Binary Block Argument Binary Data % 2 bytes wave-form values 8 bits 16-bit binary value; checksum	"%" followed by a 2-byte (16-bit) binary integer specifying the number of data bytes plus a checksum byte which follows the data bytes. The checksum is two's complement of the modulo 256 sum of the preceding binary data bytes. This includes the two bytes comprising the 16-bit integer specification.	Used to transfer large arrays of numeric data such as waveforms in a binary format.
Link Argument NR.PT:1024	Character argument (label) followed by ":" and a value represented in any of the above argument types.	Used to attach a name or label to another argument.
End Block Argument @ ABCDEFGHIJKL E O I	"@" followed by a block of data with EOI set concurrent with the last data byte. End block can only be the last argument in a message and cannot be followed by a message unit delimiter.	Used when a block of data must be sent and neither the amount of data nor its format is known.

Table 8-2 Device-dependent message structure

Format	Examples	Notes
NR1	375 +8960 –328 +0000	Value of "0" must not contain a minus sign.
NR2	+12.589 1.37592 –00037.5 0.000	Radix point should be preceded by at least one digit. Value of "0" must not contain a minus sign.
NR3	–1.51E + 03 +51.2E – 07 +00.0E + 00	Value of "0" must contain a NR2 zero followed by a zero exponent.

Table 8-3 Number formats (ANSI X3.42)

Note, however, that while the number format has been defined, its use has not. The Codes and Formats standard places no restriction on the use of the number. It does not matter whether the number is sent from a multimeter, a counter, or a spectrum analyzer — or whether it represents voltage, current, or frequency. In all cases, the syntax structure of the number is identical.

BINARY BLOCK TRANSFERS. When large amounts of data, such as waveforms, are to be transferred over the GPIB, it is often more efficient to code the data in binary rather than ASCII. Binary data can usually be transmitted much more quickly than equivalent ASCII data because an ASCII number requires one byte for each digit in the number.

For example, transmitting the number 237 in ASCII requires first the code for 2, followed by the code for 3 and the code for 7, for a total of three bytes. Transmitting the same number in binary requires only one byte, the binary equivalent of 237 (11101101). Values up to 65,535 can be transmitted in two binary bytes, and values up to 16,777,225 can be expressed using only three binary bytes. To transmit the same numbers in ASCII takes five and eight bytes, respectively. In the realm of everyday instrument values, binary coding can provide about a 2:1 reduction in the number of bytes required to transfer data over the GPIB bus, as a rule. Obviously, the larger the value, the greater the gain.

Binary numbers are transmitted in Codes and Formats standard using the *binary block argument*. The binary block argument can best be described as a frame within a frame. All binary block arguments begin with a percentage (%) symbol followed by a 2-byte field called the *integer count*. This first

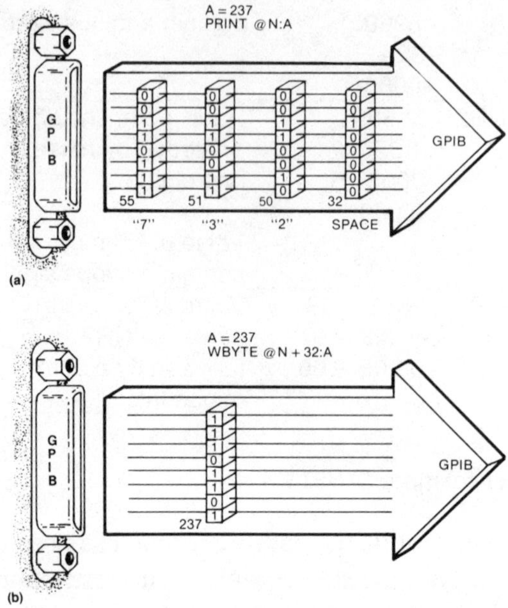

Figure 8-5 Transmitting a number in ASCII might require several bytes, while the same number can be transmitted in one binary byte. The sketch in (a) shows the process of transmitting the value 237 in ASCII, and (b) outlines transmission of the same value in binary. In both cases, the addressing bytes are omitted for clarity.

field contains a binary number that defines the size of the binary block frame, in bytes, excluding this field. This value can be used by the controller to prepare its memory area for the data packet. Following the integer count is the data itself in binary format, most-significant byte first.

At the end of the frame is an 8-bit *checksum*. The checksum is the total sum of all the bytes in the frame preceding it, including the integer count. It is expressed as the two's complement of the modulo 256 sum of the preceding binary data bits, and always occupies an area one byte wide. When the binary block is received, a similar calculation is done by the listener. This value is compared to the value in the checksum. If the two values agree, it is assumed that the data was transferred from sender to receiver without error.

When the system controller or assigned listener receives a binary block, it is usually obliged to accept the entire packet and verify the two error-checking sequences (integer count and checksum), even though only the data bytes are of any value. Not to do so is to invite errors.

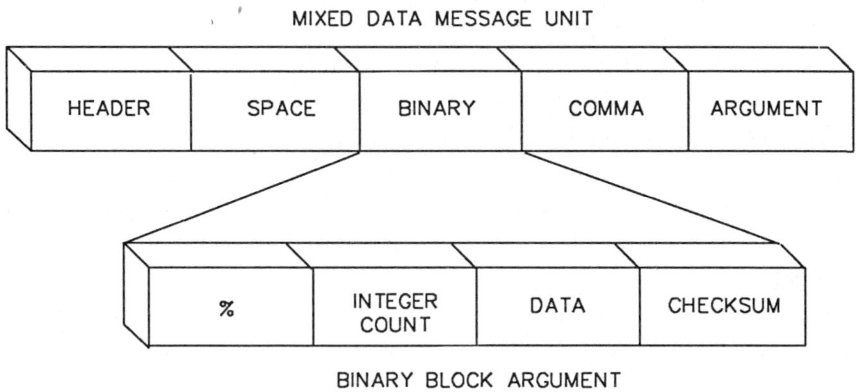

Figure 8-6 Binary block argument

MESSAGE TERMINATION. Because there is no stipulated length for a message, there needs to be a means for identifying the end of a transmission. The IEEE-488 standard defines three methods by which the end of a message may be recognized.

The first involves the use of the printer formatting ASCII line feed character (ASCII 10). Each time a message is completed, the talker appends a line feed character to the message unit. The second method uses a dedicated GPIB control line contained within the GPIB cable. It is called the *End or Identify* line and is labeled EOI. The third IEEE-488 message-termination standard allows the EOI line to be asserted concurrently with a line feed. Any of these three methods can terminate a message, but not necessarily within every system. Some instruments recognize only one as a legal termination, leading to inconsistency in some cases. Most instruments and peripherals allow the user to choose the type of terminator via a switch setting located at the rear of the instrument. The choices usually involve the exclusion of the line feed character, and are described as EOI-only or line feed and EOI together.

One must be wary, though, of arbitrarily selecting a common termination configuration. Even in a system in which all instruments recognize the same termination sequence, problems can arise. This is particularly true of systems that employ binary block communications, such as are commonly found among waveform digitizers. Because a binary-coded byte can configure itself to any form, it is possible for a data byte to coincidentally assume the sequence of the line feed ASCII character (00001010). When the listener detects this sequence, it automatically terminates the conversation, dumping those data bytes that follow and giving an erroneous reading. By setting the instrument to recognize an EOI-only termination signal, the problem is eliminated.

Another problem encountered with message terminations is the use of the carriage return (ASCII 13) terminator used by some instruments. It has been common practice for many an instrument to include a carriage return at the end of the data message just before the line feed character, mostly as a courtesy for not-so-smart printers who need to be told to return the print head to the left-hand margin in addition to advancing the page. That seems simple enough today, but in the past many printers (generally of the TeleType class) were not capable of relating the two events. That was fine until personal computers started finding their way into GPIB systems as controllers. Most personal computers recognize the carriage return as the statement terminator rather than a page advance, as would be the case when entering a keyboard command. This obviously leads to confusion.

Suppose, for example, that a computer controller has requested a voltage reading from a voltmeter over the GPIB. The voltmeter sends the number requested and, as a courtesy, ends the transmission with the carriage return followed by a line feed. The computer, however, interprets the carriage return to be the message terminator, and simply "hangs up" after receiving the CR character, leaving the line feed character in the voltmeter unsent. The next time the computer asks for a voltage reading, the voltmeter responds with the end-of-message line feed character left over from the previous reading before it begins its new cycle. This the computer does not understand, and an error message is issued. The IEEE-P981 amendment resolves the problem by not permitting GPIB instruments to generate carriage returns for any reason.

Designing a GPIB System

Instrumentation that can operate via the IEEE-488 bus ranges from simple switches and power supplies to highly advanced network analyzers and complex waveform digitizers. Because of this broad range, selecting the right system components can be complicated and difficult. Not all IEEE-488 devices have the same bus capabilities. Some permit only bus data transfers but not GPIB setting of the controls. Some can operate as data processors, acting on data secured from another instrument and passing that information on to a third device, while others may not operate together at all.

Basic to the implementation of any GPIB bus is an understanding of the system's purpose. Success with a GPIB system does not depend as heavily on a knowledge of all the details of the IEEE-488 interface as it does on a clear understanding of what the system is expected to do. Knowing what questions to ask when selecting GPIB components can get a system up and running faster and at greater efficiency than all the advice in the world.

After clearly defining the system's job, a list of required instruments can be compiled. A place to start is with the GPIB controller. To adequately direct an instrument system, a controller must have one or more GPIB interfaces and a powerful but convenient software package for handling GPIB operations. Ideally, the controller must be able to address individual instruments, send device-dependent and interface messages, transmit and receive device-dependent data, and handle service-request (SRQ) interrupts using a simple software addressing scheme and high-level command language. When designing a GPIB system, one must realize that a significant part of a system's total cost lies in writing the application programs. Consequently the more powerful and flexible the controller, the less the time and money that must be invested in software.

When searching for the ultimate controller, also keep in mind that the controller can sometimes eliminate the need for costly, specialized instruments by computing results or parameters instead of directly measuring them. If the system includes a waveform digitizer, for example, the controller can compute rms and peak values as well as other parameters from the acquired waveform. Knowing this could result in the acquisition of a cheaper — and generally more reliable — waveform digitizer.

The recent influx of cheap and powerful personal workstations, such as the IBM PC AT, has made the personal computer a strong alternative to dedicated GPIB controllers. The issues of using personal computers in computer assisted test (CAT) systems is discussed at length in Chapter 7.

The GPIB instruments themselves fall into three basic categories. One type uses the GPIB solely for transferring data. These instruments are used strictly as monitoring devices and suit applications in which programming is not required. The second type includes instruments that permit partial or full control of their settings over the bus. Instruments of this class are most often found in modular test setups that require frequent reconfiguration.

The third class of GPIB instruments can store and execute programs. These devices perform processing on board and can send either raw data or computed results to the controller via the GPIB interface. Such instruments might also use the GPIB to transfer programs to and from a peripheral device or controller. This class of GPIB instruments is the most versatile and the most costly.

GPIB INTERFACE FUNCTION SUBSETS. GPIB instruments are rated according to their IEEE-488 interface functions. The IEEE-488 standard

SERVICE REQUEST		**SR0**	**SR1**	
Full Capability	Allows an instrument to request service from the controller with the SRQ line		X	
No Capability		X		
REMOTE-LOCAL		**RL0**	**RL1**	**RL2**
Basic Remote-Local	Allows the instrument to switch between manual (local) control and programmable (remote) operation		X	X
Local Lock-out	Allows the return to local function to be disabled		X	
No Capability		X		
PARALLEL POLL		**PP0**	**PP1**	**PP2**
Basic Parallel Poll	Allows an instrument to report a single status bit to the controller on one of the data lines (D107-D108)		X	X
Remote Configuration	Allows the instrument to be configured for parallel poll by the controller		X	
No Capability		X		
DEVICE CLEAR		**DC0**	**DC1**	**DC2**
Basic Device Clear	Allows all instruments on the bus to be initialized to a predefined cleared state		X	X
Selective Device Clear	Allows individual instruments to be cleared selectively		X	
No Capability		X		
DEVICE TRIGGER		**DT0**	**DT1**	
Full Capability	Allows an instrument or group of instruments to be triggered or some action started upon receipt of the group executive trigger (GET) message		X	
No Capability		X		
SOURCE HANDSHAKE		**SH0**	**SH1**	
Full Capability	Allows a device to generate the handshake cycle for transmitting data		X	
No Capability		X		
ACCEPTOR HANDSHAKE		**AH0**	**AH1**	
Full Capability	Allows a device to generate the handshake cycle for receiving data		X	
No Capability		X		

Table 8-4 IEEE-488 subsets

TALKER (EXTENDED TALKER)*		T0 (TE0)	T1 (TE1)	T2 (TE2)	T3 (TE3)	T4 (TE4)	T5 (TE5)	T6 (TE6)	T7 (TE7)	T8 (TE8)
Basic Talker (Basic Extended Talker)	Allows an instrument to transmit data		X	X	X	X	X	X	X	X
Talk Only Mode	Allows an instrument to transmit data without a controller on the bus		X		X		X		X	
Unaddressed if My Listen Address (MLA)	Prevents an instrument from being a talker and a listener at the same time						X	X	X	X
Serial Poll	Allows an instrument to send a status byte in response to a serial poll		X	X			X	X		
No Capability		X								

LISTENER (EXTENDED LISTENER)*		L0 (LE0)	L1 (LE1)	L2 (LE2)	L3 (LE3)	L4 (LE4)
Basic Listener (Basic Extended Listener)	Allows an instrument to receive data		X	X	X	X
Listen Only Mode	Allows an instrument to receive data with a controller on the bus		X		X	
Unaddressed if My Talk Address (MTA)	Prevents an instrument from being a talker and a listener at the same time				X	X
No Capability		X				

*Extended talkers and listeners use secondary addresses; other talkers and listeners do not.

Table 8-4 (continued) IEEE-488 subsets

divides the interface into ten basic interface functions (see Table 8-1), each of which has a set of options that can be used to implement different subsets of these capabilities. An instrument may have all, some, or none of the ten basic interface functions and/or subsets implemented.

The specifications for a GPIB device should list all the subset functions it supports. Some interface functions, such as Device Trigger, have only two options — that is, either it is available or it is not. Others, such as the Controller function, have multiple options. Each option is listed as a subset of the interface function by an assigned numeric code defined by IEEE-488 standards. An instrument with a T7 rating, for instance, is capable of being a talker, has a talk-only mode available, and has a My Listen Address option that prevents the instrument from being a talker and listener at the same time. A complete summary of interface subsets is provided in Table 8-4 .

When selecting subset ratings for a system, keep in mind the compatibility of the subsets among the various instruments. In some cases, the subset capability of one device may be of no value because other devices in the system cannot provide the same service. When using the Device Trigger function to synchronize system operations, for example, it is imperative that all instruments in the system have the ability to respond to the command; otherwise the feature is wasted. While the GPIB interface provides a rich selection of interface function features, most instruments do not need all the capabilities afforded by IEEE-488. Here is where the engineer makes the decision between performance, design, and cost.

Programming the GPIB

In addition to accommodating the hardware considerations mentioned above, one must deal with the software aspects of the GPIB system. There are two levels of software necessary to operate a GPIB system: the operating system software and user-written application software programs.

The operating system software provides a set of commands and functions that the user combines into application software programs. The application software works through the controller to tell the instruments what signals to generate and what measurements to make, and to tell the controller what to do with the results.

A typical application, for example, might require a digitizer to transfer its data to an X-Y plotter for hardcopy output. The program would likely begin with a call from the controller for all network devices to stop and listen to the upcoming command by placing an attention signal on the GPIB. At this point, the address of the X-Y plotter would be specified, thus allowing the remaining devices to return to whatever it was they were doing, even if it was nothing. Next, a series of commands would be issued to prepare the plotter for the upcoming data. The sequence of commands may be long or short, depending on the plotter involved. Some plotters require a string of separate commands to set the machine, such as lift pen, locate 0,0, clear registers, etc., while another plotter may require only a single command to accomplish the same result.

Following the plotter's acknowledged willingness to cooperate, another attention call is made and the digitizer is addressed. This time the application software programs the digitizer with the proper codes for the required measurements. A final listener (L) receiver-enable command to the plotter

4050 BASIC	PRINT @ 1:"FREQ 1E6"
4040 BASIC	PRINT #1:"FREQ 1E6"
SPS BASIC (LOW-LEVEL)	PUT "FREQ 1E6" INTO @ 0.33
SPS BASIC (HIGH-LEVEL)	PUT "FREQ 1E6" INTO #1
HP 9825 (HPL)	wrt 701, "FREQ 1E6"
HP 9845 (BASIC)	OUTPUT 701; "FREQ 1E6"
FLUKE 1720A	PRINT @ 1%, "FREQ 1E6"

Figure 8-7 A message to a GPIB device from any controller involves at least three major components: the output command, the device's address, and the message itself.

followed by a talker (T) permission-to-transmit command to the digitizer completes the cycle, and the tasks of measuring and plotting begin.

The above scenario is not the ultimate programming solution, nor is it meant to be. Each instrument, each device, and each system has its own peculiarities, and each issue must be addressed individually. The issues of interrupt management, data processing, and a host of other housekeeping chores are also lacking from this simple example.

GPIB application programs are written in a controller language. Unfortunately, each GPIB controller to date has its own programming language. For the National Instruments controller card, it is Microsoft Basic, while instrument makers like Hewlett-Packard, Fluke, and Tektronix have developed their own programming languages. Unfortunately, none of the program statements from one language are transportable to another.

The chart in Figure 8-7, for example, shows the same command expressed seven different ways using seven different GPIB programming languages. All of the statements shown send the same message (FREQ 1E6); the difference lies in the syntax of the statement required by your particular controller. The two programs shown in Figure 8-8 are also identical in performance; that is, they both do the same thing. However, the program written in Microsoft Basic (a) is much more complex than the same program written in Tektronix 4041 Basic (b).

Despite their syntax differences, all GPIB programming languages are designed to do the same thing: move data from point A to point B in the most efficient manner possible. These issues are discussed below.

COMMAND STRUCTURE. A message to a GPIB device from the controller involves at least three major components: the output command, the device

```
1    CLEAR,60000!
2    IBINIT = 60000!
3    BLOAD"a:BIB.M",IBINIT
4    CALL IBINIT (IBRD%,IBWRT%,IBCMD%,IBWAIT%,
         IBRPP%,IBONL%,IBRSC%,IBSIC%,IBSRE%,IBRTL%,IBRSV%,
         IBLPE%,IBPAD%,IBSAD%,IBIST%,IBDMA%,IBEOS%,IBTMO%,
         IBEOT%,IBGTS%,IBCAC%,IBDIAG%,IBERR%,IBCNT%)
5    ONE% = 1
6    ZERO% = 0
7    CALL IBDMA% (ZERO%,ONE%)
8    CALL IBTMO%(ZERO%,ONE%)
9    CALL IBEOT%(ZERO%,ONE%)
10   REM send interface clear
11   CALL IBSIC% (ZERO%)
12   REM set controller in charge
13   CALL IBCAC% (ZERO%,ONE%)
100  REM assert remote line
110  CALL IBSRE% (ZERO%,ONE%)
120  REM send untalk,send unlisten,listen address 16,
       controller talk address
130  ADDRESS$="-?0@"
140  CALL IBCMD%(ZERO%,ADDRESS$)
150  REM send instrument message
160  MESSAGE$="INIT"
170  CALL IBWRT%(ZERO%,MESSAGE$)
180  REM unassert remote line
190  UNT$="-?"
200  CALL IBCMD%(ZERO%,UNT$)
210  REM unassert remote line
220  CALL IBSRE%(ZERO%,ZERO%)
230  END
```

(a)

```
10 PRINT #16:"INIT"
```

(b)

Figure 8-8 Program comparison

address, and the message. The output command is relative to the command language involved. We see seven examples in Figure 8-7, for instance, that all say the same thing, yet we see four different output command statements. PRINT, PUT, wrt, and OUTPUT all mean the same thing after interpretation by their native language. Also notice that each language has a unique way of identifying a GPIB address.

The message conveyed by each language, however, must remain explicitly the same at all times. In all languages, the message is the same, "FREQ

1E6." The reason is that the instrument to which this message is being directed recognizes only one format: its own. The message is not an interpreted value, as are the output and address statements, and must be transmitted over the GPIB exactly as stated. As a rule, messages are enclosed in parentheses to distinguish them from language statements.

When executed, the program organizes the instructions and bus addresses into IEEE-488 format and transmits the message to the addressed device, where it is acted upon. Program statements can also be used to program the GPIB controller, as well as peripheral instruments.

CONTROLLER TIMEOUT PERIOD. Among the many needs of the GPIB controller is a specified timeout value. The *timeout value* tells the controller how long it should wait for a bus handshake cycle to complete. If a handshake with a device is not completed within the allotted timeout period, the requested operation is aborted, thus preventing the bus from becoming gridlocked by an instrument malfunction. For some controllers, interface timeout is fixed at infinity, which is obviously not desirable for all applications. Most controllers, on the other hand, have a programmable timeout value. This value is normally set in software using the set interface timeout command. A typical timeout value is about 1 second.

SERVICE REQUEST INTERRUPTS. The controller must also be told how to handle system interrupts, or *Service Request* (SRQ) signals. SRQ signals are used by GPIB instruments to alert the controller to a change in status

		Decimal	
Abnormal Conditions	**Binary**	**X = 0**	**X = 1**
ERR query requested	011X 0000	96	112
Command error	011X 0001	97	113
Execution error	011X 0010	98	114
Internal error	011X 0011	99	115
Power fail	011X 0100	100	116
Execution error warning	011X 0101	101	117
Internal error warning	011X 0110	102	118
Normal Conditions			
No status to report out of the ordinary	000X 0000	0	16
SRQ query request	010X 0000	64	80
Power on	010X 0001	65	81
Operation complete	010X 0010	66	82

Table 8-5 Status-bytes definitions

ERR query request—This states that a device is reporting an error but is not identifying the error. The controller should send ERR? to find out about the error.

Command error—Results when a message can't be parsed or lexically analyzed.

Execution error—Exists when a message is parsed and analyzed but can't be executed (e.g., setting a device out of its range).

Internal error—Results from an out of calibration error or a malfunction in the device (may be discovered by an instrument self-check).

Power fail—Some instruments can detect a power fail long before the instrument is affected. In those cases a power fail is sent to the controller to save important information or to flag suspect operation.

Execution error warning—The device indicates that it has received and is executing a command but that a potential problem might exist. For example, an instrument may be out of range but is sending a reading anyway.

Internal error warning—The device indicates that it has an internal error but is continuing to function.

No status to report out of the ordinary—When the controller does a serial poll and a device has nothing to say, this byte is reported.

SRQ query request—This is similar to the ERR query request. It simply says "I am requesting service but am not saying why." To find out why, the controller should send SRQ?

Power on—When a device has finished its "power on" sequence, it may send a "power on" service request. This will let the controller know that a device has just come up on the bus. It should be noted that on a "power on" service request, RSV can not be cleared by a device clear. The only thing that can clear the RSV is handshaking out the status byte to the controller. This eliminates the problem of a device powering up and receiving a device clear before the controller knows it is on the bus. After every power on, this status byte should be the first reported.

Operation complete—This tells the controller some task has been completed. This is not the same thing as the BUSY bit, DIOS, described earlier. The BUSY bit going false indicates that processing and execution of a command has been completed. In a multi-task environment, the execution of a comand may cause a long task to start, e.g., data log 10,000 points. At this point the device is ready for another command. When the task is done, the "operation complete" status byte is sent to the controller.

Table 8-5 (continued) Status-bytes definitions

or to report a problem. The reason for the interrupt can be found by reading the instrument's *Status Byte*.

The IEEE-488 standard defines the Status Byte as a means for an instrument to send a byte of data to the controller for management purposes. But for bit 7, IEEE-488 has left the meaning of the Status Byte word undefined. Consequently, most instrument manufacturers have taken it upon themselves to define their own Status Byte codes. The Tektronix Codes and Formats standard Status Byte definitions can be seen in Table 8-5.

SRQ signals are sent over a dedicated wire in the GPIB bus in the same way Attention signals are sent through the cable. Because the SRQ wire is a parallel line, it is up to the controller to sort through the various instruments connected to the GPIB when this line goes high to determine which device made the SRQ call. The process is called *polling*. In polling, each device is queried as to its status to ascertain the source of the interrupt.

The IEEE-488 standard provides two versions of polling: the parallel poll and the serial poll. In the parallel poll, the controller assigns one of the eight GPIB data-bus lines to each of as many as eight instruments on the bus. Then, when the controller asks for a status bit, all devices place their current status on their assigned lines. Usually, the status bit is set if the instrument is asserting SRQ and clear if it is not. If more than eight instruments are involved, the instruments get polled in two or more groups.

In a serial poll, the controller polls the instruments one at a time. In this mode, an instrument can report fairly detailed information because it has eight bits in which to encode its status. Serial polling is normally accomplished by listing the addresses of all devices that are capable of generating an SRQ interrupt in a single POLL command. Secondary addresses may also be used in the poll to query functions within an instrument when necessary. An example of a typical poll of three devices might read:

POLL N,S;3;5,1;1

representing bus addresses 3, 5, and 1, respectively. Notice that the device at address 5 uses a secondary address. The POLL command, in this case, causes the controller to read the Status Byte from the devices in the order in which they are listed. That is, the controller first queries the instrument at bus address 3, then 5, and finally address number 1. If the instrument at address 3 is asserting the SRQ, the Status Byte is returned in S and the position of the device in the address list (number 3 in this case) is returned in N. If device 3 is not the perpetrator of the SRQ interrupt, the controller polls number 5. The process continues until all devices have been polled or the controller finds a device asserting SRQ. The polling process stops as soon as a device is found asserting SRQ. Unlike parallel polling, serial polling automatically clears a device of SRQ data.

Notice that the polling sequence of the above example is not in numerical order. Address number 1 is polled last, not first. This allows the programmer to sequence the polling process so that the GPIB devices most likely to cause an interrupt are polled first, thus reducing bus latency.

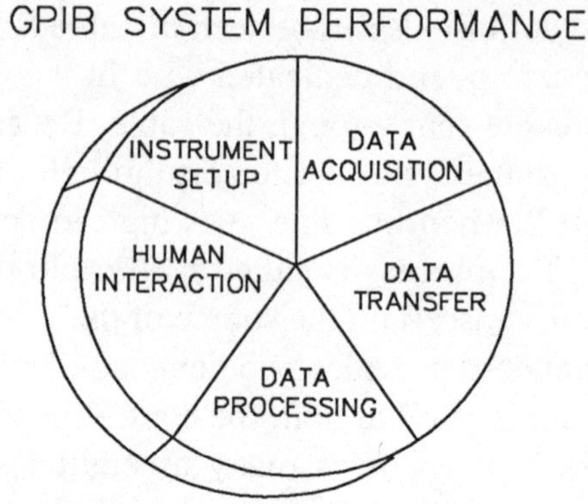

Figure 8-9 GPIB system performance

PRIMARY ADDRESSES. A GPIB device actually has two program addresses: One for talking and one for listening. The primary address of a GPIB device is only its locating address. An instrument with a primary address of 10, for example, resides at number 10 GPIB boulevard. The talk and listen addresses are subsets of the primary address, and must be addressed separately when writing an application program.

When the controller wants a device to listen to a message, it sends the device's listen address. When the controller wants the device to transmit a message, it sends the device's talk address. The listen address is equal to the primary address plus 32, while the talk address is the primary address plus 64. An instrument with a primary address of 10, for example, has its listen address located at 42 and its talk address at 74.

Many controllers generate talk and listen addresses automatically by adding the appropriate offsets to the primary address. In other cases, the programmer must explicitly provide the talk or listen address in the command, as in the case of:

```
PUT "VPOS 5" INTO @42
```

where 42 is the listen address of a device at primary address 10. To request data from the same instrument would require the command:

```
OUTPUT @74, "FREQ 1E6"
```

Whether the process of address offset is done automatically or by the programmer depends on the software package. Bear in mind that the programming commands used in the examples above are for illustration only and do not refer to any particular programming language or format.

Increasing System Performance

Fine tuning a GPIB applications program for top-notch performance generally means wringing every ounce of speed from the system. Five major processes compose the overall picture of IEEE-488 performance, as illustrated in Figure 8-9. The sum of these processes is the total time required for the system to execute its intended function.

The amount of time required for each process varies from application to application, and not all systems require all these activities. A data-logging system, for example, generally spends little time setting up the instruments, and probably does not require operator intervention. Most of its time is devoted to acquiring and transferring data. In contrast, a production-test system spends little time acquiring data and devotes most of its time to data processing with a considerable amount of human interaction involved. Two systems, two needs.

A closer look at each activity as it relates to the tasks at hand helps the programmer evaluate how much time the system spends on each function. This evaluation will locate those areas in which speed improvements are most profitable.

INSTRUMENT SETUP. The first task for the GPIB controller is instrument setup, and a major part of that is controlling the instrument's settings. The time required for instruments to process and execute setting commands and for those new settings to stabilize is called *setup time*. Setup time is divided into two sequences: program interpretation and settling time.

In most cases, the time required for a programmable instrument to decode and execute a command received from the IEEE-488 bus is small compared to other operations. Typical decode and execute times range from a few hundred microseconds to several milliseconds.

The second part of setup time is *settling time*, or step-response time. This is the time required for the instrument to stabilize to a specified accuracy after an input or change in dynamic range. For example, a programmable power supply takes a finite amount of time to interpret a voltage-setting command, such as "VPOS 10," and an additional amount of time before the output voltage actually stabilizes to the value requested. The stabilizing period is called the settling time. In many instruments, there is no direct indication of when the reading or output has settled, and the programmer is forced into programming wait states into the application program to guarantee accurate results.

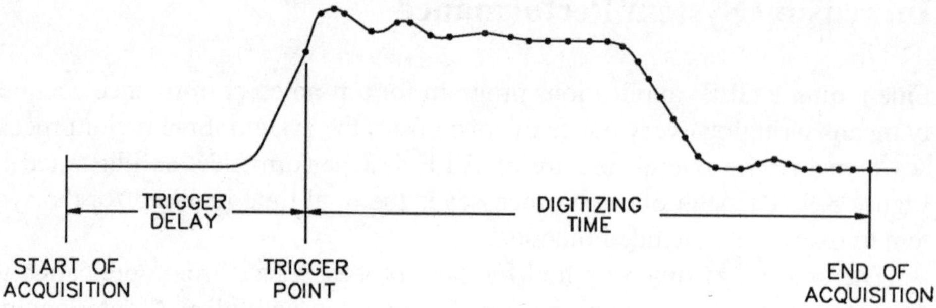

Figure 8-10 Data acquisition

The first, and most obvious, way to reduce instrument setup time is to minimize the number of setting changes required during a test. Grouping together tests that use a common instrument setting is a beginning. Autoranging should be avoided whenever possible. In autoranging mode, the instrument must capture at least one reading to implement the under- or over-range condition. Then the instrument scales up or down, accordingly, and takes another sample reading. Using a typical DMM conversion rate of 300 msec per sample, autoranging could require as much as 1.5 seconds (four ranges, five samples x 300 msec) to arrive at the correct range.

DATA ACQUISITION. The second major component in IEEE-488 system performance is the time required to acquire data. There are two factors that must be included in estimates of data-acquisition time: trigger delay and digitizing time (for digital instruments).

Trigger delay is the time between the point at which the instrument is ready to begin acquisition and the point at which the trigger actually occurs, as detailed in Figure 8-10. A bus command or an external event can initiate the trigger. Trigger delay also includes holdoff time and the time required to capture pretrigger data, if required. If a trigger occurs while the pretrigger acquisition is in progress, that trigger is ignored, and another full cycle is required before acquisition begins.

The second factor in data acquisition is *digitizing time,* the amount of time required to actually sample and digitize or latch the data. For a DMM or another single-value A/D converter, the digitizing time is simply the reciprocal of the conversion time. If the instrument performs signal averaging, on the other hand, each average requires one complete acquisition, adding substantially to the conversion time. If the instrument performs 32 averages per output value, 32 acquisitions are required.

Improving data-acquisition time requires careful attention to detail in setting up the parameters. While it is usually impossible to change the nature of the input signal, acquisition can gain speed by reducing the delay and digitizing time. A balance must be found between the resolution time afforded by higher sweep rates and the faster acquisition that comes with lower sweep rates.

DATA TRANSFER. More basic to the bus itself is the *data-transfer time.* Data-transfer time is the time required to move data over the GPIB from one location to another, such as from instrument-to-instrument or instrument-to-controller. The data-transfer time is affected by two factors: the number of bytes that are transferred and the time required to transfer each byte.

The maximum data-transfer rate for IEEE-488 is specified as 250 kilobytes per second, or roughly 250K baud. This is not, however, necessarily the rate at which the data is actually transferred. The GPIB is an asynchronous bus, and data is transferred at a rate determined entirely by the devices involved in the transfer; that is to say, the slowest device involved in the data exchange sets the limits for the transfer rate. A 9600-baud instrument communicating with a 110-baud peripheral, for example, limits the faster instrument to about 1/100th its true capacity. As a result, the maximum transfer rate of an instrument is often not fully realized in a particular system environment.

If one device involved in a transfer is significantly slower than the others, the effective transfer rate can be roughly estimated by using the rate of the slowest device. As the GPIB devices become more carefully matched in speed, however, this rule-of-thumb principle does not apply. Due to interaction of the devices involved, particularly when the speeds are similarly matched, the actual transfer rate can be considerably slower than the rate of the slowest device involved.

IEEE-488 data-transfer rates can be improved in two basic areas. First, the system can be configured to maximize bus performance. In considering bus configuration, one obvious way to improve the data-transfer rate is to use instruments that are faster on the bus. System components that are matched in performance often produce superior results. Be aware, however, that because of device interaction, one should not lean too heavily on the rated performance. Rule-of-thumb engineering dictates that one can expect one-third the rated speed performance from a system with matched specs.

In systems in which the controller supports more than one IEEE-488 port, avoid separating instruments that normally have to communicate with each other. If the instruments are placed on the same bus, they can often be programmed to communicate with one another directly, thus saving on controller setup time. When separating the same two instruments on different buses, data must pass through the controller, thus slowing the transfer rate.

Once a bus configuration is selected, performance can be further fine-tuned in software. Significant data-transfer-rate improvements can be made in software, often with very simple changes in the I/O commands. The first step in writing efficient GPIB data-transfer programs is to become intimately familiar with the IEEE-488 I/O statements in your controller's language. Many languages provide both high-level statements, such as PRINT and INPUT, and low-level statements that give the programmer more direct control of the bus.

Bus performance can also be improved by reducing the number of byte transfers required to perform a function. There are several ways to reduce bus traffic without degrading the integrity of the messages sent on the bus. First, eliminate extra spaces and zeros from the program whenever possible. Some controllers add these characters to the message, thus lengthening the amount of time it takes to complete a message. Bus traffic can also be reduced by sending messages more efficiently. When identical messages are sent to several instruments, it is more efficient to address all the devices to listen and send the message once, instead of sending individual messages. When two instruments must frequently exchange data, it is faster to set up a direct transfer between the two devices than it is to constantly funnel the data through the controller.

Other tricks for increasing bus speed include using string variables in the program in place of numeric values and working in binary code whenever the opportunity permits. String variables have a significant speed advantage over numeric variables because they need less data formatting for input and output. Data sent in binary code improves bus throughput.

DATA PROCESSING. Another significant factor in IEEE-488 system performance is *data-processing time*. The time required to process data depends entirely on the complexity of the task, the speed of the computer, and the number of other tasks assigned to the computer. The best way to estimate data-processing time is by direct measurement. Measuring data-processing time is done by running test data through the program and recording the elapsed time with the real-time clock available on most controllers.

In many GPIB systems, the system controller and intelligent instruments share the burden of data processing. By allocating different data-processing chores to intelligent instruments, one can gain a significant improvement in data-processing time. This kind of multiprocessing allows several operations to be processed at the same time. Processing data in an instrument and sending results instead of raw data has another hidden advantage: It is much faster to send one or two results to the controller than to send a complete, unprocessed waveform or measurement. Distributed processing can save data-transfer time as well as processing time.

Another ploy is to queue processing events. Instead of wasting time waiting for these events to occur, some operating systems allow for the stacking of processing tasks. Tasks are actually processed one at a time, but a task can be interrupted by an event or another higher-priority task, thus placing the present application into the background to be returned to when controller time permits. An easy way to queue processing chores is to assign priorities to the SRQ interrupts.

HUMAN INTERACTION. Human interaction is by and large the slowest part of any application. While user intervention is unavoidable in all but the most mundane systems, it can be reduced by using automated parts handlers and programmable switchers.

When procedures require the human interface, simple graphics and screen menus are the fastest way around this unavoidable bottleneck. A small keypad or digitizer device is also a more efficient form of input than typing letters into a keyboard. Above all, make the human input as idiot-proof as possible. The time wasted restarting the system when operators make errors justifies the few extra steps required to check user input and handle error conditions.

Glossary

Ac Voltmeter: Instrument for measuring ac voltage

Ac: Alternating current.

ADC: Analog-to-digital converter.

AGC: Automatic gain control.

Algorithm: A procedure for solving a mathematical problem.

AM: Amplitude modulation.

Ammeter: Instrument for measuring current flow.

Amplitude modulation: A method for the transfer of data by varying the amplitude of a carrier signal.

Analog meter: Mechanical meter movement.

Analog multiplier: Generates an output voltage equal to the product of its input voltages.

Analog-to-digital converter: Converts analog voltage or current into binary code.

Anode: Positively-charged element of an electron tube.

ANSI X3.42: Standard numerical format for GPIB programming.

Aquadag: A colloidal suspension of fine graphite particles in water; registered trademark.

ASCII: A standard binary code for alphanumeric characters.

Astable multivibrator: Logic circuit with no stable state.

ATE: Automatic test equipment

Attenuator probe: A test probe with a built-in attenuator circuit.

Attenuator: Circuit which reduces the level of an input signal using resistors and/or amplifiers.

Auto-ranging: A circuit which samples the input voltage to an instrument and automatically sets the scale factor to the proper range.

Automatic gain control: A circuit which controls the gain of an amplifier using the input signal as a reference.

Automatic test equipment: Instrument which performs measurements with no human intervention; generally an automated testing system.

Average value: The value derived from measurements taken over a period of time and mathematically averaged to eliminate random error.

Average voltage: The total area under the peak voltage curve divided by the period in radians.

Backward diode: A heavily-doped semiconductor diode with very little forward voltage drop.

Bandwidth: The range within a band of wavelengths, frequencies, or energies.

Base: The control element of a bipolar transistor.

BASIC: A computer programming language.

Baud: A variable unit of data transmission speed, usually equal to one bit per second.

Binary data: Any quantity or character expressed in binary format.

Bipolar transistor: A solid-state amplifying device with two semiconductor junctions.

Bit: A logic one or zero.

Bubble memory: Binary storage device which uses magnetic bubbles to store data.

Buffer amplifier: An amplifier usually with unity gain used to isolate one circuit from another.

Byte: A binary word composed of 8 bits.

Capacitive reactance: The effective resistance to a change in current flow in a capacitor; expressed in ohms.

Carriage Return: ASCII 13

Cartesian-coordinate: A grid of horizontal and vertical lines used to locate points on a plane.

CAT: Computer-aided testing

Cathode-ray tube: An electron tube in which a beam of electrons are projected on a fluorescent screen to produce a luminous spot.

Cathode: Negatively-charged element of a vacuum tube.

Checksum: A byte or bytes containing the arithmetic sum of a data frame; used for error detection.

CMRR: Common mode rejection ratio

Coaxial cable: A transmission line that consists of a tube of electrically conducting material surrounding a central conductor held in place by insulators to form a line with a constant impedance.

Cold junction: The temperature reference junction of a thermocouple.

Collector: The reverse-biased element of a bipolar transistor.

Common mode rejection ratio: The ability of an instrument or amplifier to reject signals which are common to both inputs.

Computer hardware: The mechanical components of a computer system.

Computer software: Instructions which control the flow of logic within a computer system.

Computer-aided testing: Measurements performed with the aid of a computer.

Constant current source: A power source with current regulation.

Control grid: The controlling element of an electron tube; controls screen brightness of a CRT.

Controller: Any device which possesses logical intelligence for the control of a system.

CR: Carriage return ASCII character

CRT: Cathode-ray tube

Crystal oscillator: Oscillator whose frequency is determined by a resonant quartz crystal.

Current shunt: Resistance used to bypass current from a device.

D'Arsonval movement: Type of mechanical meter movement.

DAC: Digital-to-analog converter.

Data-transfer rate: The speed at which data is moved between two points.

DB: Decibel.

Dc voltmeter: Instrument for measuring dc voltage.

Decade counter: Binary circuit which counts up to 10 then resets itself to zero.

Decibel: Electrical power expressed as a log of 10.

Deflection plates: Charged plates placed in the neck of a CRT used to deflect the direction of the electron beam.

Deflection pointer: The indicator of an analog meter.

Differential amplifier: A balanced amplifier which produces a differential output voltage in response to an input signal.

Digital counter: An instrument which counts pulses using digital logic circuits.

Digital display: Any device which produces a visual, numerical representation of an analog or digital value.

Digital multimeter: An instrument which measures voltage, current, and resistance using digital logic

Digital voltmeter: An instrument which displays voltage using digital logic circuits.

Digitizer: An instrument which converts analog values into binary data.

Digitizing rate: The rate at which a quantity is sampled and converted into a binary value.

Digitizing time: The total time it takes for a digital process to convert an analog value into an appropriate binary word at the specified digitizing rate.

Discriminator: A circuit which accepts or rejects signals of different characteristics, such as frequency or amplitude.

DMM: Digital multimeter.

DOS: Disk operating system; the IBM personal computer operating system.

Drain: The electron collecting element of a field-effect transistor.

Dual-beam CRT: A CRT which has two separate electron guns for displaying two traces on the screen.

Dual-channel: An instrument which has two input channels; applies to dual-trace oscilloscopes.

Dual-trace oscilloscope: An oscilloscope which can display two separate input waveforms on a single screen.

Duty cycle: The ratio of on time to off time of a square wave.

DVM: Digital voltmeter

Electrode: Any element within an electron vacuum tube; a metal contact associated with a voltage and an electrical device.

Electroluminescence: Light resulting from a high-frequency discharge through a gas or from an alternating current to a layer of phosphor.

Electron beam: A tightly-focused stream of electrons emanating from an electron source (cathode) and accelerated towards a collector plate or ring (anode).

Electron gun: Emits electrons by heating a metal or chemical compound to luminescence.

Electron lens: A series of CRT elements used to focus the electron beam.

Electron tube: An electronic device in which conduction by electrons takes place through a vacuum within a sealed gas or metal container and which has various common uses based on the controlled flow of electrons.

Emf: Electromotive force; voltage.

EMI: Electromagnetic interference; noise.

Emitter follower: An amplifier circuit which obtains its output from the emitter of a bipolar transistor.

Emitter: The forward-biased element of a bipolar transistor.

EOI: End or Identify; GPIB frame terminator.

Equivalent-time sampling: Digital sampling technique which builds a waveform over a period of time by sampling several consecutive waveforms.

Fall time: The time it takes the trailing edge of a waveform, usually a square wave, to go from 90% to 10% of its peak value.

Feedback: The return to the input of a part of the output for producing changes in an electrical circuit that improves performance or provides self-corrective action.

FET: Field-effect transistor.

Field-effect transistor: A semiconductor amplifying device that uses the electric field of an attached (gate) electrode to pinch off the flow of electrons through a bar or channel of doped semiconductor material.

Filament: A tenuous metal conductor made incandescent by the passage of an electric current.

Flash converter: An analog-to-digital converter that converts in one clock cycle.

Flip-Flop: Logic circuit with two stable states.

Floppy disk: Magnetic medium, similar in appearance and operation to a record disk, housed in a protective jacket and used by computers for the permanent storage of data.

Fluorescent: Being of fluorescence; emission of or the property of emitting visible light resulting from the absorption of radiation from some other source.

Form factor: The ratio of rms voltage with relationship to another waveform.

Forward voltage drop: The voltage differential across a forward-biased semiconductor junction at a specified current flow.

Fourier analysis: The fitting of terms to an infinite series in which the terms are constants multiplied by sine or cosine functions of integer multiples of the variable.

Frequency counter: A modified pulse counter with a time gate that permits the counting of pulses only during a precisely defined period, which is subsequently converted into a frequency measurement.

Frequency modulation: A method for the transfer of data by varying the frequency of a carrier signal.

Frequency multiplier: A circuit which uses the natural harmonics of a fundamental frequency to output a higher frequency.

Frequency synthesizer: An instrument which generates an output frequency as a result of successively mixing reference frequencies to obtain the desired frequency output.

Full scale deflection: The travel of an analog pointer from one extreme of the scale to the other.

Full-wave rectification: The conversion of both halves of an alternating waveform into dc voltage.

Function generator: An instrument for generating waveforms of varying types, usually sine, square, and triangular.

Gain factor: Amplification factor.

Galvanometer: A sensitive ammeter.

Gas discharge plasma tube: A tube filled with a rarefied gas that glows when ionized by a high voltage potential.

Glitch: A transient waveform or pulse that causes erratic operation or false measurement.

GPIB: General purpose interface bus.

Graticule: A scale on transparent material at the focal plane of an optical instrument used for the location and measurement of objects.

Half-wave rectification: The conversion of only one half of an alternating waveform into dc voltage.

Hall Effect device: Semiconductor device which generates a dc voltage proportional to the strength of a surrounding magnetic field.

Harmonic generator: A signal source which derives its output frequency as the result of harmonic resonance.

Harmonic: An overtone of a more fundamental frequency.

Header: A binary word or words attached to the beginning of a binary data string which defines the meaning of the data.

Heterodyne converter: A circuit which uses the heterodyne method to generate an output frequency.

Heterodyne: To combine one frequency with a different frequency so that a beat frequency is produced.

Holding capacitor: A storage capacitor used in a sampling circuit that retains a sampled voltage.

Holdoff period: A period of time in which the normal sequence of events are prevented from happening; used to synchronize multiple events.

HP-IB: Hewlett-Packard Interface-Bus; forerunner to the GPIB.

Hysteresis: A lagging in the values of a device when a change to the input takes place; generally surrounded by a hysteresis band, of which the input must exceed the limit of before a change is recognized.

I/O: Input/output; applies to the input/output features of a computer.

IEEE-488: The official nomenclature for the GPIB standard.

IEEE-P981: A proposed amendment to the IEEE-488 standard.

IF: Intermediate frequency.

Incandescent lamp: A light source derived from a glowing hot wire.

Input capacitance: The shunt capacitance of a circuit.

Input impedance: The total effect of input resistance, capacitance, and inductance on a signal.

Input resistance: The series-shunt resistance of a circuit.

Integrated circuit: A monolithic device containing up to thousands of interconnected transistors, resistors, etc.; usually perform a complete circuit function.

Integrator: A circuit for arriving at the result of a mathematical integration.

Interface: The point where two devices connect to each other for the purpose of communications.

Interpolation: To estimate values of a function between two known values.

Isolation transformer: Device which isolates its input from its output using magnetic coupling.

Keyboard: Typewriter-like device used for the entry of data.

LCD: Liquid crystal display.

LED: Light emitting diode.

Light emitting diode: Solid-state indicator lamp.

Liquid crystal display: A readout device using liquid crystal to reflect ambient light.

Lissajous figure: The pattern produced on a plane when two sine waves sweep against each other.

Local oscillator: Oscillator used as a reference frequency for heterodyning purposes.

Logarithmic amplifier: An amplifier which exhibits logarithmic gain in response to an input signal; gain is highest when the input signal is lowest, progressively decreasing in gain as the input voltage is increased.

LSB: Least-significant bit.

Magnetic field: The portion of space near a magnetic or current-carrying body in which the surrounding forces are magnetic nature.

Magnetic flux: Lines of force used to represent magnetic induction.

Maximum value: Peak value.

Memory address: A number used to locate data within computer memory.

Microammeter: Instrument for measuring currents down to 10-6 amps.

Microwave: A comparatively short electromagnetic wave usually between 100 cm and 1 cm in length.

Miller integrator: A sawtooth waveform generator.

Milliammeter: Instrument for measuring currents down to 10-3 amps.

Mixer: A circuit which heterodynes two frequencies to produce a beat frequency.

Monostable multivibrator: Logic circuit with one stable state.

MOSFET: Metal-oxide field-effect transistor.

Mouse: An input device used with personal computers to enter distance and directional data.

MSB: Most-significant bit.

Multimeter: Any instrument which combines the functions of voltage, current, and resistance measurement is a single unit.

Multiplier resistor: A resistor used to increase the voltage range of an analog voltmeter; any resistance device which extends the range of a measuring instrument.

Multirange voltmeter: An instrument with more than one voltage scale.

Nixie tube: A neon tube with cathodes shaped as numerals which glow when ionized.

Noise: Unwanted electrical signals or radiation.

Null: A minimum or zero value of an electrical current or voltage.

Nyquist criterion: A guideline for the faithful reproduction of a waveform using dot imagery.

Ohm's law: A mathematical relationship between volts, amps, and resistance; states that one amp of current will flow through a one-ohm resistor with a force of one volt.

Ohmmeter: Instrument for measuring resistance.

Ohms per volt: Sensitivity figure of merit for analog voltmeter.

Op amp: Operational amplifier.

Operational amplifier: An electronic device which performs the analog operations of integration, multiplication, differentiation, addition, and subtraction on an input signal; commonly used as a straightforward signal amplification device.

Optoisolator: Device which isolates its input from its output through optical means.

Oscilloscope: An instrument in which the variations in a fluctuating electrical quantity appear temporarily as a visible wave form on the fluorescent screen of a cathode-ray tube.

Overshoot: Amplitude perturbation immediately following transition time, expressed as a percentage of pulse amplitude.

Parallax: The apparent displacement of an object as seen from two different points not on a straight line with the object.

PC instrument: A measuring instrument which works in conjunction with a personal computer.

PC: Personal computer

Peak voltage: The absolute value between zero volts and the largest differential voltage attained by an ac waveform.

Peak-to-peak value: The absolute value between the most-negative voltage peak to the most positive-voltage peak of a waveform.

Perceptual aliasing: Optical illusions created by dot pattern images.

Personal computer: A small, desktop computer for personal use, such as the IBM PC.

Phase comparator: A circuit which measures relative phase angle between two signals to produce an output.

Phase-locked loop: Oscillator which uses phase a comparator in its feedback loop.

Phosphor: A substance which emits light when excited by radiation.

PIV: Peak inverse volts.

PLL: Phase-locked loop.

Potentiometer: Variable resistor.

Preamplifier: A low-level amplifier which precedes a driver amplifier.

Precision rectifier: A rectifier circuit built around an operational amplifier.

Prescaler: A circuit which reduces the frequency of an input signal by division.

Probe attenuation: The loss of signal induced by the attenuation of a test probe.

Programming language: A convention of instructions used to steer the logic of a computer.

Propagation delay: The time it takes a signal to travel from the input of a device to its output.

Protocol: A formal set of conventions governing the format and relative timing of message exchange between two communicating processes.

Pulse width: The amount of time between the 50% point on the leading edge of a waveform and the 50% point on the trailing edge.

Radian: A unit of plane angular measurement that is equal to the angle at the center of a circle subtended by an arc equal in length to the radius.

RAM: Random access memory.

RC oscillator: Oscillator whose frequency is determined by a resistor-capacitor feedback network.

Reactance: The effective resistance to a change in current flow in a capacitor or inductor; expressed in ohms.

Real-time sampling: Digital sampling done in real-time.

Record length: The number of digital samples a PC-based instrument is able to contain in a single measurement.

Rectifier diode: A one-way electronic valve that allows electrons to flow in one direction but not the other.

Register: An temporary accumulator used in digital and computer design for the containment of data bits pertinent to the task at hand; very often used as a pointer to addresses of data in RAM.

Regulated power supply: A power source for maintaining the output at a constant voltage or current.

RF: Radio frequency

RFI: Radio frequency interference; noise.

Ringing: Damped oscillation that decreases in amplitude as a function of time; similar to the ringing volume of a bell.

Rise time: The time it takes the leading edge of a waveform, usually a square wave, to go from 10% to 90% of its peak value.

Rms converter: Converts any waveform into its rms equivalent using mechanical devices (thermocouple) or signal processing methods.

Rms: Root mean square.

Roll-off: The linear attenuation of a signal with respect to decreasing frequency.

Root mean square: The square root of the arithmetic mean of th squares of set of numbers; indicates the equivalent dc-voltage heating value of an ac waveform.

RS-232: A duplex serial communications link.

RTXO: Room temperature crystal oscillator.

S/N: Signal-to-noise ratio.

Sampling bridge: A rectifier-type sampling device.

Sampling period: A predetermined time slot within which measurement samples are taken.

SAR: Successive approximation converter.

Sawtooth: A waveform with a profile similar to the teeth of a wood-cutting saw.

Scale factor: A constant used to relate an input value to an indicated measurement.

Schmitt trigger: An logic circuit with hysteresis; the gate will not generate an output until the input signal has exceeded the hysteresis range of the input voltage comparator.

Schottky diode: A rectifier fabricated from a semiconductor-to-metal junction.

Secondary electron: Electrons which are dislodged from a metal or material by high-energy electrons impinging on its surface.

Settling time: The time it takes for a voltage or current transition to stabilize at its specified amplitude.

Seven-segment display: An arrangement of seven lamps or other luminescent devices which displays the range of numerals from 0 through 10.

Signal generator: An instrument for generating ac signals.

Signal processing: The electronic manipulation of ac signals.

Signal-to-noise ratio: The ratio of signal amplitude to background noise.

Sine wave: Waveform with a sinusoidal shape.

Sinusoidal: Waveform with a sinusoidal shape; sine wave.

Source follower: An amplifier which obtains its output signal from the source of an FET transistor.

Source: The electron supplying element of a field-effect transistor.

Split-beam CRT: A dual-trace CRT with one electron gun and a beam splitter.

Square wave: Waveform with a square shape.

SRQ: Service request interrupt.

Staircase generator: A circuit which generates sawtooth waveforms by dumping quantized charges into a holding capacitor.

Storage oscilloscope: An analog or digital oscilloscope which can continue to display a trace on the screen after the event has ceased to occur.

Stray capacitance: Capacitance not deliberately introduced into a circuit; normally attributed to wiring.

Successive approximation converter: An analog-to-digital converter that converts using several clock cycles and a voltage comparator.

Sweep generator: An FM modulated generator that starts at a low frequency and sweeps up to an upper frequency limit.

Sweep rate: The speed at which a sweep progresses.

Syntax: The sequence in which computer commands are put together to form an instruction.

TCXO: Temperature compensated crystal oscillator.

Temperature coefficient: A change in operating parameters as a function of temperature.

Test probe: Device for connecting a point under test to a measuring instrument.

Thermocouple: Generates voltage as a function of temperature.

Thermopile converter: A device which uses a thermopile to convert an input current into a thermovoltage; commonly used to convert varying ac waveforms into rms values.

Thermopile: An apparatus that consists of a number of thermocouples combined so as to multiply the thermovoltage effect and is used for generating electric currents or for determining intensities of radiation.

Thermostatic control: A temperature-controlled environment.

Thermovoltage: Voltage as a result of temperature differential.

Timebase: An oscillator whose output is used as a reference for time.

Transient: A non-repetitious waveform or voltage spike passing quickly into and out of existence; generally created by external disturbances, such as EMI and RFI.

Transistor: Solid-state amplifying device.

Transition time: The time it takes a signal to switch from one voltage level to another.

Trigger: A voltage or current used to commence an event.

Trimmer capacitor: Small, variable capacitor.

Tunnel diode: Heavily doped semiconductor junction with negative resistance characteristics; may be used as an amplifier or oscillator.

Unity: Being a state of one.

USB: Useful storage bandwidth; digital sampling.

USR: Useful storage rise time; digital sampling.

Vacuum tube voltmeter: An instrument which uses electron tubes for the active element.

Vacuum tube: An electron tube evacuated to a high degree of vacuum.

Varactor: A voltage-dependent diode whose junction capacitance changes in relationship to bias voltage; commonly used to control the frequency of an oscillator.

Variable frequency oscillator: An oscillator whose output frequency is adjustable either by mechanical means or through the application of an external voltage.

VBE: The voltage drop between the base and emitter junction of a bipolar transistor when the junction is forward biased.

VCC: Positive supply voltage.

Vector: A quantity that has magnitude and direction and that is commonly represented by a directed line segment whose length represents the magnitude and whose orientation in space represents the direction.

VEE: Negative supply voltage.

Vernier: A short, expanded scale made to travel along the divisions of a graduated instrument for smaller, incremental adjustment; fine tuning.

VFO: Variable frequency oscillator.

Volt-ohm-milliammeter: An instrument that measures voltage, current, and resistance.

Voltage comparator: Compares a reference voltage to an input voltage to produce an output signal.

Voltage-to-current converter: Converts voltage to a current.

Voltmeter loading effect: Measurement errors produced by a voltmeter.

Voltmeter: Instrument for measuring voltage potential.

VOM: Volt-ohm-milliammeter; measures voltage, current, and resistance.

VTVM: Vacuum tube voltmeter.

Watts: The product of voltage and current; power.

Waveform generator: An instrument for generating waveforms of varying types, usually sine, square, and triangular.

Waveform: Shape of an ac signal.

Wein bridge: Low-frequency, sine wave oscillator.

Workstation: A computer-based measurement system based on a whole-testing concept, where the computer is the central component of the system; all operations must pass through the workstation for final approval and/or execution.

X-axis: The horizontal Cartesian coordinate component.

Y-axis: The vertical Cartesian coordinate component.

Z-axis: External screen brightness input to an oscilloscope.

Zero crossing detector: Circuit which triggers only when a signal passes through zero volts to reduce EMI/RFI.

References

Adams, R., "Using IEEE 488 Execute-only Controllers as Distributed Processors in ATE," *Electronics Test*, November 1984.

Artwick, B., *Microcomputer Interfacing*, Prentice-Hall, Englewood Cliffs, 1980.

Bell, D.A., *Electronic Instruments and Measurements*, Reston Publishing Co., Reston, VA, 1983.

Coan, S., "Test Bus Instruments Effectively to Pinpoint GPIB Noncompliance," *EDN*, October 13, 1983.

Downs, Millard, "GaAs, Silicon ICs Team Up in 1 Gsample/s Digital Scope," *Electronic Design*, October 30, 1986, pp. 167-172.

Gifford, T., "Many Choices Complicate IEEE-488 Instrument Selection," *Electronic Products*, March 1, 1985.

Goodman, R.L., *Practical Troubleshooting with Modern Test Instruments*, Tab Books, Blue Ridge Summit, PA, 1984.

Gottlieb, I., *Basic Electronic Test Procedures*, Tab Books, Blue Ridge Summit, 1985.

Healy, J.T., *Automatic Testing and Evaluation of Digital Integrated Circuits*, Reston Publishing Co., Inc., Reston, VA, 1981.

Hewlett-Packard, *PC Instruments Applications Handbook*, Publication 5952-4164, October 12, 1986.

Hughes, Luehman, "PCIB: A Low-cost, Flexible Instrument Control Interface For Personal Computers, *Hewlett-Packard Journal*," May 1986, pp. 11-16.

Hurlock, L., *The ABC's of Probes*, Tektronix, Vancouver, 1986.

Jumonville, S., "Developing a Reconfigurable Computer-aided Test System," *Electronics Test*, January 1983.

Kantrowitz, Kousourou, Zucker, *Electronic Measurements*, Prentice-Hall, New Jersey, 1979.

Kogge, P., *The Architecture of Pipelined Computer*, Hemisphere Publishing Corp., 1981.

Landee, Davis, Albrecht, *Electronics Designers' Handbook*, McGraw-Hill, New York, 1977.

Miller, Walston, *Transistor Circuit Design*, McGraw-Hill, New York, 1963.

Milne, R., "Counter-timers Combine More Power, Ease of Use, *Electronic Design*," March 31, 1983.

Milne, R.,"Digital Oscilloscopes," *Electronic Design*, February 6, 1986, pp. 98-106.

Peterson, S., "Friendly Software for Test and Measurement, *Computer Design*," October 1983.

Radio Shack, *Electronics Data Book*, 1978.

Ramirez, Tilden, "GPIB Software Configuration Determines System Performance," *EDN*, August 4, 1982.

Ramirez, Tilden, "Personal Computer Analyzes ATE Measurements," *EDN*, February 7, 1985.

Ramirez, Tilden, "Understanding IEEE-488 Basics Simplifies System Integration," *EDN*, June 9, 1982.

Rothschild, Sismilich, Walker, "Low-cost Automated Instruments for Personal Computers," *Hewlett-Packard Journal*, May 1986, pp. 4-10.

Tektronix, *2213A Oscilloscope Service Manual*, Publication No. 070-4733-00, January 1984.

Tektronix, *An Introduction to Digital Storage*, Literature No. 46W-6051, April 1984.

Tektronix, *The XYZs of Using a Scope*, Literature No. TTA-613, May 1985.

Tilden, M., "Programming Techniques Speed IEEE-488 System Execution, *EDN*," October 27, 1983.

Tilden, M., *Tektronix Codes and Formats for GPIB Instruments*, Tektronix, 1982.

Zwick, G., *The Oscilloscope*, Gernsback Library, Inc., New York, 1963.

Index